MANJIT KUMAR

QUANTUM

EINSTEIN, BOHR
AND THE GREAT DEBATE ABOUT
THE NATURE OF REALITY

D1341887

ICON BOOKS

Published in the UK in 2008 by
Icon Books Ltd, The Old Dairy, Brook Road,
Thriplow, Cambridge SG8 7RG
email: info@iconbooks.co.uk
www.iconbooks.co.uk

Sold in the UK, Europe, South Africa and Asia
by Faber & Faber Ltd, 3 Queen Square,
London WC1N 3AU or their agents

Distributed in the UK, Europe, South Africa and Asia
by TBS Ltd, TBS Distribution Centre, Colchester Road
Frating Green, Colchester CO7 7DW

Published in Australia in 2008
by Allen & Unwin Pty Ltd, PO Box 8500,
83 Alexander Street, Crows Nest, NSW 2065

ISBN: 978-184831-029-2

Text copyright © 2008 Manjit Kumar
The author has asserted his moral rights.

Line drawings by Nicholas Halliday

No part of this book may be reproduced in any form, or by any
means, without prior permission in writing from the publisher.

Typeset in 11.5 on 16pt Minion by Marie Doherty

Printed and bound in the UK by
Clays of Bungay

For
Lahmber Ram and Gurmit Kaur
Pandora, Ravinder, and Jasvinder

CONTENTS

LIST OF ILLUSTRATIONS

PLATE SECTION

About the author

Manjit Kumar has degrees in physics and philosophy. He was the founding editor of *Prometheus*, an interdisciplinary journal that covered the arts and sciences, described by one reviewer as 'perhaps the finest magazine that I've ever read'. He is the co-author of *Science and the Retreat from Reason*, which introduced key areas of modern science while defending the Enlightenment notions of social progress and scientific advance against the loss of faith in progress and science. Published in 1995, it was critically acclaimed as a 'corrective to the hype', 'thought-provoking', and 'undoubtedly one of the best introductions one can find to the crisis of confidence within science itself'. He has written and reviewed for various publications including *The Guardian*, the *Times Literary Supplement*, and the *Irish Times*. He lives in north London with his wife and two sons.

www.manjitkumar.com

THE MEETING OF MINDS

Paul Ehrenfest was in tears. He had made his decision. Soon he would attend the week-long gathering where many of those responsible for the quantum revolution would try to understand the meaning of what they had wrought. There he would have to tell his old friend Albert Einstein that he had chosen to side with Niels Bohr. Ehrenfest, the 34-year-old Austrian professor of theoretical physics at Leiden University in Holland, was convinced that the atomic realm was as strange and ethereal as Bohr argued.[1]

In a note to Einstein as they sat around the conference table, Ehrenfest scribbled: 'Don't laugh! There is a special section in purgatory for professors of quantum theory, where they will be obliged to listen to lectures on classical physics ten hours every day.'[2] 'I laugh only at their naiveté,' Einstein replied.[3] 'Who knows who would have the [last] laugh in a few years?' For him it was no laughing matter, for at stake was the very nature of reality and the soul of physics.

The photograph of those gathered at the fifth Solvay conference on 'Electrons and Photons', held in Brussels from 24 to 29 October 1927, encapsulates the story of the most dramatic period in the history of physics. With seventeen of the 29 invited eventually earning a Nobel Prize, the conference was one of the most spectacular meetings of minds ever held.[4] It marked the end of a golden age of physics, an era of scientific creativity unparalleled since the scientific revolution in the seventeenth century led by Galileo and Newton.

Paul Ehrenfest is standing, slightly hunched forward, in the back row, third from the left. There are nine seated in the front row. Eight men and one woman; six have Nobel Prizes in either physics or chemistry. The woman has two, one for physics awarded in 1903 and another for chemistry in 1911. Her name: Marie Curie. In the centre, the place of honour, sits another Nobel laureate, the most celebrated scientist since the age of Newton: Albert Einstein. Looking straight ahead, gripping the chair with his right hand, he seems ill at ease. Is it the winged collar and tie that are causing him discomfort, or what he has heard during the preceding week? At the end of the second row, on the right, is Niels Bohr, looking relaxed with a half-whimsical smile. It had been a good conference for him. Nevertheless, Bohr would be returning to Denmark disappointed that he had failed to convince Einstein to adopt his 'Copenhagen interpretation' of what quantum mechanics revealed about the nature of reality.

Instead of yielding, Einstein had spent the week attempting to show that quantum mechanics was inconsistent, that Bohr's Copenhagen interpretation was flawed. Einstein said years later that 'this theory reminds me a little of the system of delusions of an exceedingly intelligent paranoic, concocted of incoherent elements of thoughts'.[5]

It was Max Planck, sitting on Marie Curie's right, holding his hat and cigar, who discovered the quantum. In 1900 he was forced to accept that the energy of light and all other forms of electromagnetic radiation could only be emitted or absorbed by matter in bits, bundled up in various sizes. 'Quantum' was the name Planck gave to an individual packet of energy, with 'quanta' being the plural. The quantum of energy was a radical break with the long-established idea that energy was emitted or absorbed continuously, like water flowing from a tap. In the everyday world of the macroscopic where the physics of Newton ruled supreme, water could drip from a tap, but energy was not exchanged in droplets of varying size. However, the atomic and subatomic level of reality was the domain of the quantum.

In time it was discovered that the energy of an electron inside an atom was 'quantised'; it could possess only certain amounts of energy and not others. The same was true of other physical properties, as the microscopic realm was found to be lumpy and discontinuous and not some shrunken version of the large-scale world that humans inhabit, where physical properties vary smoothly and continuously, where going from A to C means passing through B. Quantum physics, however, revealed that an electron in an atom can be in one place, and then, as if by magic, reappear in another without ever being anywhere in between, by emitting or absorbing a quantum of energy. This was a phenomenon beyond the ken of classical, non-quantum physics. It was as bizarre as an object mysteriously disappearing in London and an instant later suddenly reappearing in Paris, New York or Moscow.

By the early 1920s it had long been apparent that the advance of quantum physics on an ad hoc, piecemeal basis had left it without solid foundations or a logical structure. Out of this state of confusion and crisis emerged a bold new theory known as quantum mechanics. The picture of the atom as a tiny solar system with electrons orbiting a nucleus, still taught in schools today, was abandoned and replaced with an atom that was impossible to visualise. Then, in 1927, Werner Heisenberg made a discovery that was so at odds with common sense that even he, the German wunderkind of quantum mechanics, initially struggled to grasp its significance. The uncertainty principle said that if you want to know the exact velocity of a particle, then you cannot know its exact location, and vice versa.

No one knew how to interpret the equations of quantum mechanics, what the theory was saying about the nature of reality at the quantum level. Questions about cause and effect, or whether the moon exists when no one is looking at it, had been the preserve of philosophers since the time of Plato and Aristotle, but after the emergence of quantum mechanics they were being discussed by the twentieth century's greatest physicists.

With all the basic components of quantum physics in place, the fifth Solvay conference opened a new chapter in the story of the quantum. For the debate that the conference sparked between Einstein and Bohr raised

issues that continue to preoccupy many eminent physicists and philosophers to this day: what is the nature of reality, and what kind of description
of reality should be regarded as meaningful? 'No more profound intellectual debate has ever been conducted', claimed the scientist and novelist
C.P. Snow. 'It is a pity that the debate, because of its nature, can't be common currency.'[6]

Of the two main protagonists, Einstein is a twentieth-century icon. He
was once asked to stage his own three-week show at the London Palladium.
Women fainted in his presence. Young girls mobbed him in Geneva. Today
this sort of adulation is reserved for pop singers and movie stars. But in
the aftermath of the First World War, Einstein became the first superstar
of science when in 1919 the bending of light predicted by his theory of
general relativity was confirmed. Little had changed when in January
1931, during a lecture tour of America, Einstein attended the premiere of
Charlie Chaplin's movie *City Limits* in Los Angeles. A large crowd cheered
wildly when they saw Chaplin and Einstein. 'They cheer me because they
all understand me,' Chaplin told Einstein, 'and they cheer you because no
one understands you.'[7]

Whereas the name Einstein is a byword for scientific genius, Niels Bohr
was, and remains, less well known. Yet to his contemporaries he was every
inch the scientific giant. In 1923 Max Born, who played a pivotal part in the
development of quantum mechanics, wrote that Bohr's 'influence on theoretical and experimental research of our time is greater than that of any
other physicist'.[8] Forty years later, in 1963, Werner Heisenberg maintained
that 'Bohr's influence on the physics and the physicists of our century was
stronger than that of anyone else, even than that of Albert Einstein'.[9]

When Einstein and Bohr first met in Berlin in 1920, each found an
intellectual sparring partner who would, without bitterness or rancour,
push and prod the other into refining and sharpening his thinking about
the quantum. It is through them and some of those gathered at Solvay
1927 that we capture the pioneering years of quantum physics. 'It was a
heroic time', recalled the American physicist Robert Oppenheimer, who
was a student in the 1920s.[10] 'It was a period of patient work in the labo-

ratory, of crucial experiments and daring action, of many false starts and many untenable conjectures. It was a time of earnest correspondence and hurried conferences, of debate, criticism and brilliant mathematical improvisation. For those who participated it was a time of creation.' But for Oppenheimer, the father of the atom bomb: 'There was terror as well as exaltation in their new insight.'

Without the quantum, the world we live in would be very different. Yet for most of the twentieth century, physicists accepted that quantum mechanics denied the existence of a reality beyond what was measured in their experiments. It was a state of affairs that led the American Nobel Prize-winning physicist Murray Gell-Mann to describe quantum mechanics as 'that mysterious, confusing discipline which none of us really understands but which we know how to use'.[11] And use it we have. Quantum mechanics drives and shapes the modern world by making possible everything from computers to washing machines, from mobile phones to nuclear weapons.

The story of the quantum begins at the end of the nineteenth century when, despite the recent discoveries of the electron, X-rays, and radioactivity, and the ongoing dispute about whether or not atoms existed, many physicists were confident that nothing major was left to uncover. 'The more important fundamental laws and facts of physical science have all been discovered, and these are now so firmly established that the possibility of their ever being supplanted in consequence of new discoveries is exceedingly remote', said the American physicist Albert Michelson in 1899. 'Our future discoveries,' he argued, 'must be looked for in the sixth place of decimals.'[12] Many shared Michelson's view of a physics of decimal places, believing that any unsolved problems represented little challenge to established physics and would sooner or later yield to time-honoured theories and principles.

James Clerk Maxwell, the nineteenth century's greatest theoretical physicist, had warned as early as 1871 against such complacency: 'This characteristic of modern experiments – that they consist principally of measurements – is so prominent, that the opinion seems to have got abroad that in a few years all the great physical constants will have been approximately estimated, and that the only occupation which will be left to men of science will be to carry on these measurements to another place of decimals.'[13] Maxwell pointed out that the real reward for the 'labour of careful measurement' was not greater accuracy but the 'discovery of new fields of research' and 'the development of new scientific ideas'.[14] The discovery of the quantum was the result of just such a 'labour of careful measurement'.

In the 1890s some of Germany's leading physicists were obsessively pursuing a problem that had long vexed them: what was the relationship between the temperature, the range of colours, and the intensity of light emitted by a hot iron poker? It seemed a trivial problem compared to the mystery of X-rays and radioactivity that had physicists rushing to their laboratories and reaching for their notebooks. But for a nation forged only in 1871, the quest for the solution to the hot iron poker, or what became known as 'the blackbody problem', was intimately bound up with the need to give the German lighting industry a competitive edge against its British and American competitors. But try as they might, Germany's finest physicists could not solve it. In 1896 they thought they had, only to find within a few short years that new experimental data proved that they had not. It was Max Planck who solved the blackbody problem, at a cost. The price was the quantum.

PART I

THE QUANTUM

*'Briefly summarized, what I did can be described as simply
an act of desperation.'*
—MAX PLANCK

*'It was as if the ground had been pulled out from under one, with no firm
foundation to be seen anywhere, upon which one could have built.'*
—ALBERT EINSTEIN

*'For those who are not shocked when they first come across quantum theory
cannot possibly have understood it.'*
—NIELS BOHR

Chapter 1

THE RELUCTANT REVOLUTIONARY

'A new scientific truth does not triumph by convincing its opponents and making them see the light, but rather because its opponents eventually die, and a new generation grows up that is familiar with it', wrote Max Planck towards the end of his long life.[1] Bordering on cliché, it could easily have served as his own scientific obituary had he not as an 'act of desperation' abandoned ideas that he had long held dear.[2] Wearing his dark suit, starched white shirt and black bow tie, Planck looked the archetypal late nineteenth-century Prussian civil servant but 'for the penetrating eyes under the huge dome of his bald head'.[3] In characteristic mandarin fashion he exercised extreme caution before committing himself on matters of science or anything else. 'My maxim is always this,' he once told a student, 'consider every step carefully in advance, but then, if you believe you can take responsibility for it, let nothing stop you.'[4] Planck was not a man to change his mind easily.

His manner and appearance had hardly changed when to students in the 1920s, as one recalled later, 'it seemed inconceivable that this was the man who had ushered in the revolution'.[5] The reluctant revolutionary could scarcely believe it himself. By his own admission he was 'peacefully inclined' and avoided 'all doubtful adventures'.[6] He confessed that he lacked 'the capacity to react quickly to intellectual stimulation'.[7] It often took him years to reconcile new ideas with his deep-rooted conservatism. Yet at the age of 42, it was Planck who unwittingly started the quantum revolution in December 1900 when he discovered the equation for the distribution of radiation emitted by a blackbody.

———

All objects, if hot enough, radiate a mixture of heat and light, with the intensity and colour changing with the temperature. The tip of an iron poker left in a fire will start to glow a faint dull red; as its temperature rises it becomes a cherry red, then a bright yellowish-orange, and finally a bluish-white. Once taken out of the fire the poker cools down, running through this spectrum of colours backwards until it is no longer hot enough to emit any visible light. Even then it still gives off an invisible glow of heat radiation. After a time this too stops as the poker continues to cool and finally becomes cold enough to touch.

It was the 23-year-old Isaac Newton who, in 1666, showed that a beam of white light was woven from different threads of coloured light and that passing it through a prism simply unpicked the seven separate strands: red, orange, yellow, green, blue, indigo, and violet.[8] Whether red and violet represented the limits of the light spectrum or just those of the human eye was answered in 1800. It was only then, with the advent of sufficiently sensitive and accurate mercury thermometers, that the astronomer William Herschel placed one in front of a spectrum of light and found that as he moved it across the bands of different colours from violet to red, the temperature rose. To his surprise it continued to rise when he accidentally left the thermometer up to an inch past the region of red light. Herschel had detected what was later called infrared radiation, light that was invisible to human eyes from the heat that it generated.[9] In 1801, using the fact that silver nitrate darkens when exposed to light, Johann Ritter discovered invisible light at the other end of the spectrum beyond the violet: ultraviolet radiation.

The fact that all heated objects emit light of the same colour at the same temperature was well known to potters long before 1859, the year that Gustav Kirchhoff, a 34-year-old German physicist at Heidelberg University, started his theoretical investigations into the nature of this correlation. To help simplify his analysis, Kirchhoff developed the concept of a perfect absorber and emitter of radiation called a blackbody. His choice of name was apt. A body that was a perfect absorber would reflect no radiation and therefore appear black. However, as a perfect emitter its appearance would be anything but black if its temperature was high enough for it to radiate at wavelengths from the visible part of the spectrum.

Kirchhoff envisaged his imaginary blackbody as a simple hollow container with a tiny hole in one of its walls. Since any radiation, visible or invisible light, entering the container does so through the hole, it is actually the hole that mimics a perfect absorber and acts like a blackbody. Once inside, the radiation is reflected back and forth between the walls of the cavity until it is completely absorbed. Imagining the outside of his blackbody to be insulated, Kirchhoff knew that if heated, only the interior surface of the walls would emit radiation that filled the cavity.

At first the walls, just like a hot iron poker, glow a deep cherry-red even though they still radiate predominantly in the infrared. Then, as the temperature climbs ever higher, the walls glow a bluish-white as they radiate at wavelengths from across the spectrum from the far infrared to the ultra-violet. The hole acts as a perfect emitter since any radiation that escapes through it will be a sample of *all* the wavelengths present inside the cavity at that temperature.

Kirchhoff proved mathematically what potters had long observed in their kilns. Kirchhoff's law said that the range and intensity of the radiation inside the cavity did not depend on the material that a real blackbody could be made of, or on its shape and size, but only on its temperature. Kirchhoff had ingeniously reduced the problem of the hot iron poker: what was the exact relationship between the range and intensity of the colours it emitted at a certain temperature to how much energy is radiated by a blackbody at that temperature? The task that Kirchhoff set himself and his colleagues became known as the blackbody problem: measure the spectral energy distribution of blackbody radiation, the amount of energy at each wavelength from the infrared to the ultraviolet, at a given temperature and derive a formula to reproduce the distribution at any temperature.

Unable to go further theoretically without experiments with a real blackbody to guide him, Kirchhoff nevertheless pointed physicists in the right direction. He told them that the distribution being independent of the material from which a blackbody was made meant that the formula should contain only two variables: the temperature of the blackbody and the wavelength of the emitted radiation. Since light was thought to be a wave, any particular colour and hue was distinguished from every other by its defining characteristic: its wavelength, the distance between two successive peaks or troughs of the wave. Inversely proportional to the wavelength is the frequency of the wave – the number of peaks, or troughs, that pass a fixed point in one second. The longer the wavelength, the lower the frequency and vice versa. But there was also a different but equivalent way of measuring the frequency of a wave: the number of times it jiggled up and down, 'waved', per second.[10]

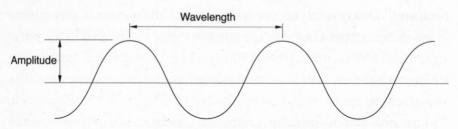

Figure 1: The characteristics of a wave

The technical obstacles in constructing a real blackbody and the precision instruments needed to detect and measure the radiation ensured that no significant progress was made for almost 40 years. It was in the 1880s, when German companies tried to develop more efficient light bulbs and lamps than their American and British rivals, that measuring the blackbody spectrum and finding Kirchhoff's fabled equation became a priority.

The incandescent light bulb was the latest in a series of inventions, including the arc lamp, dynamo, electric motor, and telegraphy, fuelling the rapid expansion of the electrical industry. With each innovation the need for a globally agreed set of units and standards of electrical measurement became increasingly urgent.

Two hundred and fifty delegates from 22 countries gathered in Paris, in 1881, for the first International Conference for the Determination of Electrical Units. Although the volt, amp and other units were defined and named, no agreement was reached on a standard for luminosity and it began to hamper the development of the most energy-efficient means of producing artificial light. As a perfect emitter at any given temperature, a blackbody emits the maximum amount of heat, infrared radiation. The blackbody spectrum would serve as a benchmark in calibrating and producing a bulb that emitted as much light as possible while keeping the heat it generated to a minimum.

'In the competition between nations, presently waged so actively, the country that first sets foot on new paths and first develops them into established branches of industry has a decisive upper hand', wrote the industrialist and inventor of the electrical dynamo, Werner von Siemens.[11] Determined to be first, in 1887 the German government founded the Physikalisch-Technische Reichsanstalt (PTR), the Imperial Institute of Physics and Technology. Located on the outskirts of Berlin in Charlottenburg, on land donated by Siemens, the PTR was conceived as an institute fit for an empire determined to challenge Britain and America. The construction of the entire complex lasted more than a decade, as the PTR became the best-equipped and most expensive research facility in the world. Its mission was to give Germany the edge in the appliance of science by developing standards and testing new products. Among its list of priorities was to devise an internationally recognised unit of luminosity. The need to make a better light bulb was the driving force behind the PTR blackbody research programme in the 1890s. It would lead to the accidental discovery of the quantum as Planck turned out to be the right man, in the right place, at the right time.

Max Karl Ernst Ludwig Planck was born in Kiel, then a part of Danish Holstein, on 23 April 1858 into a family devoted to the service of Church and State. Excellence in scholarship was almost his birthright. Both his paternal great-grandfather and grandfather had been distinguished theologians, while his father became professor of constitutional law at Munich University. Venerating the laws of God and Man, these duty-bound men of probity were also steadfast and patriotic. Max was to be no exception.

Planck attended the most renowned secondary school in Munich, the Maximilian Gymnasium. Always near the top of his class, but never first, he excelled through hard work and self-discipline. These were just the qualities demanded by an educational system with a curriculum founded on the retention of enormous quantities of factual knowledge through rote learning. A school report noted that 'despite all his childishness' Planck at ten already possessed 'a very clear, logical mind' and promised 'to be something right'.[12] By the time he was sixteen it was not Munich's famous taverns but its opera houses and concert halls that attracted the young Planck. A talented pianist, he toyed with the idea of pursuing a career as a professional musician. Unsure, he sought advice and was bluntly told: 'If you have to *ask*, you'd better study something else!'[13]

In October 1874, aged sixteen, Planck enrolled at Munich University and opted to study physics because of a burgeoning desire to understand the workings of nature. In contrast to the near-militaristic regime of the Gymnasiums, German universities allowed their students almost total freedom. With hardly any academic supervision and no fixed requirements, it was a system that enabled students to move from one university to another, taking courses as they pleased. Sooner or later those wishing to pursue an academic career took the courses by the pre-eminent professors at the most prestigious universities. After three years at Munich, where he was told 'it is hardly worth entering physics anymore' because there was nothing important left to discover, Planck moved to the leading university in the German-speaking world, Berlin.[14]

With the creation of a unified Germany in the wake of the Prussian-led victory over France in the war of 1870–71, Berlin became the capital of a mighty new European nation. Situated at the confluence of the Havel and the Spree rivers, French war reparations allowed its rapid redevelopment as it sought to make itself the equal of London and Paris. A population of 865,000 in 1871 swelled to nearly 2 million by 1900, making Berlin the third-largest city in Europe.[15] Among the new arrivals were Jews fleeing persecution in Eastern Europe, especially the pogroms in Tsarist Russia. Inevitably the cost of housing and living soared, leaving many homeless and destitute. Manufacturers of cardboard boxes advertised 'good and cheap boxes for habitation' as shanty towns sprung up in parts of the city.[16]

Despite the bleak reality that many found on arriving in Berlin, Germany was entering a period of unprecedented industrial growth, technological progress, and economic prosperity. Driven largely by the abolition of internal tariffs after unification and French war compensation, by the outbreak of the First World War Germany's industrial output and economic power would be second only to the United States. By then it was producing over two-thirds of continental Europe's steel, half its coal, and was generating more electricity than Britain, France and Italy combined. Even the recession and anxiety that affected Europe after the stock market crash of 1873 only slowed the pace of German development for a few years.

With unification came the desire to ensure that Berlin, the epitome of the new Reich, had a university second to none. Germany's most renowned physicist, Herman von Helmholtz, was enticed from Heidelberg. A trained surgeon, Helmholtz was also a celebrated physiologist who had made fundamental contributions to understanding the workings of the human eye after his invention of the ophthalmoscope. The 50-year-old polymath knew his worth. Apart from a salary several times the norm, Helmholtz demanded a magnificent new physics institute. It was still being built in 1877 when Planck arrived in Berlin and began attending lectures in the university's main building, a former palace on Unter den Linden opposite the Opera House.

As a teacher, Helmholtz was a severe disappointment. 'It was obvious,' Planck said later, 'that Helmholtz never prepared his lectures properly.'[17] Gustav Kirchhoff, who had also transferred from Heidelberg to become the professor of theoretical physics, was so well prepared that he delivered his lectures 'like a memorized text, dry and monotonous'.[18] Expecting to be inspired, Planck admitted 'that the lectures of these men netted me no perceptible gain'.[19] Seeking to quench his 'thirst for advanced scientific knowledge', he stumbled across the work of Rudolf Clausius, a 56-year-old German physicist at Bonn University.[20]

In stark contrast to the lacklustre teaching of his two esteemed professors, Planck was immediately enthralled by Clausius' 'lucid style and enlightening clarity of reasoning'.[21] His enthusiasm for physics returned as he read Clausius' papers on thermodynamics. Dealing with heat and its relationship to different forms of energy, the fundamentals of thermodynamics were at the time encapsulated in just two laws.[22] The first was a rigorous formulation of the fact that energy, in whatever guise, possessed the special property of being conserved. Energy could neither be created nor destroyed but only converted from one form to another. An apple hanging from a tree possesses potential energy by virtue of its position in the earth's gravitational field, its height above the ground. When it falls, the apple's potential energy is converted into kinetic energy, the energy of motion.

Planck was a schoolboy when he first encountered the law of the conservation of energy. It struck him 'like a revelation' he said later, because it possessed 'absolute, universal validity, independently from all human agency'.[23] It was the moment he caught a glimpse of the eternal, and from then on he considered the search for absolute or fundamental laws of nature 'as the most sublime scientific pursuit in life'.[24] Now Planck was just as spellbound reading Clausius' formulation of the second law of thermodynamics: 'Heat will not pass *spontaneously* from a colder to a hotter body.'[25] The later invention of the refrigerator illustrated what Clausius meant by 'spontaneously'. A refrigerator needed to be plugged into an external supply of energy, in this case electrical, so that heat could be made to flow from a colder to a hotter body.

Planck understood that Clausius was not simply stating the obvious, but something of deep significance. Heat, the transfer of energy from A to B due to a temperature difference, explained such everyday occurrences as a hot cup of coffee getting cold and an ice cube in a glass of water melting. But left undisturbed, the reverse never happened. Why not? The law of conservation of energy did not forbid a cup of coffee from getting hotter and the surrounding air colder, or the glass of water becoming warmer and the ice cooler. It did not outlaw heat flowing from a cold to a hot body spontaneously. Yet something was preventing this from happening. Clausius discovered that something and called it entropy. It lay at the heart of why some processes occur in nature and others do not.

When a hot cup of coffee cools down, the surrounding air gets warmer as energy is dissipated and irretrievably lost, ensuring that the reverse cannot happen. If the conservation of energy was nature's way of balancing the books in any *possible* physical transaction, then nature also demanded a price for every transaction that *actually* occurred. According to Clausius, entropy was the price for whether something happened or not. In any isolated system only those processes, transactions, in which entropy either stayed the same or increased were allowed. Any that led to a decrease of entropy were strictly forbidden.

Clausius defined entropy as the amount of heat in or out of a body or a system divided by the temperature at which it takes place. If a hot body at 500 degrees loses 1000 units of energy to a colder body at 250 degrees, then its entropy has decreased by $-1000/500 = -2$. The colder body at 250 degrees has gained 1000 units of energy, $+1000/250$, and its entropy has increased by 4. The overall entropy of the system, the hot and cold bodies combined, has increased by 2 units of energy per degree. All real, actual processes are irreversible because they result in an increase in entropy. It is nature's way of stopping heat from passing spontaneously, of its own accord, from something cold to something hot. Only ideal processes in which entropy remains unchanged can be reversed. They, however, never occur in practice, only in the mind of the physicist. The entropy of the universe tends towards a maximum.

Alongside energy, Planck believed that entropy was 'the most important property of physical systems'.[26] Returning to Munich University after his year-long sojourn in Berlin, he devoted his doctoral thesis to an exploration of the concept of irreversibility. It would be his calling card. To his dismay, he 'found no interest, let alone approval, even among the very physicists who were closely concerned with the topic'.[27] Helmholtz did not read it; Kirchhoff did, but disagreed with it. Clausius, who had such a profound influence on him, did not even answer his letter. 'The effect of my dissertation on the physicists of those days was nil', Planck recalled with some bitterness even 70 years later. But driven by 'an inner compulsion', he was undeterred.[28] Thermodynamics, particularly the second law, became the focus of Planck's research as he began his academic career.[29]

German universities were state institutions. Extraordinary (assistant) and ordinary (full) professors were civil servants appointed and employed by the ministry of education. In 1880 Planck became a *privatdozent*, an unpaid lecturer, at Munich University. Employed neither by the state nor the university, he had simply gained the right to teach in exchange for fees paid by students attending his courses. Five years passed as he waited in vain for an appointment as an extraordinary professor. As a theorist uninterested in conducting experiments, Planck's chances for promotion were slim, as theoretical physics was not yet a firmly established distinct discipline. Even in 1900 there were only sixteen professors of theoretical physics in Germany.

If his career was to progress, Planck knew that he had 'to win, somehow, a reputation in the field of science'.[30] His chance came when Göttingen University announced that the subject for its prestigious essay competition was 'The Nature of Energy'. As he worked on his paper, in May 1885, 'a message of deliverance' arrived.[31] Planck, aged 27, was offered an extraordinary professorship at the University of Kiel. He suspected it was his father's friendship with Kiel's head of physics that had led to the offer. Planck knew there were others, more established than he, who would have expected advancement. Nevertheless, he accepted and finished his entry for the Göttingen competition shortly after arriving in the city of his birth.

Even though only three papers were submitted in search of the prize, an astonishing two years passed before it was announced that there would be no winner. Planck was awarded second place and denied the prize by the judges because of his support for Helmholtz in a scientific dispute with a member of the Göttingen faculty. The behaviour of the judges drew the attention of Helmholtz to Planck and his work. After a little more than three years at Kiel, in November 1888, Planck received an unexpected honour. He had not been first, or even second choice. But after others had turned it down, Planck, with Helmholtz's backing, was asked to succeed Gustav Kirchhoff at Berlin University as professor of theoretical physics.

In the spring of 1889, the capital was not the city Planck had left eleven years earlier. The stench that always shocked visitors had disappeared as a new sewer system replaced the old open drains, and at night the main streets were lit by modern electric lamps. Helmholtz was no longer head of the university's physics institute but running the PTR, the majestic new research facility three miles away. August Kundt, his successor, had played no part in Planck's appointment, but welcomed him as 'an excellent acquisition' and 'a splendid man'.[32]

In 1894 Helmholtz, aged 73, and Kundt, only 55, both died within months of each other. Planck, only two years after finally being promoted to the rank of ordinary professor, found himself as the senior physicist at Germany's foremost university at just 36. He had no choice but to bear the weight of added responsibilities, including that of adviser on theoretical physics for *Annalen der Physik*. It was a position of immense influence that gave him the right of veto on all theoretical papers submitted to the premier German physics journal. Feeling the pressure of his newly elevated position and a deep sense of loss at the deaths of his two colleagues, Planck sought solace in his work.

As a leading member of Berlin's close-knit community of physicists, he was well aware of the ongoing, industry-driven blackbody research programme of the PTR. Although thermodynamics was central to a theoretical analysis of the light and heat radiated by a blackbody, the lack of reliable experimental data had stopped Planck from trying to derive the exact form of Kirchhoff's unknown equation. Then a breakthrough by an old friend at PTR meant that he could no longer avoid the blackbody problem.

In February 1893, 29-year-old Wilhelm Wien discovered a simple mathematical relationship that described the effect of a change in temperature on the distribution of blackbody radiation.[33] Wien found that as the temperature of a blackbody increases, the wavelength at which it emits radiation with the greatest intensity becomes ever shorter.[34] It was already known that the rise in temperature would result in an increase in the total amount of energy radiated, but Wien's 'displacement law' revealed something very precise: the wavelength at which the maximum amount of radiation is emitted multiplied by the temperature of a blackbody is always a constant. If the temperature is doubled, then the 'peak' wavelength will be half the previous length.

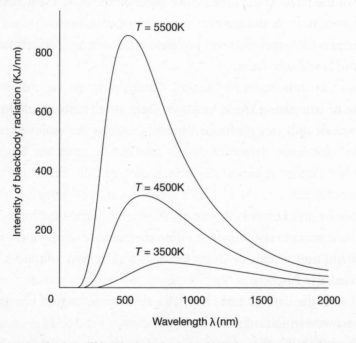

Figure 2: Distribution of blackbody radiation which shows Wien's displacement law

Wien's discovery meant that once the numerical constant was calculated by measuring the peak wavelength – the wavelength that radiates most strongly, at a certain temperature – then the peak wavelength could be calculated for any temperature.[35] It also explained the changing colours of a hot iron poker. Starting at low temperatures, the poker emits predominantly long-wavelength radiation from the infrared part of the spectrum. As the temperature increases, more energy is radiated in each region and the peak wavelength decreases. It is 'displaced' towards the shorter wavelengths. Consequently the colour of the emitted light changes from red to orange, then yellow and finally a bluish-white as the quantity of radiation from the ultraviolet end of the spectrum increases.

Wien had quickly established himself as a member of that endangered breed of physicist, one who was both an accomplished theorist and a skilled experimenter. He found the displacement law in his spare time and was forced to publish it as a 'private communication' without the imprimatur of the PTR. At the time he was working as an assistant in the PTR's optics laboratory under the leadership of Otto Lummer. Wien's day job was the practical work that was a prerequisite for an experimental investigation of blackbody radiation.

Their first task was to construct a better photometer, an instrument capable of comparing the intensity of light – the amount of energy in a given wavelength range – from different sources such as gas lamps and electric bulbs. It was the autumn of 1895 before Lummer and Wien devised a new and improved hollow blackbody capable of being heated to a uniform temperature.

While he and Lummer developed their new blackbody during the day, Wien continued to spend his evenings searching for Kirchhoff's equation for distribution of blackbody radiation. In 1896, Wien found a formula that Friedrich Paschen, at the University of Hanover, quickly confirmed agreed with the data he had collected on the allocation of energy among the short wavelengths of blackbody radiation.

In June that year, the very month the 'distribution law' appeared in print, Wien left the PTR for an extraordinary professorship at the Technische Hochschule in Aachen. He would win the Nobel Prize for physics in 1911 for his work on blackbody radiation, but left Lummer to put his distribution law through a rigorous test. To do so required measurements over a greater range and at higher temperatures than ever before. Working with

Ferdinand Kurlbaum and then Ernst Pringsheim, it took Lummer two long years of refinements and modifications but in 1898 he had a state-of-the-art electrically heated blackbody. Capable of reaching temperatures as high as 1500°C, it was the culmination of more than a decade of painstaking work at the PTR.

Plotting the intensity of radiation along the vertical axis of a graph against the wavelength of the radiation along the horizontal axis, Lummer and Pringsheim found that the intensity rose as the wavelength of radiation increased until it peaked and then began to drop. The spectral energy distribution of blackbody radiation was almost a bell-shaped curve, resembling a shark's dorsal fin. The higher the temperature, the more pronounced the shape as the intensity of radiation emitted increased. Taking readings and plotting curves with the blackbody heated to different temperatures showed that the peak wavelength that radiated with maximum intensity was displaced towards the ultraviolet end of the spectrum with increasing temperature.

Lummer and Pringsheim reported their results at a meeting of the German Physical Society held in Berlin on 3 February 1899.[36] Lummer told the assembled physicists, among them Planck, that their findings confirmed Wien's displacement law. However, the situation regarding the distribution law was unclear. Although the data was in broad agreement with Wien's theoretical predictions, there were some discrepancies in the infrared region of the spectrum.[37] In all likelihood these were due to experimental errors, but it was an issue, they argued, that could be settled only once 'other experiments spread over a greater range of wavelengths and over a greater interval of temperature can be arranged'.[38]

Within three months Friedrich Paschen announced that his measurements, though conducted at a lower temperature than those of Lummer and Pringsheim, were in complete harmony with the predictions of Wien's distribution law. Planck breathed a sigh of relief and read out Paschen's paper at a session of the Prussian Academy of Sciences. Such a law appealed deeply to him. For Planck the theoretical quest for the spectral energy distribution of blackbody radiation was nothing less than the search for the absolute, and 'since I had always regarded the search for the absolute as the loftiest goal of all scientific activity, I eagerly set to work'.[39]

Soon after Wien published his distribution law, in 1896, Planck set about trying to place the law on rock-solid foundations by deriving it from first principles. Three years later, in May 1899, he thought he had succeeded by using the power and authority of the second law of thermodynamics. Others agreed and started calling Wien's law by a new name, Wien-Planck, despite the claims and counter-claims of the experimentalists. Planck remained confident enough to assert that 'the limits of validity of this law, in case there are any at all, coincide with those of the second fundamental law of the theory of heat'.[40] He advocated further testing of the distribution law as a matter of urgency, since for him it would be a simultaneous examination of the second law. He got his wish.

At the beginning of November 1899, after spending nine months extending the range of their measurements as they eliminated possible sources of experimental error, Lummer and Pringsheim reported that they had found 'discrepancies of a systematic nature between theory and experiment'.[41] Although in perfect agreement for short wavelengths, they discovered that Wien's law consistently overestimated the intensity of radiation at long wavelengths. However, within weeks Paschen contradicted Lummer and Pringsheim. He presented another set of new data and claimed that the distribution law 'appears to be a rigorously valid law of nature'.[42]

With most of the leading experts living and working in Berlin, the meetings of the German Physical Society held in the capital became the main forum for discussions concerning blackbody radiation and the status of Wien's law. It was the subject that again dominated the proceedings of the society at its fortnightly meeting on 2 February 1900 when Lummer and Pringsheim disclosed their latest measurements. They had found systematic discrepancies between their measurements and the predictions of Wien's law in the infrared region that could not be the result of experimental error.

This breakdown of Wien's law led to a scramble to find a replacement. But these makeshift alternatives proved unsatisfactory, prompting calls for further testing at even longer wavelengths to unequivocally establish the extent of any failure of Wien's law. It did, after all, agree with the available data covering the shorter wavelengths, and all other experiments bar those of Lummer-Pringsheim had found in its favour.

As Planck was only too well aware, any theory is at the mercy of hard experimental facts, but he strongly believed that 'a conflict between observation and theory can only be confirmed as valid beyond all doubt if the figures of various observers substantially agree with each other'.[43] Nevertheless, the disagreement between the experimentalists forced him to reconsider the soundness of his ideas. In late September 1900, as he continued to review his derivation, the failure of Wien's law in the deep infrared was confirmed.

The question was finally settled by Heinrich Rubens, a close friend of Planck's, and Ferdinand Kurlbaum. Based at the Technische Hochschule on Berlinerstrasse, where at the age of 35 he had recently been promoted to ordinary professor, Rubens spent most of his time as a guest worker at the nearby PTR. It was there, with Kurlbaum, that he built a blackbody that allowed measurements of the uncharted territory deep within the infrared region of the spectrum. During the summer they tested Wien's law between wavelengths of 0.03mm and 0.06mm at temperatures ranging from 200 to 1500°C. At these longer wavelengths, they found the difference between theory and observation was so marked that it could be evidence of only one thing, the breakdown of Wien's law.

Rubens and Kurlbaum wanted to announce their results in a paper to the German Physical Society. The next meeting was on Friday, 5 October. With little time to write a paper, they decided to wait until the following meeting two weeks later. In the meantime, Rubens knew that Planck would be eager to hear the latest results.

It was among the elegant villas of bankers, lawyers, and other professors in the affluent suburb of Grunewald in west Berlin that Planck lived for 50 years in a large house with an enormous garden. On Sunday, 7 October,

Rubens and his wife came for lunch. Inevitably the talk between the two friends soon turned to physics and the blackbody problem. Rubens explained that his latest measurements left no room for doubt: Wien's law failed at long wavelengths and high temperatures. Those measurements, Planck learnt, revealed that at such wavelengths the intensity of blackbody radiation was proportional to the temperature.

That evening Planck decided to have a go at constructing the formula that would reproduce the energy spectrum of blackbody radiation. He now had three crucial pieces of information to help him. First, Wien's law accounted for the intensity of radiation at short wavelengths. Second, it failed in the infrared where Rubens and Kurlbaum had found that intensity was proportional to the temperature. Third, Wien's displacement law was correct. Planck had to find a way to assemble these three pieces of the blackbody jigsaw together to build the formula. His years of hard-won experience were quickly put into practice as he set about manipulating the various mathematical symbols of the equations at his disposal.

After a few unsuccessful attempts, through a combination of inspired scientific guesswork and intuition, Planck had a formula. It looked promising. But was it Kirchhoff's long-sought-after equation? Was it valid at any given temperature for the entire spectrum? Planck hurriedly penned a note to Rubens and went out in the middle of the night to post it. After a couple of days, Rubens arrived at Planck's home with the answer. He had checked Planck's formula against the data and found an almost perfect match.

On Friday, 19 October at the meeting of the German Physical Society, with Rubens and Planck sitting among the audience, it was Ferdinand Kurlbaum who made the formal announcement that Wien's law was valid only at short wavelengths and failed at the longer wavelengths of the infrared. After Kurlbaum sat down, Planck rose to deliver a short 'comment' billed as 'An Improvement of Wien's Equation for the Spectrum'. He began by admitting that he had believed 'Wien's law must necessarily be true', and had said so at a previous meeting.[44] As he continued, it quickly became clear that Planck was not simply proposing 'an improvement', some minor tinkering with Wien's law, but a completely new law of his own.

After speaking for less than ten minutes, Planck wrote his equation for the blackbody spectrum on the blackboard. Turning around to look at the familiar faces of his colleagues, he told them that this equation 'as far as I can see at the moment, fits the observational data, published up to now'.[45] As he sat down, Planck received polite nods of approval. The muted response was understandable. After all, what Planck had just proposed was another ad hoc formula manufactured to explain the experimental results. There were others who had already put forward equations of their own in the hope of filling the void, should the suspected failure of Wien's law at long wavelengths be confirmed.

The next day Rubens visited Planck to reassure him. 'He came to tell me that after the conclusion of the meeting he had that very night checked my formula against the results of his measurements,' Planck remembered, 'and found satisfactory concordance at every point.'[46] Less than a week later, Rubens and Kurlbaum announced that they had compared their measurements with the predictions of five different formulae and found Planck's to be much more accurate than any of the others. Paschen too confirmed that Planck's equation matched his data. Yet despite this rapid corroboration by the experimentalists of the superiority of his formula, Planck was troubled.

He had his formula, but what did it mean? What was the underlying physics? Without an answer, Planck knew that it would, at best, be just an 'improvement' on Wien's law and have 'merely the standing of a law disclosed by a lucky intuition' that possessed no more 'than a formal significance'.[47] 'For this reason, on the very first day when I formulated this law,' Planck said later, 'I began to devote myself to the task of investing it with true physical meaning.'[48] He could achieve this only by deriving his equation step by step using the principles of physics. Planck knew his destination, but he had to find a way of getting there. He possessed a priceless guide, the equation itself. But what price was he prepared to pay for such a journey?

The next six weeks were, Planck recalled, 'the most strenuous work of my life', after which 'the darkness lifted and an unexpected vista began to appear'.[49] On 13 November he wrote to Wien: 'My new formula is well satisfied; I now have also obtained a theory for it, which I shall present in four weeks at the Physical Society here [in Berlin].'[50] Planck said nothing to Wien either of the intense intellectual struggle that had led to his theory or the theory itself. He had strived long and hard during those weeks to reconcile his equation with the two grand theories of nineteenth-century physics: thermodynamics and electromagnetism. He failed.

'A theoretical interpretation therefore had to be found at any cost,' he accepted, 'no matter how high.'[51] He 'was ready to sacrifice every one of my previous convictions about physical laws'.[52] Planck no longer cared what it cost him, as long as he could 'bring about a positive result'.[53] For such an emotionally restrained man, who only truly expressed himself freely at the piano, this was highly charged language. Pushed to the limit in the struggle to understand his new formula, Planck was forced into 'an act of desperation' that led to the discovery of the quantum.[54]

As the walls of a blackbody are heated they emit infrared, visible, and ultraviolet radiation into the heart of the cavity. In his search for a theoretically consistent derivation of his law, Planck had to come up with a physical model that reproduced the spectral energy distribution of blackbody radiation. He had already been toying with an idea. It did not matter if the model failed to capture what was really going on; all Planck needed was a way of getting the right mix of frequencies, and therefore wavelengths, of the radiation present inside the cavity. He used the fact that this distribution depends only on the temperature of the blackbody and not on the material from which it is made to conjure up the simplest model he could.

'Despite the great success that the atomic theory has so far enjoyed,' Planck wrote in 1882, 'ultimately it will have to be abandoned in favour of the assumption of continuous matter.'[55] Eighteen years later, in the absence of indisputable proof of their existence, he still did not believe in atoms. Planck knew from the theory of electromagnetism that an electric charge oscillating at a certain frequency emits and absorbs radiation only of that frequency. He therefore chose to represent the walls of the blackbody as an enormous array of oscillators. Although each oscillator emits only a single frequency, collectively they emit the entire range of frequencies found within the blackbody.

A pendulum is an oscillator and its frequency is the number of swings per second, a single oscillation being one complete to and fro swing that returns the pendulum to its starting point. Another oscillator is a weight hanging from a spring. Its frequency is the number of times per second the weight bounces up and down after being pulled from its stationary position and released. The physics of such oscillations had long been understood and given the name 'simple harmonic motion' by the time Planck used oscillators, as he called them, in his theoretical model.

Planck envisaged his collection of oscillators as massless springs of varying stiffness, so as to reproduce the different frequencies, each with an electric charge attached. Heating the walls of the blackbody provided the energy needed to set the oscillators in motion. Whether an oscillator was active or not would depend only upon the temperature. If it were, then it would emit radiation into, and absorb radiation from, the cavity. In time, if the temperature is held constant, this dynamic give and take of radiation energy between the oscillators and the radiation in the cavity comes into balance and a state of thermal equilibrium is achieved.

Since the spectral energy distribution of blackbody radiation represents how the total energy is shared among the different frequencies, Planck assumed that it was the number of oscillators at each given frequency that determined the allocation. After setting up his hypothetical model, he had to devise a way to share out the available energy among the oscillators. In the weeks following its announcement, Planck discovered the hard way that he could not derive his formula using physics that he had long accepted as dogma. In desperation he turned to the ideas of an Austrian physicist, Ludwig Boltzmann, who was the foremost advocate of the atom. On the road to his blackbody formula, Planck became a convert as he accepted that atoms were more than just a convenient fiction, after years of being openly 'hostile to the atomic theory'.[56]

The son of a tax collector, Ludwig Boltzmann was short and stout with an impressive late nineteenth-century beard. Born in Vienna on 20 February 1844, he was, for a while, taught the piano by the composer Anton Bruckner. A better physicist than a pianist, Boltzmann obtained his doctorate from the University of Vienna in 1866. He quickly made his reputation with fundamental contributions to the kinetic theory of gases, so called because its proponents believed that gases were made up of atoms or molecules in a state of continual motion. Later, in 1884, Boltzmann provided the theoretical justification for the discovery by Josef Stefan, his former mentor, that the total energy radiated by a blackbody is proportional to the temperature raised to the fourth power, T^4 or $T{\times}T{\times}T{\times}T$. It meant that doubling the temperature of a blackbody increased the energy it radiated by a factor of sixteen.

Boltzmann was a renowned teacher and, although a theorist, a very capable experimentalist despite being severely shortsighted. Whenever a vacancy arose at one of Europe's leading universities his name was usually on the list of potential candidates. It was only after he turned down the professorship at Berlin University left vacant by the death of Gustav Kirchhoff that a downgraded version was offered to Planck. By 1900 a much-travelled Boltzmann was at Leipzig University and universally acknowledged as one the great theoreticians. Yet there were many, like Planck, who found his approach to thermodynamics unacceptable.

Boltzmann believed that properties of gases, such as pressure, were the macroscopic manifestations of microscopic phenomena regulated by the laws of mechanics and probability. For those whose believed in atoms, the classical physics of Newton governed the movement of each gas molecule, but using Newtonian laws of motion to determine that of each of the countless molecules of a gas was for all practical purposes impossible. It was the 28-year-old Scottish physicist James Clerk Maxwell who, in 1860, captured the motion of gas molecules without measuring the velocity of a single one. Using statistics and probability, Maxwell worked out the most likely distribution of velocities as the gas molecules underwent incessant collisions with each other and the walls of a container. The introduction of statistics and probability was bold and innovative; it allowed Maxwell to explain many of the observed properties of gases. Thirteen years younger, Boltzmann followed in Maxwell's footsteps to help shore up the kinetic theory of gases. In the 1870s he went one step further and developed a statistical interpretation of the second law of thermodynamics by linking entropy with disorder.

According to what became known as Boltzmann's principle, entropy is a measure of the probability of finding a system in a particular state. A well-shuffled pack of playing cards, for example, is a disordered system with high entropy. However, a brand-new deck with cards arranged according to suit and from ace to king is a highly ordered system with low entropy. For Boltzmann the second law of thermodynamics concerns the evolution of a system with a low probability, and therefore low entropy, into a state of higher probability and high entropy. The second law is not an absolute law. It is possible for a system to go from a disordered state to a more ordered one, just as a shuffled pack of cards may, if shuffled again, become ordered. However, the odds against that happening are so astronomical that it would require many times the age of the universe to pass for it to occur.

Planck believed that the second law of thermodynamics was absolute – entropy always increases. In Boltzmann's statistical interpretation, entropy *nearly* always increases. There was a world of difference between these two views as far as Planck was concerned. For him to turn to Boltzmann was a renunciation of everything that he held dear as a physicist, but he had no choice in his quest to derive his blackbody formula. 'Until then I had paid no attention to the relationship between entropy and probability, in which I had little interest since every probability law permits exceptions; and at that time I assumed that the second law of thermodynamics was valid without exceptions.'[57]

A state of maximum entropy, maximum disorder, is the most probable state for a system. For a blackbody that state is thermal equilibrium – just the situation that Planck faced as he tried to find the most probable distribution of energy among his oscillators. If there are 1000 oscillators in total and ten have a frequency v, it is these oscillators that determine the intensity of radiation emitted at that frequency. While the frequency of any one of Planck's electric oscillators is fixed, the amount of energy it emits and absorbs depends solely upon its amplitude, the size of its oscillation. A pendulum completing five swings in five seconds has a frequency of one oscillation per second. However, if it swings through a wide arc the pendulum has more energy than if it traces out a smaller one. The frequency remains unchanged because the length of the pendulum fixes it, but the extra energy allows it to travel faster through a wide arc. The pendulum therefore completes the same number of oscillations in the same time as an identical pendulum swinging through a narrower arc.

Applying Boltzmann's techniques, Planck discovered that he could derive his formula for the distribution of blackbody radiation only if the oscillators absorbed and emitted packets of energy that were proportional to their frequency of oscillation. It was the 'most essential point of the whole calculation', said Planck, to consider the energy at each frequency as being composed of a number of equal, indivisible 'energy elements' that he later called quanta.[58]

Guided by his formula, Planck had been forced into slicing up energy (E) into hv-sized chunks, where v is the frequency of the oscillator and h is a constant. $E=hv$ would become one of the most famous equations in the whole of science. If, for example, the frequency was 20 and h was 2, then each quantum of energy would have a magnitude of $20\times2=40$. If the total energy available at this frequency were 3600, then there would be $3600/40=90$ quanta to be distributed among the ten oscillators of that frequency. Planck learnt from Boltzmann how to determine the most probable distribution of these quanta among the oscillators.

He found that his oscillators could only have energies: o, hv, $2hv$, $3hv$, $4hv$ ... all the way up to nhv, where n is a whole number. This corresponded to either absorbing or emitting a whole number of 'energy elements' or 'quanta' of size hv. It was like a bank cashier able to receive and dispense money only in denominations of £1, £2, £5, £10, £20 and £50.

Since Planck's oscillators cannot have any other energy, the amplitude of their oscillations is constrained. The strange implications of this are manifest if scaled up to the everyday world of a spring with a weight attached.

If the weight oscillates with an amplitude of 1cm, then it has an energy of 1 (ignoring the units of measuring energy). If the weight is pulled down to 2cm and allowed to oscillate, its frequency remains the same as before. However its energy, which is proportional to the square of the amplitude, is now 4. If the restriction on Planck's oscillators applied to the weight, then between 1cm and 2cm it can oscillate only with amplitudes of 1.42cm and 1.73cm, because they have energies of 2 and 3.[59] It cannot, for example, oscillate with an amplitude of 1.5cm because the associated energy would be 2.25. A quantum of energy is indivisible. An oscillator cannot receive a fraction of a quantum of energy; it must be all or nothing. This ran counter to the physics of the day. It placed no restrictions on the size of oscillation and therefore on how much energy an oscillator can emit or absorb in a single transaction – it could have any amount.

In his desperation Planck had discovered something so remarkable and unexpected that he failed to grasp its significance. It is not possible for his oscillators to absorb or emit energy continuously like water from a tap. Instead they can only gain and lose energy discontinuously, in small, indivisible units of $E=h\nu$, where ν is the frequency with which the oscillator vibrates that exactly matches the frequency of the radiation it can absorb or emit.

The reason why large-scale oscillators are not seen to behave like Planck's atomic-sized ones is because h is equal to 0.000000000000000000000000006626 erg seconds or 6.626 divided by one thousand trillion trillion. According to Planck's formula, there could be no smaller step than h in the increase or decrease of energy, but the infinitesimal size of h makes quantum effects invisible in the world of the everyday when it comes to pendulums, children's swings and vibrating weights.

Planck's oscillators forced him to slice and dice radiation energy so as to feed them the correct bite-sized chunks of hv. He did not believe that the energy of radiation was really chopped up into quanta. It was just the way his oscillators could receive and emit energy. The problem for Planck was that Boltzmann's procedure for slicing energy required that at the end the slices be made ever thinner until mathematically their thickness was zero and they vanished, with the whole being restored. To reunite a sliced-up quantity in such a fashion was a mathematical technique at the very heart of calculus. Unfortunately for Planck, if he did the same his formula vanished too. He was stuck with quanta, but was unconcerned. He had his formula; the rest could be sorted out later.

'Gentlemen!' said Planck as he faced the members of the German Physical Society seated in the room at Berlin University's Physics Institute. He could see Rubens, Lummer and Pringsheim among them as he began his lecture, 'Zur Theorie des Gesetzes der Energieverteilung im Normalspektrum', On the Theory of the Energy Distribution Law of the Normal Spectrum. It was just after 5pm on Friday, 14 December 1900. 'Several weeks ago I had the honour of directing your attention to a new equation that seemed suitable to me for expressing the law of the distribution of radiating energy over all areas of the normal spectrum.'[60] Planck now presented the physics behind that new equation as he derived it.

At the end of the meeting his colleagues roundly congratulated him. Just as Planck regarded the introduction of the quantum, a packet of energy, as a 'purely formal assumption' to which he 'really did not give much thought', so did everyone else that day. What was important to them was that Planck had succeeded in providing a physical justification for the formula he had presented in October. To be sure, his idea of chopping up energy into quanta for the oscillators was rather strange, but it would be ironed out in time. All believed that it was nothing more than the usual theorist's sleight of hand, a neat mathematical trick on the path to getting the right answer. It had no true physical significance. What continued to impress his colleagues was the accuracy of his new radiation law. Nobody really took much notice of the quantum of energy, including Planck himself.

Early one morning Planck left home with his seven-year-old son, Erwin. Father and son were headed to nearby Grunewald Forest. Walking there was a favourite pastime of Planck's and he enjoyed taking his son along. Erwin later recalled that as the pair walked and talked, his father told him: 'Today I have made a discovery as important as that of Newton.'[61] When he recounted the tale years later, Erwin could not remember exactly when the walk took place. It was probably some time before the December lecture. Was it possible that Planck understood the full implications of the quantum after all? Or was he just trying to convey to his young son something of the importance of his new radiation law? Neither. He was simply expressing his joy at discovering not one but two new fundamental constants: k, which he called Boltzmann's constant, and h, which he called the quantum of action but which physicists would call Planck's constant. They were fixed and eternal, two of nature's absolutes.[62]

Planck acknowledged his debt to Boltzmann. Having named k after the Austrian, a constant that he had discovered in his research leading up to the blackbody formula, Planck also nominated Boltzmann for the Nobel Prize in 1905 and 1906. By then it was too late. Boltzmann had long been plagued by ill health – asthma, migraines, poor eyesight and angina. Yet none of these were as debilitating as the bouts of severe manic depression he suffered. In September 1906, while on holiday in Duino near Trieste, he hanged himself. He was 62, and though some of his friends had long feared the worst, news of his death came as a terrible shock. Boltzmann had felt increasingly isolated and unappreciated. It was untrue. He was among the most widely honoured and admired physicists of the age. But continuing disputes over the existence of atoms had left him vulnerable during periods of despair to believing that his life's work was being undermined. Boltzmann had returned to Vienna University for the third and last time in 1902. Planck was asked to succeed him. Describing Boltzmann's work as 'one of the most beautiful triumphs of theoretical research', Planck was tempted by the Viennese offer but declined.[63]

h was the axe that chopped up energy into quanta, and Planck had been the first to wield it. But what he quantised was the way his imaginary oscillators could receive and emit energy. Planck did not quantise, chop into hv-sized chunks, energy itself. There is a difference between making a discovery and fully understanding it, especially in a time of transition. There was much that Planck did that was only implicit in his derivation, and not even clear to him. He never explicitly quantised individual oscillators, as he should have done, but only groups of them.

Part of the problem was that Planck thought he could get rid of the quantum. He only realised the far-reaching consequences of what he had done much later. His deep conservative instincts compelled him to try for the best part of a decade to incorporate the quantum into the existing framework of physics. He knew that some of his colleagues saw this as bordering on a tragedy. 'But I feel differently about it', Planck wrote.[64] 'I now know for a fact that the elementary quantum of action [h] played a far more significant part in physics that I had originally been inclined to suspect.'

Years after Planck's death in 1947, at the age of 89, his former student and colleague James Franck recalled watching his hopeless struggle 'to avoid quantum theory, [to see] whether he could not at least make the influence of quantum theory as little as it could possibly be'.[65] It was clear to Franck that Planck 'was a revolutionary against his own will' who 'finally came to the conclusion, "It doesn't help. We have to live with quantum theory. And believe me, it will expand."'[66] It was a fitting epitaph for a reluctant revolutionary.

Physicists did have to learn to 'live with' the quantum. The first to do so was not one of Planck's distinguished peers, but a young man living in Bern, Switzerland. He alone realised the radical nature of the quantum. He was not a professional physicist, but a junior civil servant whom Planck credited with the discovery that energy itself is quantised. His name was Albert Einstein.

Chapter 2

THE PATENT SLAVE

Bern, Switzerland, Friday, 17 March 1905. It was nearly eight o'clock in the morning as the young man dressed in the unusual plaid suit hurried to work clutching an envelope. To a passer-by, Albert Einstein appeared to have forgotten that he was wearing a pair of worn-out green slippers with embroidered flowers.[1] At the same time six days a week, he left his wife and baby son, Hans Albert, behind in their small two-room apartment in the middle of Bern's picturesque Old Town quarter, and walked to the rather grand sandstone building ten minutes away. With its famous clock tower, the Zytloggeturm, and arcades lining both sides of the cobbled street, Kramgasse was one of the most beautiful streets in the Swiss capital. Lost in thought, Einstein hardly noticed his surroundings as he made his way to the administrative headquarters of the Federal Post and Telephone Service. Once inside he headed straight for the stairs and the third floor that housed the Federal Office of Intellectual Property, better known as the Swiss Patent Office. Here he and the dozen other technical experts, men in more sober dark suits, laboured at their desks for eight hours a day sorting out the barely viable from the fatally flawed.

Three days earlier, Einstein had celebrated his 26th birthday. He had been a 'patent slave', as he called it, for nearly three years.[2] For him the job brought to an end 'the annoying business of starving'.[3] The work itself he enjoyed for its variety, the 'many-sided thinking' it encouraged and the relaxed atmosphere of the office. It was an environment Einstein later referred to as his 'worldly monastery'. Although the post of technical expert, third class, was a humble one, it was well-paid and allowed him time enough to pursue his own research. Despite the watchful eye of his boss, the formidable Herr Haller, Einstein spent so much time between examining patents secretly doing his own calculations that his desk had become his 'office for theoretical physics'.[4]

'It was as if the ground had been pulled out from under one, with no firm foundation to be seen anywhere, upon which one could have built', was how Einstein recalled feeling after reading Planck's solution of the blackbody problem soon after it was published.[5] What he sent in the envelope to the editor of Annalen der Physik, the world's leading physics journal, on 17 March 1905 was even more radical than Planck's original introduction of the quantum. Einstein knew that his proposal of a quantum theory of light was nothing short of heresy.

Two months later, in the middle of May, Einstein wrote to his friend Conrad Habicht promising to send four papers he hoped to see published before the year's end. The first was the quantum paper. The second was his PhD dissertation in which he set out a new way to determine the sizes of atoms. The third offered an explanation of Brownian motion, the erratic dance of tiny particles, like grains of pollen, suspended in liquid. 'The fourth paper,' Einstein admitted, 'is only a rough draft at this point and is an electrodynamics of moving bodies which employs a modification of the theory of space and time.'[6] It is an extraordinary list. In the annals of science only one other scientist and one other year bears comparison with Einstein and his achievements in 1905: Isaac Newton in 1666, when the 23-year-old Englishman laid the foundations of calculus and the theory of gravity, and outlined his theory of light.

Einstein would become synonymous with the theory first sketched out in his fourth paper: relativity. Although it would change humanity's very understanding of the nature of space and time, it was the extension of Planck's quantum concept to light and radiation that he described as 'very revolutionary', not relativity.[7] Einstein regarded relativity as simply a 'modification' of ideas already developed and established by Newton and others, whereas his concept of light-quanta was something totally new, entirely his own, and represented the greatest break with the physics of the past. Even for an amateur physicist it was sacrilegious.

For more than half a century it had been universally accepted that light was a wave phenomenon. In 'On a Heuristic Point of View Concerning the Production and Transformation of Light', Einstein put forward the idea that light was not made up of waves, but particle-like quanta. In his resolution of the blackbody problem Planck had reluctantly introduced the idea that energy was absorbed or emitted as quanta, in discrete lumps. However, he, like everyone else, believed that electromagnetic radiation itself was a continuous wave phenomenon, whatever the mechanism of how it exchanged energy when it interacted with matter. Einstein's revolutionary 'point of view' was that light, indeed all electromagnetic radiation, was not wavelike at all but chopped up into little bits, light-quanta. For the next twenty years, virtually no one but he believed in his quantum of light.

From the beginning Einstein knew it would be an uphill struggle. He signalled as much by including 'On a Heuristic Point of View' in the title of his paper. 'Heuristic', as defined by *The Shorter Oxford English Dictionary*, means 'serving to find out'. What he was offering physicists was a way to explain the unexplained when it came to light, not a fully worked-out theory derived from first principles. His paper was a signpost towards such a theory, but even that proved too much for those unprepared to travel to a destination in the opposite direction to the long-established wave theory of light.

Received by the *Annalen der Physik* between 18 March and 30 June, Einstein's four papers would transform physics in the years ahead. Remarkably, he also found the time and energy to write 21 book reviews for the journal during the course of the year. Almost as an afterthought, since he did not tell Habicht about it, he wrote a fifth paper. It contained the one equation that almost everyone would come to know, $E=mc^2$. 'A storm broke loose in my mind', was how he described the surge of creativity that consumed him as he produced his breathtaking succession of papers during that glorious Bern spring and summer of 1905.[8]

Max Planck, the adviser on theoretical physics for the *Annalen der Physik*, was among the first to read 'On the Electrodynamics of Moving Bodies'. Planck was immediately won over by what he, and not Einstein, later called the theory of relativity. As for the quantum of light, though he profoundly disagreed with it, Planck allowed Einstein's paper to be published. As he did so he must have wondered about the identity of this physicist capable of the sublime and the ridiculous.

'The people of Ulm are mathematicians' was the unusual medieval motto of the city on the banks of the Danube in the south-western corner of Germany where Albert Einstein was born. It was an apt birthplace on 14 March 1879 for the man who would become the epitome of scientific genius. The back of his head was so large and distorted, his mother feared her newborn son was deformed. Later he took so long to speak that his parents worried he never would. Not long after the birth of his sister, and only sibling, Maja in November 1881, Einstein adopted the rather strange ritual of softly repeating every sentence he wanted to say until satisfied it was word-perfect before uttering it aloud. At seven, to the relief of his parents, Hermann and Pauline, he began to speak normally. By then the family had lived in Munich for six years, having moved so Hermann could open an electrical business in partnership with his younger brother Jakob.

In October 1885, with the last of the private Jewish schools in Munich closed for more than a decade, the six-year-old Einstein was sent to the nearest school. Not surprisingly in the heartland of German Catholicism, religious education formed an integral part of the curriculum, but the teachers, he recalled many years later, 'were liberal and did not make any denominational distinctions'.[9] However liberal and accommodating his teachers may have been, the anti-Semitism that permeated German society was never buried too far beneath the surface, even in the schoolroom. Einstein never forgot the lesson in which his religious studies teacher told the class how the Jews had nailed Christ to the cross. 'Among the children,' Einstein recalled years later, 'anti-Semitism was alive especially in elementary school.'[10] Not surprisingly, he had few, if any, school friends. 'I am truly a lone traveller and have never belonged to my country, my home, my friends, or even my immediate family, with my whole heart', he wrote in 1930. He called himself an Einspänner, a one-horse cart.

As a schoolboy he preferred solitary pursuits and enjoyed nothing more than constructing ever-taller houses of cards. He had the patience and tenacity, even as a ten-year-old, to build them as high as fourteen storeys. These traits, already such a fundamental part of his make-up, would allow him to pursue his own scientific ideas when others might have given up. 'God gave me the stubbornness of a mule,' he said later, 'and a fairly keen scent.'[11] Though others disagreed, Einstein maintained he possessed no special talents, only a passionate curiosity. This quality that others had, however, coupled with his stubbornness, meant that he continued to seek the answer to almost childlike questions long after his peers were taught to stop even asking them. What would it be like to ride on a beam of light? It was trying to answer this question that set him on his decade-long path to the theory of relativity.

In 1888, aged nine, Einstein started at the Luitpold Gymnasium, and he later spoke bitterly of his days there. Whereas young Max Planck enjoyed and thrived under a strict, militaristic discipline focused on rote learning, Einstein did not. Despite resenting his teachers and their autocratic methods, he excelled academically even though the curriculum was orientated towards the humanities. He scored top marks in Latin and did well in Greek, even after being told by his teacher 'that nothing would ever become of him'.[12]

The stifling emphasis on mechanical learning at school, and during music lessons with tutors at home, was in stark contrast to the nurturing influence of a penniless Polish medical student. Max Talmud was 21, and Albert ten, when every Thursday he began dining with the Einsteins as they adopted their own version of an old Jewish tradition of inviting a poor religious scholar to lunch on the Sabbath. Talmud quickly recognised the inquisitive young boy as a kindred spirit. Before long the two would spend hours discussing the books that Talmud had given him to read or had recommended. They began with books on popular science that brought to an end what Einstein called his 'religious paradise of youth'.[13]

The years at a Catholic school and instruction at home by a relative on Judaism had left their mark. Einstein, to the surprise of his secular parents, had developed what he described as 'a deep religiosity'. He stopped eating pork, sang religious songs on the way to school, and accepted the biblical story of creation as an established fact. Then, as he devoured one book after another on science, came the realisation that much of the Bible could not be true. It unleashed what he called 'a fanatic freethinking coupled with the impression that youth is intentionally being deceived by the State through lies; it was a crushing impression'.[14] It sowed the seeds of a lifelong suspicion of every kind of authority. He came to view the loss of his 'religious paradise' as the first attempt to free himself from 'the chains of the "merely personal", from an existence which is dominated by wishes, hopes and primitive feelings'.[15]

As he lost faith in the teachings of one sacred book, he began to experience the wonder of his sacred little geometry book. He was still at primary school when his Uncle Jakob introduced him to the rudiments of algebra and began posing problems for him to solve. By the time Talmud gave him a book on Euclid's geometry, Einstein was already well versed in mathematics not normally expected of a boy of twelve. Talmud was surprised at the speed with which Einstein worked through the book, proving the theorems and completing the exercises. Such was his zeal that during the summer vacation he mastered the mathematics to be taught the following year at school.

With a father and an uncle in the electrical industry, Einstein not only learnt about science through reading but was surrounded by the technology that its application could produce. It was his father who unwittingly introduced Einstein to the wonder and mystery of science. One day, as his son lay ill in bed with a fever, Hermann showed him a compass. The movement of the needle appeared so miraculous that the five-year-old trembled and grew cold at the thought that 'Something deeply hidden had to be behind things.'[16]

The Einstein brothers' electrical business initially prospered. They went from manufacturing electric devices to installing power and lighting networks. The future seemed bright as the Einsteins notched up one success after another, including the contract to provide the first electric lighting for Munich's famous Oktoberfest.[17] But in the end the brothers were simply outgunned by the likes of Siemens and AEG. There were many small electrical firms that prospered and survived in the shadow of these giants, but Jakob was over-ambitious and Hermann too indecisive for their company to be one of them. Beaten but not bowed, the brothers decided that Italy, where electrification was just beginning, was the place to start afresh. So in June 1894 the Einsteins relocated to Milan. All except fifteen-year-old Albert who was left behind in the care of distant relatives to complete the three remaining years to graduation from the school he detested.

For the sake of his parents he pretended that everything was fine in Munich. However, he was increasingly troubled by the thought of compulsory military service. Under German law, if he remained in the country until his seventeenth birthday, Einstein would have no choice but to report for duty when the time came or be declared a deserter. Alone and depressed, he had to think of a way out, when suddenly the perfect opportunity arose.

Dr Degenhart, the teacher of Greek who thought Einstein would never amount to anything, was now also his form tutor. During a heated argument, Degenhart told Einstein he should leave the school. Requiring no further encouragement, he did just that after obtaining a medical certificate stating that he was suffering from exhaustion and required complete rest to recover. At the same time, Einstein secured a testimonial from his mathematics teacher that he had mastered the subject to a level required to graduate. It had taken him just six months to follow in the footsteps of his family and cross the Alps into Italy.

His parents tried to reason with him, but Einstein refused to go back to Munich. He had an alternative plan. He would stay in Milan and prepare for the entrance exams, the following October, of the Federal Polytechnikum in Zurich. Established in 1854, and renamed Eidgenossische Technische Hochschule (ETH) in 1911, the 'Poly' was not as prestigious as Germany's leading universities. However, it did not require graduation from a gymnasium as a precondition for entry. To be accepted, he explained to his parents, he just needed to pass its entrance exams.

They soon discovered the second part of their son's plan. He wanted to renounce his German nationality and thereby remove the possibility of ever being called up for military service by the Reich. Too young to do it himself, Einstein needed his father's consent. Hermann duly gave it and formally applied to the authorities for his son's release. It was January 1896 before they received official notification that Albert, at the cost of three marks, was no longer a German citizen. For the next five years he was legally stateless until he became a Swiss citizen. A renowned pacifist later in life, once he was granted his new nationality Einstein turned up for his Swiss army medical, on 13 March 1901, the day before his 22nd birthday. Fortunately, he was found unfit for military service because of sweaty flat feet and varicose veins.[18] As a teenager back in Munich, it was not the thought of serving in the army that bothered him, but the prospect of donning a grey uniform on behalf of the militarism of the German Reich which he hated.

'The happy months of my sojourn in Italy are my most beautiful memories' was how Einstein, even after 50 years, recalled his new carefree existence.[19] He helped his father and uncle with their electrical business and travelled here and there visiting friends and family. In the spring of 1895 the family moved to Pavia, just south of Milan, where the brothers opened a new factory that lasted little more than a year before it too closed. Although amid the upheaval he worked hard to prepare, Einstein failed the Poly entrance exams. Yet his mathematics and physics results were so impressive that the professor of physics invited him to attend his lectures. It was a tantalising offer, but for once Einstein took some sound advice. He had done so badly in languages, literature and history that the director of the Poly urged him to go back to school for another year and recommended one in Switzerland.

By the end of October Einstein was in Aarau, a town 30 miles west of Zurich. With its liberal ethos, the Aargau canton school provided a stimulating environment that enabled Einstein to thrive. The experience of boarding with the classics teacher and his family was to leave an indelible mark. Jost Winteler and his wife Pauline encouraged freethinking among their three daughters and four sons, and dinner each evening was always a lively and noisy affair. Before long the Wintelers became surrogate parents and he even referred to them as 'Papa Winteler' and 'Mama Winteler'. Whatever the old Einstein said later about being a lone traveller, the young Einstein needed people who cared about him and he for them. Soon it was September 1896 and exam time. Einstein passed easily and headed to Zurich and the Federal Polytechnikum.[20]

'A happy man is too satisfied with the present to dwell too much upon the future', Einstein had written at the start of a short essay called 'My Future Plans', during his two-hour French exam. But an inclination for abstract thinking and the lack of practical sense had led him to decide on a future as a teacher of mathematics and physics.[21] So it was that Einstein found himself, in October 1896, the youngest of eleven new students entering the Poly's School for Specialised Teachers in the Mathematical and Science Subjects. He was one of the five seeking to qualify to teach maths and physics. The only woman among them was to be his future wife.

None of Albert's friends could understand why he was attracted to Mileva Maric. A Hungarian Serb, she was four years older and a bout of childhood tuberculosis had left her with a slight limp. During the first year they sat through the five compulsory maths courses and mechanics – the single physics course offered. Although he had devoured his little sacred book of geometry in Munich, Einstein was no longer interested in mathematics for its own sake. Hermann Minkowski, his maths professor at the Poly, recalled that Einstein had been a 'lazy dog'. It was not apathy but a failure to grasp, as Einstein later confessed, 'that the approach to a more profound knowledge of the basic principles of physics is tied up with the most intricate mathematical methods'.[22] It was something he learnt the hard way in the years of research that followed. He regretted not having tried harder to get 'a sound mathematical education'.[23]

Fortunately, Marcel Grossmann, one of the other three besides Einstein and Mileva enrolled on the course, was a better mathematician and more studious than either of them. It would be to Grossmann that Einstein later turned for help as he struggled with the mathematics needed to formulate the general theory of relativity. The two quickly became friends as they talked 'about anything that might interest young people whose eyes were open'.[24] Only a year older, Grossmann must have been an astute judge of character, for he was so impressed by his classmate that he took him home to meet his parents. 'This Einstein,' he told them, 'will one day be a very great man.'[25]

It was only by using Grossmann's excellent set of notes that he passed the intermediate exams in October 1898. In old age, Einstein could barely bring himself to contemplate what might have happened without Grossmann's help after he began skipping lectures. It had all been so different at the beginning of Heinrich Weber's physics course, when Einstein looked 'forward from one of his lectures to the next'.[26] Weber, who was in his mid-fifties, could make physics come alive for his students, and Einstein conceded that he lectured on thermodynamics with 'great mastery'. But he became disenchanted because Weber did not teach Maxwell's theory of electromagnetism or any of the latest developments. Soon Einstein's independent streak and contemptuous manner began to alienate his professors. 'You're a smart boy', Weber told him. 'But you have one great fault: you do not let yourself be told anything.'[27]

When the final exams took place in July 1900 he came fourth out of five. Einstein felt coerced by the exams, and they had such a deterring effect upon him that afterwards he found 'the consideration of any scientific problems distasteful to me for an entire year'.[28] Mileva was last, and the only one to fail. It was a bitter blow for the couple who were now affectionately calling each other 'Johonzel' (Johnny) and 'Doxerl' (Dollie). Another soon followed.

A future as a schoolteacher no longer appealed to Einstein. Four years in Zurich had given rise to a new ambition. He wanted to be a physicist. The chances of getting a full-time job at a university were slim even for the best students. The first step was an assistant's position with one of the professors at the Poly. None wanted him and Einstein began searching further afield. 'Soon I will have honoured all physicists from the North Sea to the Southern tip of Italy with my offer!' he wrote to Mileva in April 1901 while visiting his parents.[29]

One of those honoured was Wilhelm Ostwald, a chemist at the Leipzig University. Einstein wrote to him twice; both letters went unanswered. It must have been distressing for his father to watch his son's growing despair. Hermann, unknown to Albert then or later, took it upon himself to intervene. 'Please forgive a father who is so bold as to turn to you, esteemed Herr professor, in the interest of his son', he wrote to Ostwald.[30] 'All those in position to give a judgement in the matter, praise his talents; in any case, I can assure you that he is extraordinarily studious and diligent and clings with great love to his science.'[31] The heartfelt plea went unanswered. Later Ostwald would be the first to nominate Einstein for the Nobel Prize.

Although anti-Semitism may have played a part, Einstein was convinced that it was Weber's poor references that were behind his failure to secure an assistantship. As he grew increasingly despondent, a letter from Grossmann held out the possibility of a decent, well-paying job. Grossmann senior had learnt of Einstein's desperate situation and wanted to help the young man whom his son held in such high regard. He strongly recommended Einstein for the next vacancy that arose to his friend Friedrich Haller, the director of the Swiss Patent Office in Bern. 'When I found your letter yesterday,' Einstein wrote to Marcel, 'I was deeply moved by your devotion and compassion which did not let you forget your old luckless friend.'[32] After five years of being stateless, Einstein had recently acquired Swiss citizenship and was certain it would help when applying for the job.

Maybe his luck had changed at last. He was offered and accepted a temporary teaching job at the school in Winterthur, a small town less than twenty miles from Zurich. The five or six classes Einstein taught each morning left him free to pursue physics in the afternoon. 'I cannot tell you how happy I would feel in such a job', he wrote to Papa Winteler shortly before his time in Winterthur ended. 'I have completely given up my ambition to get a position at a university, since I see that even as it is, I have enough strength and desire left for scientific endeavour.'[33] Soon that strength was put to the test when Mileva announced she was pregnant.

After failing the Poly exams a second time, Mileva returned to her parents in Hungary to await the arrival of the baby. Einstein took the news of the pregnancy in his stride. He had already entertained thoughts of becoming an insurance clerk and now vowed to find any job, no matter how humble, so that they could marry. When their daughter was born, Einstein was in Bern. He never saw Lieserl. What happened to her, whether she was given up for adoption or died in infancy, remains a mystery.

In December 1901, Friedrich Haller wrote to Einstein asking him to apply for a vacancy at the Patent Office that was about to be advertised.[34] The long search for a permanent job seemed at an end as Einstein sent off his application before Christmas. 'All the time I rejoice in the fine prospects which are in store for us in the near future', he wrote to Mileva. 'Have I already told you how rich we will be in Bern?'[35] Convinced that everything would be settled quickly, Einstein quit a year-long tutoring job at a private boarding school in Schaffhausen after only a few months.

Bern was home to some 60,000 people when Einstein arrived during the first week of February 1902. The medieval elegance of the Old Town quarter had changed little in the 500 years since it had been rebuilt following a fire that destroyed half the city. It was here that Einstein found a room on Gerechtigkeitgasse, not far from the city's famous bear pit.[36] Costing 23 francs a month, it was anything but the 'large, beautiful room' he described to Mileva.[37] Not long after he unpacked his bags, Einstein went down to the local newspaper to place an advert offering his services as a private tutor of mathematics and physics. It appeared on Wednesday, 5 February and offered a free trial lesson. Within days it paid off. One of the students described his new tutor as 'about five foot ten, broad-shouldered, slightly stooped, a pale brown skin, a sensuous mouth, black moustache, nose slightly aquiline, radiant brown eyes, a pleasant voice, speaking French correctly but with a slight accent'.[38]

A young Romanian Jew, Maurice Solovine, also came across the advert as he read his newspaper walking down the street. A philosophy student at Bern University, Solovine was also interested in physics. Frustrated that a lack of mathematics was preventing him from gaining a deeper understanding of physics, he immediately made his way to the address given in the newspaper. When Solovine rang the bell, Einstein had found a kindred spirit. The student and tutor talked for two hours. They shared many of the same interests and after spending another half hour chatting in the street, they agreed to meet the following day. When they did, all thoughts of a structured lesson were forgotten amid a shared enthusiasm for exploring ideas. 'As a matter of fact, you don't have to be tutored in physics', Einstein told him on the third day.[39] What Solovine liked about Einstein, as the two quickly became friends, was the care with which he outlined a topic or problem as lucidly as possible.

Before long, Solovine suggested that they read a particular book and then discuss it. Having done the same with Max Talmud in Munich as a schoolboy, Einstein thought it an excellent idea. Soon Conrad Habicht joined them. A friend from Einstein's aborted stint teaching at the boarding school in Schaffhausen, Habicht had moved to Bern to complete a mathematics thesis at the university. United by their enthusiasm for studying and clarifying the problems of physics and philosophy for their own satisfaction, the three men started calling themselves the 'Akademie Olympia'.

Even though Einstein came highly recommended by a friend, Haller had to make sure he was capable of doing the job. The ever-growing number of patent applications for all manner of electrical devices had made the hiring of a competent physicist to work alongside his engineers a necessity rather than a favour for a friend. Einstein impressed Haller sufficiently to be provisionally appointed a 'Technical Expert, Third Class' with a salary of 3,500 Swiss francs. At eight o'clock in the morning on 23 June 1902, Einstein reported for his first day as a 'respectable Federal ink pisser'.[40]

'As a physicist,' Haller told Einstein, 'you haven't a clue about blueprints.'[41] Until he could read and assess technical drawings, there would be no permanent contract. Haller took it upon himself to teach Einstein what he needed to know, including the art of expressing himself clearly, concisely, and correctly. Although he had never taken kindly to being instructed as a schoolboy or student, he knew that he needed to learn all he could from Haller, 'a splendid character and a clever mind'.[42] 'One soon gets used to his rough manner', Einstein wrote. 'I hold him in very high regard.'[43] As he proved his worth, Haller likewise came to respect his young protégé as a prized member of staff.

In October 1902, aged only 55, his father fell seriously ill. Einstein travelled to Italy to see him one last time. It was then, as he lay dying, that Hermann gave Albert his permission to marry Mileva – a prospect that he and Pauline had long opposed. With only Solovine and Habicht as witnesses, Einstein and Mileva married the following January in a civil ceremony at the Bern registrar's office. 'Marriage is,' Einstein said later, 'the unsuccessful attempt to make something lasting out of an incident.'[44] But in 1903 he was just happy to have a wife that cooked, cleaned, and simply looked after him.[45] Mileva had hoped for more.

The Patent Office took up 48 hours a week. From Monday to Saturday Einstein started at eight o'clock and worked until noon. Then it was lunch either at home or with a friend at a nearby café. He was back in the office from two until six. It left 'eight hours for fooling around' each day, and 'then there's also Sunday', he told Habicht.[46] It was September 1904 before Einstein's 'provisional' position was made permanent with a pay rise of 400 francs. By the spring of 1906 Haller was so impressed with Einstein's ability to 'tackle technically very difficult patent applications' that he rated him as 'one of the valued experts at the office'.[47] He was promoted to technical expert, second class.

'I will be grateful to Haller for as long as I live', Einstein had written to Mileva soon after moving to Bern in the expectation that a job at the

Patent Office would sooner or later be his.[48] And he was. But it was only much later that he recognised the extent of the influence that Haller and the Patent Office exerted on him: 'I might not have died, but I would have been intellectually stunted.'[49] Haller demanded that every patent application be evaluated rigorously enough to withstand any legal challenge. 'When you pick up an application, think that anything the inventor says is wrong,' he advised Einstein, or else 'you will follow the inventor's way of thinking, and that will prejudice you. You have to remain critically vigilant.'[50] Accidentally, Einstein had found a job that suited his temperament and honed his abilities. The critical vigilance he exercised in assessing an inventor's hopes and dreams, often on the basis of unreliable drawings and inadequate technical specifications, Einstein brought to bear on the physics that occupied him. The 'many-sided thinking' his job entailed he described as a 'veritable blessing'.[51]

'He had the gift of seeing a meaning behind inconspicuous, well-known facts which had escaped everyone else', recalled Einstein's friend and fellow theoretical physicist Max Born. 'It was this uncanny insight into the working of nature which distinguished him from all of us, not his mathematical skill.'[52] Einstein knew that his mathematical intuition was not strong enough to differentiate what was really basic 'from the rest of the more or less dispensable erudition'.[53] But when it came to physics, his nose was second to none. Einstein said he 'learned to scent out that which was able to lead to fundamentals and to turn aside from everything else, from the multitude of things which clutter up the mind and divert it from the essential'.[54]

His years at the Patent Office only heightened his sense of smell. As with the patents that inventors submitted, Einstein looked for subtle flaws and inconsistencies in the blueprints of the workings of nature put forward by physicists. When he found such a contradiction in a theory, Einstein probed it ceaselessly until it yielded a new insight resulting in its elimination or an alternative where none had existed before. His 'heuristic' principle that light behaved in certain instances as if it was made up of a stream of particles, light-quanta, was Einstein's solution to a contradiction at the very heart of physics.

Einstein had long accepted that everything was composed of atoms and that these discrete, discontinuous bits of matter possessed energy. The energy of a gas, for example, was the sum total of the energies of the individual atoms of which it was made up. The situation was entirely different when it came to light. According to Maxwell's theory of electromagnetism, or any wave theory, the energy of a light ray continuously spreads out over an ever-increasing volume like the waves radiating outwards from the point where a stone hits the surface of a pond. Einstein called it a 'profound formal difference' and it made him uneasy while stimulating his 'many-sided thinking'.[55] He realised that the dichotomy between the discontinuity of matter and the continuity of electromagnetic waves would dissolve if light was also discontinuous, made up of quanta.[56]

The quantum of light emerged out of Einstein's review of Planck's derivation of the blackbody radiation law. He accepted that Planck's formula was correct, but his analysis revealed what Einstein had always suspected. Planck should have arrived at an entirely different formula. However, since he knew the equation he was looking for, Planck fashioned his derivation to get it. Einstein worked out exactly where Planck had gone astray. In his desperation to justify his equation that he knew to be in perfect agreement with experiments, Planck had failed to consistently apply the ideas and techniques he used or that were available to him. If he had done so, Einstein realised that Planck would have obtained an equation that did not agree with the data.

Lord Rayleigh had originally proposed this other formula in June 1900, but Planck had taken little, if any, notice of it. At the time he did not believe in the existence of atoms and therefore disapproved of Rayleigh's use of the equipartition theorem. Atoms are free to move in only three ways: up and down, back and forth, and side to side. Called a 'degree of freedom', each is an independent way in which an atom can receive and store energy. In addition to these three kinds of 'translational' motion, a molecule made up of two or more atoms has three types of rotational motion about the imaginary axes joining the atoms, giving a total of six degrees of freedom. According to the equipartition theorem, the energy of a gas should be distributed equally among its molecules and then divided equally among the different ways in which a molecule can move.

Rayleigh employed the equipartition theorem to divide up the energy of blackbody radiation among the different wavelengths of radiation present inside a cavity. It had been a flawless application of the physics of Newton, Maxwell and Boltzmann. Aside from a numerical error that was later corrected by James Jeans, there was a problem with what became known as the Rayleigh-Jeans law. It predicted a build-up of an infinite amount of energy in the ultraviolet region of the spectrum. It was a breakdown of classical physics that many years later, in 1911, was dubbed 'the ultraviolet catastrophe'. Thankfully it did not actually happen, for a universe bathed in a sea of ultraviolet radiation would have made human life impossible.

Einstein had derived the Rayleigh-Jeans law on his own and knew that the distribution of blackbody radiation that it forecast contradicted the experimental data and led to the absurdity of an infinite energy in the ultraviolet. Given that the Rayleigh-Jeans law tallied with the behaviour of blackbody radiation only at long wavelengths (very low frequencies), Einstein's point of departure was Wilhelm Wien's earlier blackbody radiation law. It was the only safe choice, even though Wien's law managed to replicate the behaviour of blackbody radiation only at short wavelengths (high frequencies) and failed at longer wavelengths (lower frequencies) of the infrared. Yet it had certain advantages that appealed to Einstein. He had no doubts about the soundness of its derivation, and it perfectly described at least a portion of the blackbody spectrum to which he would restrict his argument.

Einstein devised a simple but ingenious plan. A gas is just a collection of particles, and in thermodynamic equilibrium it is the properties of these particles that determine, for example, the pressure exerted by the gas at a given temperature. If there were similarities between the properties of blackbody radiation and the properties of a gas, then he could argue that electromagnetic radiation is itself particle-like. Einstein began his analysis with an imaginary blackbody that was empty. But unlike Planck, he filled it with gas particles and electrons. The atoms in the walls of the blackbody, however, contained other electrons. As the blackbody is heated, they oscillate with a broad range of frequencies resulting in the emission and absorption of radiation. Soon the interior of the blackbody is teeming with speeding gas particles and electrons, and the radiation emitted by the oscillating electrons. After a while, thermal equilibrium is reached when the cavity and everything inside it is at the same temperature T.

The first law of thermodynamics, that energy is conserved, can be translated to connect the entropy of a system to its energy, temperature and volume. It was now that Einstein used this law, Wien's law and Boltzmann's ideas to analyse how the entropy of blackbody radiation depended on the volume it occupied 'without establishing any model for the emission or propagation of radiation'.[57] What he found was a formula that looked exactly like one describing how the entropy of a gas, made up of atoms, is dependent on the volume it occupies. Blackbody radiation behaved as if it was made up of individual particle-like bits of energy.

Einstein had discovered the quantum of light without having to use either Planck's blackbody radiation law or his method. In keeping Planck at arm's length, Einstein wrote the formula slightly differently but it meant and encoded the same information as $E=h\nu$, that energy is quantised, that it comes only in units of $h\nu$. Whereas Planck had only quantised the emission and absorption of electromagnetic radiation so that his imaginary oscillators would produce the correct spectral distribution of blackbody radiation, Einstein had quantised electromagnetic radiation, and therefore light, itself. The energy of a quantum of yellow light was just Planck's constant multiplied by the frequency of yellow light.

By showing that electromagnetic radiation sometimes behaves like the particles of a gas, Einstein knew that he had smuggled his light-quanta in through the back door, by analogy. To convince others of the 'heuristic' value of his new 'point of view' concerning the nature of light, he used it to explain a little-understood phenomenon.[58]

The German physicist Heinrich Hertz first observed the photoelectric effect in 1887 while in the middle of performing a series of experiments that demonstrated the existence of electromagnetic waves. By chance he noticed that the spark between two metal spheres became brighter when one of them was illuminated by ultraviolet light. After months of investigating the 'completely new and very puzzling phenomenon' he could offer no explanation, but believed, incorrectly, that it was confined to the use of ultraviolet light.[59]

'Naturally, it would be nice if it were less puzzling,' Hertz admitted, 'however, there is some hope that when this puzzle is solved, more new facts will be clarified than if it were easy to solve.'[60] It was a prophetic statement, but one that he never lived to see fulfilled. He died tragically young at the age of 36 in 1894.

It was Hertz's former assistant, Philipp Lenard, who in 1902 deepened the mystery surrounding the photoelectric effect when he discovered that it also occurred in a vacuum when he placed two metal plates in a glass tube and removed the air. Connecting the wires from each plate to a battery, Lenard found that a current flowed when one of the plates was irradiated with ultraviolet light. The photoelectric effect was explained as the emission of electrons from the illuminated metal surface. Shining ultraviolet light onto the plate gave some electrons enough energy to escape from the metal and cross the gap to the other plate, thereby completing the circuit to produce a 'photoelectric current'. However, Lenard also found facts that contradicted established physics. Enter Einstein and his quantum of light.

It was expected that increasing the intensity of a light beam, by making it brighter, would yield the same number of electrons from the metal surface, but with each having more energy. Lenard, however, found the exact opposite: a greater number of electrons were emitted with no change in their individual energy. Einstein's quantum solution was simple and elegant: if light is made up of quanta, then increasing the intensity of the beam means that it is now made up of a greater number of quanta. When a more intense beam strikes the metal plate, the increase in the number of light-quanta leads to a corresponding increase in the number of electrons being emitted.

Lenard's second curious discovery was that the energy of the emitted electrons was not governed by the intensity of the light beam, but by its frequency. Einstein had a ready answer. Since the energy of a light-quantum is proportional to the frequency of the light, a quantum of red light (low frequency) has less energy than one of blue light (high frequency). Changing the colour (frequency) of light does not alter the number of quanta in beams of the same intensity. So, no matter what the colour of light, the same number of electrons will be emitted since the same numbers of quanta strike the metal plate. However, since different frequencies of light are made up of quanta of different energies, the electrons that are emitted will have more or less energy depending on the light used. Ultraviolet light will yield electrons with a greater maximum kinetic energy than those emitted by quanta of red light.

There was another intriguing feature. For any particular metal there was a minimum or 'threshold frequency' below which no electrons were emitted at all, no matter how long or intensively the metal was illuminated. However, once this threshold was crossed, electrons were emitted no matter how dim the beam of light. Einstein's quantum of light supplied the answer once again as he introduced a new concept, the work function.

Einstein envisaged the photoelectric effect as the result of an electron acquiring enough energy from a quantum of light to overcome the forces holding it within the metal surface and to escape. The work function, as Einstein labelled it, was the minimum energy an electron needed to escape from the surface, and it varied from metal to metal. If the frequency of light is too low, then the light-quanta will not possess enough energy to allow an electron to break the bonds that keep it bound within the metal.

Einstein encoded all this in a simple equation: the maximum kinetic energy of an electron emitted from a metal surface was equal to the energy of the light-quanta it absorbed minus the work function. Using this equation, Einstein predicted that a graph of the maximum kinetic energy of the electrons versus the frequency of light used would be a straight line, beginning at the threshold frequency of the metal. The gradient of the line, irrespective of the metal used, would always be exactly equal to Planck's constant, h.

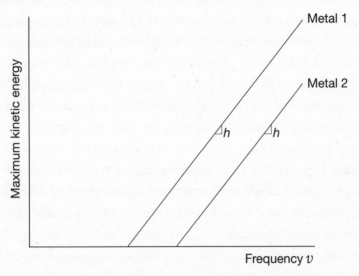

Figure 3: The photoelectric effect – maximum kinetic energy of emitted electrons versus the frequency of light striking the metal surface

'I spent ten years of my life testing that 1905 equation of Einstein's and contrary to all my expectations,' complained the American experimental physicist Robert Millikan, 'I was compelled to assert its unambiguous verification in spite of its unreasonableness, since it seemed to violate everything we knew about the interference of light.'[61] Although Millikan won the 1923 Nobel Prize partly in recognition of this work, even in the face of his own data he balked at the underlying quantum hypothesis: 'the physical theory upon which the equation is based is totally untenable.'[62] From the very beginning, physicists at large had greeted Einstein's light-quanta with similar disbelief and cynicism. A handful wondered if light-quanta existed at all or whether they were simply a useful fictional contrivance of practical value in calculations. At best some thought that light, and therefore all electromagnetic radiation, did not consist of quanta, but only behaved as such when exchanging energy with matter.[63] Foremost among them was Planck.

When in 1913 he and three others nominated Einstein for membership of the Prussian Academy of Sciences, they concluded their testimonial by trying to excuse his light-quanta proposal: 'In sum, it can be said that among the important problems, which are so abundant in modern physics, there is hardly one in which Einstein did not take a position in a remarkable manner. That he might sometimes have overshot the target in his speculations, as for example in his light-quantum hypothesis, should not be counted against him too much. Because without taking a risk from time to time it is impossible, even in the most exact natural science, to introduce real innovations.'[64]

Two years later, Millikan's painstaking experiments made it difficult to ignore the validity of Einstein's photoelectric equation. By 1922 it was becoming almost impossible, as Einstein was belatedly awarded the 1921 Nobel Prize for physics explicitly for his photoelectric effect law, described by his formula, and not for his underlying explanation using light-quanta. No longer the unknown patent clerk in Bern, he was by then world-famous for his theories of relativity and widely acknowledged as the greatest scientist since Newton. Yet his quantum theory of light was just too radical for physicists to accept.

The stubborn opposition to Einstein's idea of light-quanta rested on the overwhelming evidence in support of a wave theory of light. However, whether light was a particle or a wave had been hotly disputed before. During the eighteenth century and in the early years of the nineteenth, it was Isaac Newton's particle theory that had triumphed. 'My Design in this Book is not to explain the Properties of Light by Hypotheses,' Newton wrote at the beginning of *Opticks*, published in 1704, 'but to propose and prove them by Reason and Experiments.'[65] Those first experiments were conducted in 1666, when he split light into the colours of the rainbow with a prism and wove them back together into white light using a second prism. Newton believed that rays of light were composed of particles or, as he called them, 'corpuscles', the 'very small Bodies emitted from shining Substances'.[66] With the particles of light travelling in straight lines, such a theory would, according to Newton, explain the everyday fact that while a person can be heard talking around a corner, they cannot be seen, since light cannot not bend around corners.

Newton was able to give a detailed mathematical account for a host of optical observations, including reflection and refraction – the bending of light as it passes from a less to a more dense medium. However, there were other properties of light that Newton could not explain. For example, when a beam of light hit a glass surface, part of it passed through and the rest was reflected. The question Newton had to address was why some particles of light were reflected and others not? To answer it, he was forced to adapt his theory. Light particles caused wavelike disturbances in the ether. These 'Fits of easy Reflexion and easy Transmission', as he called them, were the mechanism by which some of the beam of light was transmitted through the glass and the remainder reflected.[67] He linked the 'bigness' of these disturbances to colour. The biggest disturbances, those having the longest wavelength, in the terminology that came later, were responsible for producing red. The smallest, those having the shortest wavelength, produced violet.

The Dutch physicist Christiaan Huygens argued that there was no Newtonian particle of light. Thirteen years older than Newton, by 1678 Huygens had developed a wave theory of light that explained reflection and refraction. However, his book on the subject, *Traité de la Lumière*, was not published until 1690. Huygens believed that light was a wave travelling through the ether. It was akin to the ripples that fanned out across the still surface of a pond from a dropped stone. If light was really made up of particles, Huygens asked, then where was the evidence of collisions that should occur when two beams of light crossed each other? There was none, argued Huygens. Sound waves do not collide; ergo light must also be wavelike.

Although the theories of Newton and Huygens were able to explain reflection and refraction, each predicted different outcomes when it came to certain other optical phenomena. None could be tested with any degree of precision for decades. However, there was one prediction that could be observed. A beam of light made up of Newton's particles travelling in straight lines should cast sharp shadows when striking objects, whereas Huygens' waves, like water waves bending around an object they encounter, should result in shadows whose outline is slightly blurred. The Italian Jesuit and mathematician Father Francesco Grimaldi christened this bending of light around the edge of an object, or around the edges of an extremely narrow slit, diffraction. In a book published in 1665, two years after his death, he described how an opaque object placed in a narrow shaft of sunlight allowed to enter an otherwise darkened room through a very small hole in a window shutter, cast a shadow larger than expected if light consisted of particles travelling in straight lines. He also found that around the shadow were fringes of coloured light and fuzziness where there should have been a sharp, well-defined separation between light and dark.

Newton was well aware of Grimaldi's discovery and later conducted his own experiments to investigate diffraction, which seemed more readily explicable in terms of Huygens' wave theory. However, Newton argued that diffraction was the result of forces exerted on light particles and indicative of the nature of light itself. Given his pre-eminence, Newton's particle theory of light, though in truth a strange hybrid of particle and wave, was accepted as the orthodoxy. It helped that Newton outlived Huygens, who died in 1695, by 32 years. 'Nature and Nature's Laws lay hid in Night; / God said, Let Newton be! And all was Light.' Alexander Pope's famous epitaph bears witness to the awe in which Newton was held in his own day. In the years after his death in 1727, Newton's authority was undiminished and his view on the nature of light barely questioned. At the dawn of the nineteenth century the English polymath, Thomas Young, did challenge it, and in time his work led to a revival of the wave theory of light.

Born in 1773, Young was the eldest of ten children. He was reading fluently by the age of two and had read the entire Bible twice by six. A master

of more than a dozen languages, Young went on make important contributions towards the deciphering of Egyptian hieroglyphics. A trained
physician, he could indulge his myriad intellectual pursuits after a bequest
from an uncle left him financially secure. His interest in the nature of light
led Young to examine the similarities and differences between light and
sound, and ultimately to 'one or two difficulties in the Newtonian system'.[68]
Convinced that light was a wave, he devised an experiment that was to
prove the beginning of the end for Newton's particle theory.

Young shone monochromatic light onto a screen with a single slit. From
this slit a beam of light spread out to strike a second screen with two very
narrow and parallel slits close together. Like a car's headlights, these two
slits acted as new sources of light, or as Young wrote, 'as centres of diver-
gence, from whence the light diffracted in every direction'.[69] What Young
found on another screen placed some distance behind the two slits was a
central bright band surrounded on each side by a pattern of alternating
dark and bright bands.

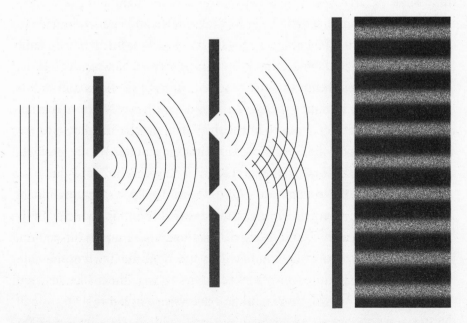

*Figure 4: Young's two-slits experiment. At far right, the resulting interference pattern on
the screen is shown*

To explain the appearance of these bright and dark 'fringes', Young used an analogy. Two stones are dropped simultaneously and close together into a still lake. Each stone produces waves that spread out across the lake. As they do so, the ripples originating from one stone encounter those from the other. At each point where two wave troughs or two wave crests meet, they coalesce to produce a new single trough or crest. This was constructive interference. But where a trough meets a crest or vice versa, they cancel each other out, leaving the water undisturbed at that point – destructive interference.

In Young's experiment, light waves originating from the two slits similarly interfere with each other before striking the screen. The bright fringes indicate constructive interference while the dark fringes are a product of destructive interference. Young recognised that only if light is a wave phenomenon could these results be explained. Newton's particles would simply produce two bright images of the slits with nothing but darkness in between. An interference pattern of bright and dark fringes was simply impossible.

When he first put forward the idea of interference and reported his early results in 1801, Young was viciously attacked in print for challenging Newton. He tried to defend himself by writing a pamphlet in which he let everyone know his feelings about Newton: 'But, much as I venerate the name of Newton, I am not therefore obliged to believe that he was infallible. I see, not with exultation, but with regret, that he was liable to err, and that his authority has, perhaps, sometimes even retarded the progress of science.'[70] Only a single copy was sold.

It was a French civil engineer who followed Young in stepping out of Newton's shadow. Augustin Fresnel, fifteen years his junior, independently rediscovered interference and much else of what Young, unknown to him, had already done. However, compared to the Englishman, Fresnel's elegantly designed experiments were more extensive, with the presentation of results and accompanying mathematical analysis so impeccably thorough that the wave theory started to gain distinguished converts by the 1820s. Fresnel convinced them that the wave theory could better explain an array of optical phenomena than Newton's particle theory. He also answered the long-standing objection to the wave theory: light cannot travel around corners. It does, he said. However, since light waves are millions of times smaller than sound waves, the bending of a beam of light from a straight path is very, very small and therefore extremely difficult to detect. A wave bends only around an obstacle not much longer than itself. Sound waves are very long and can easily move around most barriers they encounter.

One way to get opponents and sceptics to finally decide between the two rival theories was to find observations for which they predicted different results. Experiments conducted in France in 1850 revealed that the speed of light was slower in a dense medium such as glass or water than in the air. This was exactly what the wave of light predicted, while Newton's corpuscles failed to travel as fast as expected. But the question remained: if light was a wave, what were its properties? Enter James Clerk Maxwell and his theory of electromagnetism.

Born in 1831 in Edinburgh, Maxwell, the son of a Scottish landowner, was destined to become the greatest theoretical physicist of the nineteenth century. At the age of fifteen, he wrote his first published paper on a geometrical method for tracing ovals. In 1855 he won Cambridge University's Adams Prize for showing that Saturn's rings could not be solid, but had to be made of small, broken bits of matter. In 1860 he instigated the final phase of the development of the kinetic theory of gases, the properties of gases explained by maintaining that they consisted of particles in motion. But his greatest achievement was the theory of electromagnetism.

In 1819 the Danish physicist Hans Christian Oersted discovered that an electric current flowing through a wire deflected a compass needle. A year later the Frenchman François Arago found that a wire carrying an electric current acted as a magnet and could attract iron filings. Soon his compatriot André Marie Ampère demonstrated that two parallel wires were attracted towards one another if each had a current flowing through it in the same direction. However, they repelled each other if the currents flowed in the opposite directions. Intrigued by the fact that a flow of electricity could create magnetism, the great British experimentalist Michael

Faraday decided to see if he could generate electricity using magnetism. He pushed a bar magnet in and out of a helix coil of wire and found an electric current being generated. The current ceased whenever the magnet was motionless within the coil.

Just as ice, water and steam are different manifestations of H_2O, Maxwell showed in 1864 that electricity and magnetism were likewise different manifestations of the same underlying phenomenon – electromagnetism. He managed to encapsulate the disparate behaviour of electricity and magnetism into a set of four elegant mathematical equations. On seeing them, Ludwig Boltzmann immediately recognised the magnitude of Maxwell's achievement and could only quote Goethe in admiration: 'Was it a God that wrote these signs?'[71] Using these equations, Maxwell was able to make the startling prediction that electromagnetic waves travelled at the speed of light through the ether. If he was right, then light was a form of electromagnetic radiation. But did electromagnetic waves actually exist? If so, did they really travel at the speed of light? Maxwell did not live long enough to see his prediction confirmed by experiment. Aged just 48, he died from cancer in November 1879, the year Einstein was born. Less than a decade later, in 1887, Heinrich Hertz provided the experimental corroboration that ensured Maxwell's unification of electricity, magnetism and light was the crowning achievement of nineteenth-century physics.

Hertz proclaimed in his paper outlining his investigations: 'The experiments described appear to me, at any rate, eminently adapted to remove any doubt as to the identity of light, radiant heat, and electromagnetic wave motion. I believe that from now on we shall have greater confidence in making use of the advantages, which this identity enables us to derive both in the study of optics and electricity.'[72] Ironically, it was during these very experiments that Hertz discovered the photoelectric effect that provided Einstein with evidence for a case of mistaken identity. His light-quanta challenged the wave theory of light that Hertz and everyone else thought was well and truly established. Light as a form of electromagnetic radiation had proved so successful that for physicists to even contemplate discarding it in favour of Einstein's light-quanta was unthinkable. Many found light-quanta absurd. After all, the energy of a particular quantum of light was determined by the frequency of that light, but surely frequency was something associated with waves, not particle-like bits of energy travelling through space.

Einstein readily accepted that the wave theory of light had 'proved itself superbly' in explaining diffraction, interference, reflection and refraction, and that it would 'probably never be replaced by another theory'.[73] However, this success, he pointed out, rested on the vital fact that all these optical phenomena involved the behaviour of light over a period of time, and any particle-like properties would not be manifest. The situation was starkly different when it came to the virtually 'instantaneous' emission and absorption of light. This was the reason, Einstein suggested, why the wave theory faced 'especially great difficulties' explaining the photoelectric effect.[74]

A future Nobel laureate, but in 1906 a *privatdozent* at Berlin University, Max Laue wrote to Einstein that he was willing to accept that quanta may be involved during the emission and absorption of light. However, that was all. Light itself was not made up of quanta, warned Laue, but it is 'when it is exchanging energy with matter that it behaves as if it consisted of them'.[75] Few even conceded that much. Part of the problem lay with Einstein himself. In his original paper he did say that light 'behaves' as though it consisted of quanta. This was hardly a categorical endorsement of the quantum of light. This was because Einstein wanted something more than just a 'heuristic point of view': he craved a fully-fledged theory.

The photoelectric effect had proved to be a battlefield for the clash between the supposed continuity of light waves and the discontinuity of matter, atoms. But in 1905 there were still those who doubted the reality of atoms. On 11 May, less than two months after he finished his quantum paper, the *Annalen der Physik* received Einstein's second paper of the year. It was his explanation of Brownian motion and it became a key piece of evidence in support of the existence of atoms.[76]

When in 1827 the Scottish botanist Robert Brown peered through a microscope at some pollen grains suspended in water, he saw that they were in a constant state of haphazard motion as if buffeted by some unseen force. It had already been noted by others that this erratic wiggling increased as the temperature of the water rose, and it was assumed that some sort of biological explanation lay behind the phenomenon. However, Brown discovered that when he used pollen grains that were up to twenty years old they moved in exactly the same way. Intrigued, he produced fine powders of all manner of inorganic substances, from glass to a piece of the Sphinx, and suspended each of them in water. He found the same zigzagging motion in each case and realised that it could not be animated by some vital force. Brown published his research in pamphlet entitled: *A Brief Account of Microscopical Observations Made in the Months of June, July, and August 1827, on the Particles Contained in the Pollen of Plants; and on the General Existence of Active Molecules in Organic and Inorganic Bodies*. Others offered plausible explanations of 'Brownian motion', but all were sooner or later found wanting. By the end of the nineteenth century, those who believed in the existence of atoms and molecules accepted that Brownian motion was the result of collisions with water molecules.

What Einstein recognised was that the Brownian motion of a pollen grain was not caused by a single collision with a water molecule, but was the product of a large number of such collisions. At each moment, the collective effect of these collisions was the random zigzagging of the pollen grain or suspended particle. Einstein suspected that the key to understanding this unpredictable motion lay in deviations, statistical fluctuations, from the expected 'average' behaviour of water molecules. Given their relative sizes, on average, many water molecules would strike an individual pollen grain simultaneously from different directions. Even on this scale, each collision would result in an infinitesimal push in one direction, but the overall effect of all of them would leave the pollen unmoved as they cancelled each other out. Einstein realised that Brownian motion was due to water molecules regularly deviating from their 'normal' behaviour as some of them got bunched up and struck the pollen together, sending it in particular direction.

Using this insight, Einstein succeeded in calculating the average horizontal distance a particle would travel as it zigzagged along in a given time. He predicted that in water at 17°C, suspended particles with a diameter of one-thousandth of a millimetre would move on average just six-thousandths of a millimetre in one minute. Einstein had come up with a formula that offered the possibility of working out the size of atoms armed only with a thermometer, microscope and stopwatch. Three years later, in 1908, Einstein's predictions were confirmed in a delicate series of experiments conducted at the Sorbonne by Jean Perrin, for which he received the Nobel Prize in 1926.

With Planck championing the theory of relativity, and the analysis of Brownian motion recognised as a decisive breakthrough in favour of the atom, Einstein's reputation grew despite the rejection of his quantum theory of light. He received letters often addressed to him at Bern University, as few knew he was a patent clerk. 'I must tell you quite frankly that I was surprised to read that you must sit in an office for 8 hours a day,' wrote Jakob Laub from Würzburg. 'History is full of bad jokes.'[77] It was March 1908 and Einstein agreed. After almost six years he no longer wanted to be a patent slave.

He applied for a job as a mathematics teacher at a school in Zurich, stating that he would be ready and willing to teach physics as well. With his application he enclosed a copy of his thesis that had earned him, at the third attempt, a doctorate from Zurich University in 1905 and laid the groundwork for the paper on Brownian motion. Hoping it would bolster his chances, he also sent all of his published papers. Despite his impressive scientific achievements, of the 21 applicants, Einstein did not even make the short list of three.

It was at the behest of Alfred Kleiner, the professor of experimental physics at Zurich University, that Einstein tried for a third time to become a *privatdozent*, an unpaid lecturer, at the University of Bern. The first application was rejected because at the time he did not have a PhD. In June

1907, he failed a second time because he did not submit a *habilitations-schrift* – a piece of unpublished research. Kleiner wanted Einstein to fill a soon-to-be-created extraordinary professorship in theoretical physics, and being a *privatdozent* was a necessary stepping-stone to such an appointment. So he produced a *habilitationsschrift* as demanded and was duly appointed a *privatdozent* in the spring of 1908.

Only three students attended his first lecture course on the theory of heat. All three were friends. They had to be, since Einstein had been allocated Tuesdays and Saturdays between seven and eight in the morning. University students had the choice of whether or not to attend courses offered by a *privatdozent* and none were willing to get up that early. As a lecturer, then and later, Einstein was often under-prepared and made frequent mistakes. And when he did, he simply turned to the students and asked: 'Who can tell me where I went wrong?' or 'Where have I made a mistake?' If a student pointed out an error in his mathematics, Einstein would say, 'I have often told you, my mathematics have never been up to much.'[78]

The ability to teach was a vital consideration for the job earmarked for Einstein. To ensure that he was up to the task, Kleiner organised to attend one of his lectures. Annoyed at 'having-to-be-investigated', he performed poorly.[79] However, Kleiner gave him a second chance to impress and he did. 'I was lucky', Einstein wrote to his friend Jakob Laub. 'Contrary to my habit, I lectured well on that occasion – and so it came to pass.'[80] It was May 1909 and Einstein could finally boast that he was 'an official member of the guild of whores' as he accepted the Zurich post.[81] Before moving to Switzerland with Mileva and five-year-old Hans Albert, Einstein travelled to Salzburg in September to give the keynote lecture to the cream of German physics at a conference of the Gesellschaft Deutscher Naturforscher und Ärtze. He went well prepared.

It was a singular honour to be asked to deliver such a lecture. It was one usually reserved for a distinguished elder statesman of physics, not someone who had just turned 30 and was about take up his first extraordinary professorship. So all eyes were on Einstein, but he seemed oblivious as he paced the podium and delivered what would turn out to be a celebrated lecture: 'On the Development of Our Views Concerning the Nature and Constitution of Radiation'. He told the audience that 'the next stage in the development of theoretical physics will bring us a theory of light that may be conceived of as a sort of fusion of the wave and of the emission theory of light'.[82] It was not a hunch, but based on the result of an inspired thought experiment involving a mirror suspended inside a blackbody. He managed to derive an equation for the fluctuations of the energy and momentum of radiation that contained two very distinct parts. One corresponded to the wave theory of light, while the other had all the hallmarks of the radiation being composed of quanta. Both parts appeared to be indispensable, as did the two theories of light. It was the first prediction of what would later be called wave-particle duality – that light was both a particle and a wave.

Planck, who was chairing, was the first to speak after Einstein sat down. He thanked him for the lecture and then told everyone he disagreed. He reiterated his firmly held belief that quanta were necessary only in the exchange between matter and radiation. To believe as Einstein did that light was actually made up of quanta, Planck said, was 'not yet necessary'. Only Johannes Stark stood up to support Einstein. Sadly, he, like Lenard, would later become a Nazi and the two of them would attack Einstein and his work as 'Jewish Physics'.

Einstein left the Patent Office to devote more of his time to research. He was in for a rude awakening when he arrived in Zurich. The time he needed to prepare for the seven hours of lectures that he gave each week left him complaining that his 'actual free time is less than in Bern'.[83] The students were struck by the shabby appearance of their new professor, but Einstein quickly gained their respect and affection by his informal style as he encouraged them to interrupt if anything was unclear. Outside formal lectures, at least once a week he took his students along to the Café Terasse to chat and gossip until closing time. Before long he got used to his workload and turned his attention to using the quantum to solve a long-standing problem.

In 1819 two French scientists, Pierre Dulong and Alexis Petit, measured the specific heat capacity, the amount of energy needed to raise the temperature of a kilogram of a substance by one degree, for various metals from copper to gold. For the next 50 years no one who believed in atoms doubted their conclusion that 'the atoms of all simple bodies have exactly the same heat capacity'.[84] It therefore came as a great surprise when, in the 1870s, exceptions were discovered.

Imagining that the atoms of a substance oscillated when heated, Einstein adapted Planck's approach as he tackled the specific heat anomalies. Atoms could not oscillate with just any frequency, but were 'quantised' – able to oscillate only with those frequencies that were multiples of a certain 'fundamental' frequency. Einstein came up with a new theory of how solids absorb heat. Atoms are permitted to absorb energy only in discrete amounts, quanta. However, as the temperature drops, the amount of energy the substance has decreases, until there is not enough available to provide each atom with the correct-sized quantum of energy. This results in less energy being taken up by the solid and leads to a decrease in specific heat.

For three years there was hardly a murmur of interest in what Einstein had done, despite the fact that he had shown how the quantisation of energy – how at the atomic level energy comes wrapped up in bite-sized chunks – resolved a problem in a completely new area of physics. It was Walter Nernst, an eminent physicist from Berlin, who made others sit up and take note as they discovered that he had been to see Einstein in Zurich. Soon it was clear why. Nernst had succeeded in accurately measuring the specific heats of solids at low temperatures and found the results to be in total agreement with Einstein's predictions based on his quantum solution.

With each passing success his reputation soared ever higher, and Einstein was offered an ordinary professorship at the German University in Prague. It was an opportunity he could not refuse, even if it meant leaving

Switzerland after fifteen years. Einstein, Mileva and their sons Hans Albert and Eduard, who was not yet one, moved to Prague in April 1911.

'I no longer ask whether these quanta really exist', Einstein wrote to his friend Michele Besso soon after taking up his new post. 'Nor do I try to construct them any longer, for I now know that my brain cannot get through in this way.' Instead, he told Besso, he would limit himself to trying to understand the consequences of the quantum.[85] There were others who also wanted to try. Less than a month later, on 9 June, Einstein received a letter and an invitation from an unlikely correspondent. Ernst Solvay, a Belgian industrialist who had made a substantial fortune by revolutionising the manufacture of sodium carbonate, offered to pay 1,000 francs to cover his travel expenses if he agreed to attend a week-long 'Scientific Congress' to be held in Brussels later that year from 29 October to 4 November.[86] He would be one of a select group of 22 physicists from across Europe brought together to discuss 'current questions concerning the molecular and kinetic theories'. Planck, Rubens, Wien and Nernst would be attending. It was a summit meeting on the quantum.

Planck and Einstein were among the eight asked to prepare reports on a particular topic. To be written in French, German or English, they were to be sent out to the participants before the meeting and serve as the starting point for discussion during the planned sessions. Planck would discuss blackbody radiation theory, while Einstein had been assigned his quantum theory of specific heat. Although Einstein was accorded the honour of giving the final talk, a discussion of his quantum theory of light was not on the agenda.

'I find the whole undertaking extremely attractive,' Einstein wrote to Walter Nernst, 'and there is little doubt in my mind that you are its heart and soul.'[87] By 1910 Nernst believed that the time was ripe to get to grips with the quantum that he regarded as nothing more than a 'rule with most curious, indeed grotesque properties'.[88] He convinced Solvay to finance the conference and the Belgian spared no expense booking the plush Hotel Metropole as the venue. In its luxurious surroundings, with all their needs catered for, Einstein and his colleagues spent five days talking about the quantum. Whatever slim hopes he harboured for progress at what he called 'the Witches' Sabbath', Einstein returned to Prague disappointed and complained of learning nothing that he did not know before.[89]

Nevertheless, he had enjoyed getting to know some of the other 'witches'. Marie Curie, whom he found to be 'unpretentious', appreciated 'the clearness of his mind, the shrewdness with which he marshalled his facts and the depth of his knowledge'.[90] During the congress it was announced that she had been awarded the Nobel Prize for chemistry. She had become the first scientist to win two, having already won the physics prize in 1903. It was a tremendous achievement that was overshadowed by the scandal that broke around her during the congress. The French press had learned that she was having an affair with a married French physicist. Paul Langevin, a slender man with an elegant moustache, was a delegate at the conference and the papers were full of stories that the pair had eloped. Einstein, who had seen no signs of a special relationship between the two, dismissed the reports as rubbish. Despite her 'sparkling intelligence', he thought Curie was 'not attractive enough to represent a danger to anyone'.[91]

Even though at times he appeared to waver under the strain, Einstein had been the first to learn to live with the quantum, and by doing so revealed a hidden element of the true nature of light. Another young theorist also learned to live with the quantum after he used it to resurrect a flawed and neglected model of the atom.

Chapter 3

THE GOLDEN DANE

M*anchester, England, Wednesday, 19 June 1912.* 'Dear Harald, Perhaps I have found out a little about the structure of atoms,' Niels Bohr wrote to his younger brother.[1] 'Don't talk about it to anybody,' he warned, 'for otherwise I couldn't write to you so soon.' Silence was essential for Bohr, as he hoped to do what every scientist dreams of: unveiling 'a little bit of reality'. There was still work to be done and he was 'eager to finish it in a hurry, and to do that I have taken off a couple of days from the laboratory (this is also a secret)'. It would take the 26-year-old Dane much longer than he thought to turn his fledgling ideas into a trilogy of papers all entitled 'On the Constitution of Atoms and Molecules'. The first, published in July 1913, was truly revolutionary, as Bohr introduced the quantum directly into the atom.

———

It was his mother Ellen's 25th birthday when Niels Henrik David Bohr was born on 7 October 1885 in Copenhagen. She had returned to the comfort of her parents' home for the birth of her second child. Across the wide cobbled street from Christianborg Castle, the seat of the Danish parliament, Ved Stranden 14 was one of the most magnificent residences in the city. A banker and politician, her father was one of the wealthiest men in

Denmark. Although the Bohrs did not stay there long, it was to be the first of the grand and elegant homes in which Niels lived throughout his life.

Christian Bohr was the distinguished professor of physiology at Copenhagen University. He had discovered the role of carbon dioxide in the release of oxygen by haemoglobin, and together with his research on respiration it led to nominations for the Nobel Prize for physiology or medicine. From 1886 until his untimely death in 1911, at just 56, the family lived in a spacious apartment in the university's Academy of Surgery.[2] Situated in the city's most fashionable street and a ten-minute walk from the local school, it was ideal for the Bohr children: Jenny, two years older than Niels, and Harald, eighteen months younger.[3] With three maids and a nanny to look after them, they enjoyed a comfortable and privileged childhood far removed from the squalid and overcrowded conditions in which most of Copenhagen's ever-increasing inhabitants lived.

His father's academic position and his mother's social standing ensured that many of Denmark's leading scientists and scholars, writers and artists were regular visitors to the Bohr home. Three such guests were, like Bohr senior, members of the Royal Danish Academy of Sciences and Letters: the physicist Christian Christiansen, the philosopher Harald Høffding and the linguist Vilhelm Thomsen. After the Academy's weekly meeting, the discussion would continue at the home of one of the quartet. In their teens, whenever their father played host to his fellow Academicians, Niels and Harald were allowed to eavesdrop on the animated debates that took place. It was a rare opportunity to listen to the intellectual concerns of a group of such men as the mood of *fin-de-siècle* gripped Europe. They left on the boys, as Niels said later, 'some of our earliest and deepest impressions'.[4]

Bohr the schoolboy excelled at mathematics and science, but had little aptitude for languages. 'In those days,' recalled a friend, 'he was definitely not afraid to use his strength when it came to blows during the break between classes.'[5] By the time he enrolled at Copenhagen University, then Denmark's only university, to study physics in 1903, Einstein had spent more than a year at the Patent Office in Bern.[6] When he received his Master's degree in 1909, Einstein was extraordinary professor of theoretical physics at the University of Zurich and had received his first nomination for the Nobel Prize. Bohr had also distinguished himself, albeit on a far smaller stage. In 1907, aged 21, he won the Gold Medal of the Royal Danish Academy with a paper on the surface tension of water. It was the reason why his father, who had won the silver medal in 1885, often proudly proclaimed, 'I'm silver but Niels is gold'.[7]

Bohr struck gold after his father persuaded him to abandon the laboratory for a place in the countryside to finish writing his award-winning paper. Although he submitted it just hours before the deadline, Bohr still found something to add, and handed in a postscript two days later. The need to rework any piece of writing until he was satisfied that it conveyed exactly what he wanted verged on an obsession. A year before he finished his doctoral thesis, Bohr admitted that he had already written 'fourteen more or less divergent rough drafts'.[8] Even the simple act of penning a letter became a protracted affair. One day Harald, seeing a letter lying on Niels' desk, offered to post it, only to be told: 'Oh no, that is just one of the first drafts for a rough copy.'[9]

All their lives, the brothers remained the closest of friends. Apart from mathematics and physics they shared a passion for sport, particularly football. Harald, the better player, won a silver medal at the 1908 Olympics as a member of the Danish football team that lost to England in the final. Also regarded by many to be intellectually more gifted, he gained a doctorate in mathematics a year before Niels received his in physics in May 1911. Their father, however, always maintained that his eldest son was 'the special one in the family'.[10]

Dressed in white tie and tails as custom demanded, Bohr began the public defence of his doctoral thesis. It lasted just 90 minutes, the shortest on record. One of the two examiners was his father's friend Christian Christiansen. He regretted that no Danish physicist 'was well enough informed about the theory of metals to be able to judge a dissertation on the subject'.[11] Nevertheless, Bohr was awarded his doctorate and sent copies of the thesis to men like Max Planck and Hendrik Lorentz. When no one replied he knew it had been a mistake to send it without first having it translated. Instead of German or French, which many leading physicists spoke fluently, Bohr decided on an English translation and managed to convince a friend to produce one.

Whereas his father had chosen Leipzig and his brother Göttingen, German universities being the traditional place for high-flying Danes to complete their education, Bohr chose Cambridge University. The intellectual home of Newton and Maxwell was for him 'the centre of physics'.[12] The translated thesis would be his calling card. He hoped that it would lead to a dialogue with Sir Joseph John Thomson, the man he described later as 'the genius who showed the way for everybody'.[13]

After a lazy summer of sailing and hiking, Bohr arrived in England at the end of September 1911 on a one-year scholarship funded by Denmark's famous Carlsberg brewery. 'I found myself rejoicing this morning, when I stood outside a shop and by chance happened to read the address "Cambridge" over the door', he wrote to his fiancée Margrethe Nørland.[14] The letters of introduction and the Bohr name led to a warm welcome from the university's physiologists who remembered his late father. They helped him find a small two-room flat on the edge of town and he was kept 'very busy with arrangements, visits and dinner parties'.[15] But for Bohr it was his meeting with Thomson, J.J. to his friends and students alike, which soon preyed on his mind.

A bookseller's son from Manchester, Thomson had been elected the third head of the Cavendish Laboratory in 1884 within a week of his 28th birthday. He was an unlikely choice, after James Clerk Maxwell and Lord Rayleigh, to lead the prestigious experimental research facility, and not just because of his youth. 'J.J. was very awkward with his fingers,' one of his assistants later admitted, 'and I found it necessary not to encourage him to handle the instruments.'[16] Yet if the man who won the Nobel Prize for discovering the electron lacked a delicate touch, others testified to Thomson's 'intuitive ability to comprehend the inner working of intricate apparatus without the trouble of handling it'.[17]

The polite manner of the slightly dishevelled Thomson, the epitome of the absent-minded professor in his round-rimmed glasses, tweed jacket and winged collar, helped calm Bohr's nerves when they first met. Eager to impress, he had walked into the professor's office clutching his thesis and a book written by Thomson. Opening the book, Bohr pointed to an equation and said, 'This is wrong.'[18] Though not used to having his past mistakes paraded before him in such a forthright manner, J.J. promised to read Bohr's thesis. Placing it on top of a stack of papers on his overcrowded desk, he invited the young Dane to dinner the following Sunday.

Initially delighted, as the weeks passed and the thesis remained unread, Bohr became increasingly anxious. 'Thomson,' he wrote to Harald, 'has so far not been easy to deal with as I thought the first day.'[19] Yet his admiration for the 55-year-old was undiminished: 'He is an excellent man, incredibly clever and full of imagination (you should hear one of his elementary lectures) and extremely friendly; but he is so immensely busy with so many things, and he is so absorbed in his work that it is very difficult to get to talk to him.'[20] Bohr knew that his poor English did not help. So with the aid of a dictionary he began reading *The Pickwick Papers* as he fought to overcome the language barrier.

Early in November, Bohr went to see a former student of his father's who was now the professor of physiology at Manchester University. During the visit, Lorrain Smith introduced him to Ernest Rutherford, who had just returned from a physics conference in Brussels.[21] The charismatic New Zealander, he recalled years later, 'spoke with characteristic enthusiasm about the many new prospects in physical science'.[22] After being regaled with a 'vivid account of the discussions at the Solvay meeting', Bohr left Manchester charmed and impressed by Rutherford – both the man and the physicist.[23]

On his first day, in May 1907, the new head of physics at Manchester University caused a stir as he searched for his new office. 'Rutherford went up three stairs at a time, which was horrible to us, to see a Professor going up the stairs like that', remembered a laboratory assistant.[24] But within a few weeks the boundless energy and earthy no-nonsense approach of the 36-year-old had captivated his new colleagues. Rutherford was on his way to creating an exceptional research team whose success over the next decade or so would be unmatched. It was a group shaped as much by Rutherford's personality as his inspired scientific judgement and ingenuity. He was not only its head, but also its heart.

Born on 30 August 1871 in a small, single-storey wooden house in Spring Grove on New Zealand's South Island, Rutherford was the fourth of twelve children. His mother was a schoolteacher and his father ended up working in a flax mill. Given the harshness of life in the scattered rural community, James and Martha Rutherford did what they could to ensure that their children had a chance to go as far as talent and luck would carry them. For Ernest it meant a series of scholarships that took him to the other side of the world and Cambridge University.

When he arrived at the Cavendish to study under Thomson in October 1895, Rutherford was far from the exuberant and self-confident man he would become within a few years. The transformation began as he continued work started in New Zealand on the detection of 'wireless' waves, later called radio waves. In only a matter of months Rutherford developed a much-improved detector and toyed with the idea of making money from it. Just in time, he realised that exploiting research for financial gain in a scientific culture where patents were rare would harm the chances of a young man yet to make his reputation. As the Italian Guglielmo Marconi amassed a fortune that could have been his, Rutherford never regretted abandoning his detector to explore a discovery that had been front-page news around the world.

On 8 November 1895, Wilhelm Röntgen found that every time he passed a high-voltage electric current through an evacuated glass tube, some unknown radiation was causing a small paper screen coated with barium platinocyanide to glow. When Röntgen, the 50-year-old professor of physics at the University of Würzburg, was later asked what he had thought on discovering his mysterious new rays, he replied: 'I did not think;

I investigated.'[25] For nearly six weeks, he did 'the same experiment over and over again to make absolutely certain that the rays actually existed'.[26] He confirmed that the tube was the source of the strange emanation causing the fluorescence.[27]

Röntgen asked his wife Bertha to place her hand on a photographic plate while he exposed it to 'X-rays', as he called the unknown radiation. After fifteen minutes Röntgen developed the plate. Bertha was frightened when she saw the outlines of her bones, her two rings and the dark shadows of her flesh. On 1 January 1896, Röntgen mailed copies of his paper, 'A New Kind of Rays', together with photographs of weights in a box and the bones in Bertha's hand, to leading physicists in Germany and abroad. Within days, news of Röntgen's discovery and his amazing photographs spread like wildfire. The world's press latched on to the ghostly photograph revealing the bones in his wife's hand. Within a year, 49 books and over a thousand scientific and semi-popular articles on X-rays would be published.[28]

Thomson had begun studying the sinister-sounding X-rays even before an English translation of Röntgen's paper appeared in the weekly science journal *Nature* on 23 January. Engaged in investigating the conduction of electricity through gases, Thomson turned his attention to X-rays when he read that they turned a gas into a conductor. Quickly confirming the claim, he asked Rutherford to help measure the effects of passing X-rays through a gas. For Rutherford the work led to four published papers in the next two years that brought him international recognition. Thomson provided a brief note to the first, suggesting, correctly as it later proved, that X-rays, like light, were a form of electromagnetic radiation.

While Rutherford was busy conducting his experiments, in Paris the Frenchman Henri Becquerel was trying to discover whether phosphorescent substances, which glow in the dark, could also emit X-rays. Instead he found that uranium compounds emitted radiation whether they were phosphorescent or not. Becquerel's announcement of his 'uranic rays' aroused little scientific curiosity and no newspapers clamoured to report his discovery. Only a handful of physicists were interested in Becquerel's rays for, like their discoverer, most believed that only uranium compounds emitted them. However, Rutherford decided to investigate the effects of 'uranic rays' on the electrical conductivity of gases. It was a decision he later described as the most important of his life.

Testing the penetration of uranium radiation using wafer-thin layers of 'Dutch metal', a copper-zinc alloy, Rutherford found that the amount of radiation detected depended on the number of layers used. At a certain point, adding further layers had little effect in reducing the intensity of radiation, but then surprisingly it began to fall once again as more layers were added. After repeating the experiment with different materials and finding the same general pattern, Rutherford could offer only one explanation. Two types of radiation were being emitted, and he called them alpha and beta rays.

When the German physicist Gerhard Schmidt announced that thorium and its compounds also emitted radiation, Rutherford compared it with alpha and beta rays. He found the thorium radiation to be more powerful and concluded that 'rays of a more penetrative kind were present'.[29] These were later called gamma rays.[30] It was Marie Curie who introduced the term 'radioactivity' to describe the emission of radiation and who labelled substances that emitted 'Becquerel rays' as 'radioactive'. She believed that since radioactivity was not confined to uranium alone, it must be an atomic phenomenon. It set her on the path to discovering, with her husband Pierre, the radioactive elements radium and polonium.

In April 1898, as Curie's first paper was published in Paris, Rutherford learned that there was a vacant professorship at McGill University in Montreal, Canada. Although acknowledged as a pioneer in the new field of radioactivity, Rutherford put his name forward with little expectation of being appointed, despite a glowing letter of recommendation from Thomson. 'I have never had a student with more enthusiasm or ability for original research than Mr Rutherford,' wrote Thomson, 'and I am sure if elected, he would establish a distinguished school of physics at Montreal.'[31] He concluded: 'I should consider any institution fortunate that secured the services of Mr Rutherford as professor of physics.' After a stormy voyage,

Rutherford, just turned 27, arrived in Montreal at the end of September and stayed for the next nine years.

Even before he left England he knew that he was 'expected to do a lot of original work and to form a research school to knock the shine out of the Yankees!'[32] He did just that, beginning with the discovery that the radio-activity of thorium decreased by half in one minute and then by half again in the next. After three minutes it had fallen to an eighth of its original value.[33] Rutherford called this exponential reduction of radioactivity the 'half-life', the time taken for the intensity of radiation emitted to fall by half. Each radioactive element had its own characteristic half-life. Then came the discovery that would earn him the professorship in Manchester and a Nobel Prize.

In October 1901, Rutherford and Frederick Soddy, a 25-year-old British chemist at Montreal, began a joint study of thorium and its radiation and were soon faced with the possibility that it could be turning into another element. Soddy recalled how he stood stunned at the thought and let slip, 'this is transmutation'. 'For Mike's sake, Soddy, don't call it transmutation', warned Rutherford. 'They'll have our heads off as alchemists.'[34]

The pair were soon convinced that radioactivity was the transforma-tion of one element into another through the emission of radiation. Their heretical theory was met with widespread scepticism but the experimen-tal evidence quickly proved decisive. Their critics had to discard long-cherished beliefs in the immutability of matter. No longer an alchemist's dream, but a scientific fact: all radioactive elements did spontaneously transform into other elements, the half-life measuring the time it took for half the atoms to do so.

'Youthful, energetic, boisterous, he suggested anything but the scientist', is how Chaim Weizmann, later the first president of Israel but then a chem-ist at Manchester University, remembered Rutherford. 'He talked readily and vigorously on any subject under the sun, often without knowing anything about it. Going down to the refectory for lunch, I would hear the loud, friendly voice rolling up the corridor.'[35] Weizmann found Rutherford 'devoid of any political knowledge or feelings, being entirely taken up with his epoch-making scientific work'.[36] At the centre of that work lay his use of the alpha particle to probe the atom.

But what exactly was an alpha particle? It was a question that had long vexed Rutherford after he discovered that alpha rays were in fact particles with a positive charge that were deflected by strong magnetic fields. He believed that an alpha particle was a helium ion, a helium atom that had lost two electrons, but never said so publicly because the evidence was purely circumstantial. Now, almost ten years after discovering alpha rays, Rutherford hoped to find definitive proof of their true character. Beta rays had already been identified as fast-moving electrons. With the help of another young assistant, this time 25-year-old German Hans Geiger, Rutherford confirmed in the summer of 1908 what he had long suspected: an alpha particle was indeed a helium atom that had lost two electrons.

'The scattering is the devil', Rutherford had complained as he and Geiger tried to unmask the alpha particle.[37] He had first noticed the effect two years earlier in Montreal when some alpha particles that had passed through a sheet of mica were slightly deflected from their straight-line trajectory, causing fuzziness on a photographic plate. Rutherford made a mental note to follow it up. Soon after arriving in Manchester, he had drawn up a list of potential research topics. Rutherford now asked Geiger to investigate one of those items – the scattering of alpha particles.

Together they devised a simple experiment that involved counting scintillations, tiny flashes of light produced by alpha particles when they strike a paper screen coated with zinc sulphide, after passing through a thin sheet of gold foil. Counting scintillations was an arduous task, with long hours spent in total darkness. Luckily, according to Rutherford, Geiger was 'a demon at the work and could count at intervals for a whole night without disturbing his equanimity'.[38] He found that alpha particles either passed straight through the gold foil or were deflected by one or two degrees. This was as expected. However, surprisingly, Geiger also reported finding a few alpha particles 'deflected through quite an appreciable angle'.[39]

Before he could fully consider the implications, if any, of Geiger's results, Rutherford was awarded the Nobel Prize for chemistry for discovering that radioactivity was the transformation of one element into another. For a man who regarded 'all science as either physics or stamp collecting', he appreciated the funny side of his own instant transmutation from physicist to chemist.[40] After returning from Stockholm with his prize, Rutherford learnt to evaluate the probabilities associated with different degrees of alpha particle scattering. His calculations revealed that there was a very small chance, almost zero, that an alpha particle passing through gold foil would undergo multiple scatterings resulting in an overall large-angle deflection.

It was while Rutherford was preoccupied with these calculations that Geiger spoke to him about assigning a project to Ernest Marsden, a promising undergraduate. 'Why not,' said Rutherford, 'let him see if any alpha particles can be scattered through a large angle?'[41] He was surprised when Marsden did. As the search continued at ever-larger angles, there should have been none of the tell-tale flashes of light that Marsden had seen, signalling alpha particles crashing into the zinc sulphide screen.

As Rutherford struggled to make sense of 'the nature of the huge electric or magnetic forces which could turn aside or scatter a beam of alpha particles', he asked Marsden to check if any were reflected backwards.[42] Not expecting him to find anything, he was utterly astonished when Marsden discovered alpha particles bouncing off the gold foil. 'It was,' Rutherford said, 'almost as incredible as if you had fired a 15-inch shell at a piece of tissue paper and it came back and hit you.'[43]

Geiger and Marsden set about making comparative measurements using different metals. They found that gold scattered backwards almost twice as many alpha particles as silver and twenty times more than aluminium. Only one alpha particle in every 8,000 bounced off a sheet of platinum. When they published these and other results in June 1909, Geiger and Marsden simply recounted the experiments and stated the facts without further comment. A baffled Rutherford brooded for the next eighteen months as he tried to think his way through to an explanation.

The existence of atoms had been a matter of considerable scientific and philosophical debate throughout the nineteenth century, but by 1909 the reality of atoms had been established beyond any reasonable doubt. The critics of atomism were silenced by the sheer weight of evidence against them, two key pieces of which were Einstein's explanation of Brownian motion and its confirmation, and Rutherford's discovery of the radioactive transformation of elements. After decades of argument, in which many eminent physicists and chemists had denied its existence, the most favoured representation of the atom to emerge was the so-called 'plum pudding' model put forward by J.J. Thomson.

In 1903 Thomson suggested that the atom was a ball of massless, positive charge in which were embedded like plums in a pudding the negatively-charged electrons he had discovered six years earlier. The positive charge would neutralise the repulsive forces between the electrons that would otherwise tear the atom apart.[44] For any given element, Thomson envisaged these atomic electrons to be uniquely arranged in a set of concentric rings. He argued that it was the different number and distribution of electrons in gold and lead atoms, for example, which distinguished the metals from one another. Since all the mass of a Thomson atom was due to the electrons it contained, it meant there were thousands in even the lightest atoms.

Exactly one hundred years earlier, in 1803, the English chemist John Dalton first put forward the idea that atoms of every element were uniquely characterised by their weight. With no direct way of measuring atomic weights, Dalton determined their relative weights by examining the proportions in which different elements combined to form various compounds. First he needed a benchmark. Hydrogen being the lightest known element, Dalton assigned it an atomic weight of one. The atomic weights of all the other elements were then fixed relative to that of hydrogen.

Thomson knew his model was wrong after studying the results of experiments involving the scattering of X-rays and beta particles by atoms. He had overestimated the number of electrons. According to his new calculations, an atom could not have more electrons than prescribed by its atomic weight. The precise number of electrons in the atoms of the different elements was unknown, but this upper limit was quickly accepted as a first step in the right direction. The hydrogen atom with an atomic weight of one could have only one electron. However, the helium atom with an atomic weight of four could have two, three, or even four electrons, and so on for the other elements.

This drastic reduction in electron numbers revealed that most of the weight of an atom was due to the diffuse sphere of positive charge. Suddenly, what Thomson had originally invoked as nothing more than a necessary artifice to produce a stable, neutral atom took on a reality of its own. But even this new, improved model could not explain alpha particle scattering and failed to pin down the exact number of electrons in a particular atom.

Rutherford believed that alpha particles were scattered by an enormously strong electric field within the atom. But inside J.J.'s atom, with its positive charge evenly distributed throughout, there was no such intense electric field. Thomson's atom simply could not send alpha particles hurtling backwards. In December 1910, Rutherford finally managed to 'devise an atom much superior to J.J.'s'.[45] 'Now,' he told Geiger, 'I know what the atom looks like!'[46] It was nothing like Thomson's.

Rutherford's atom consisted of a tiny positively-charged central core, the nucleus, which contained virtually all the atom's mass. It was 100,000 times smaller than the atom, occupying only a minute volume, 'like a fly in a cathedral'.[47] Rutherford knew that electrons inside an atom could not be responsible for the large deflection of alpha particles, so to determine their exact configuration around the nucleus was unnecessary. His atom was no longer the 'nice hard fellow, red or grey in colour, according to taste' that he once, tongue-in-cheek, said he had been brought up to believe in.[48]

Most alpha particles would pass straight through Rutherford's atom in any 'collision', since they were too far from the tiny nucleus at its heart to suffer any deflection. Others would veer off course slightly as they encountered the electric field generated by the nucleus, resulting in a small deflection. The closer they passed to the nucleus, the stronger the effect of its electric field and the greater the deflection from their original path. But if an alpha particle approached the nucleus head-on, the repulsive force between the two would cause it to recoil straight back like a ball bouncing off a brick wall. As Geiger and Marsden had found, such direct hits were extremely rare. It was, Rutherford said, 'like trying to shoot a gnat in the Albert Hall at night'.[49]

Rutherford's model allowed him to make definite predictions, using a simple formula he had derived, about the fraction of scattered alpha particles to be found at any angle of deflection. He did not want to present his atomic model until it had been tested by a careful investigation of the angular distribution of scattered alpha particles. Geiger undertook the task and found alpha particle distribution to be in total agreement with Rutherford's theoretical estimates.

On 7 March 1911, Rutherford announced his atomic model in a paper presented at a meeting of the Manchester Literary and Philosophical Society. Four days later, he received a letter from William Henry Bragg, the professor of physics at Leeds University, informing him that 'about 5 or 6 years ago' the Japanese physicist Hantaro Nagaoka had constructed an atom with 'a big positive centre'.[50] Unknown to Bragg, Nagaoka had visited Rutherford the previous summer as part of a grand tour of Europe's leading physics laboratories. Less than two weeks after Bragg's letter, Rutherford received one from Tokyo. Nagaoka wrote offering his gratitude 'for the great kindness you showed me in Manchester' and pointing out that in 1904 he had proposed a 'Saturnian' model of the atom.[51] It consisted of a large heavy centre surrounded by rotating rings of electrons.[52]

'You will notice that the structure assumed in my atom is somewhat similar to that suggested by you in your paper some years ago', acknowledged Rutherford in his reply. Though alike in some respects, there were significant differences between the two models. In Nagaoka's the central body was positively-charged, heavy and occupied most of the flat pancake-like atom. Whereas Rutherford's spherical model had an incredibly tiny positively-charged core that contained most of the mass, leaving the atom largely empty. However, both models were fatally flawed and few physicists gave them a second thought.

An atom with stationary electrons positioned around a positive nucleus would be unstable, because the electrons with their negative charge would be irresistibly pulled towards it. If they moved around the nucleus, like planets orbiting the sun, the atom would still collapse. Newton had shown long ago that any object moving in a circle undergoes acceleration. According to Maxwell's theory of electromagnetism, if it is a charged particle, like an electron, it will continuously lose energy in the form of electromagnetic radiation as it accelerates. An orbiting electron would spiral into the nucleus within a thousandth of a billionth of a second. The very existence of the material world was compelling evidence against Rutherford's nuclear atom.

He had long been aware of what appeared to be an intractable problem. 'This necessary loss of energy from an accelerated electron,' Rutherford wrote in his 1906 book *Radioactive Transformations*, 'has been one of the greatest difficulties met with in endeavouring to deduce the constitution of a stable atom.'[53] But in 1911 he chose to ignore the difficulty: 'The question of the stability of the atom proposed need not be considered at this stage, for this will obviously depend upon the minute structure of the atom, and on the motion of the constituent charged part.'[54]

Geiger's initial testing of Rutherford's scattering formula had been quick and limited in scope. Marsden now joined him in spending most of the next year conducting a more thorough investigation. By July 1912 their results confirmed the scattering formula and the main conclusions of Rutherford's theory.[55] 'The complete check,' Marsden recalled years later, 'was a laborious but exciting task.'[56] In the process they also discovered that the charge of the nucleus, taking into account experimental error, was about half the atomic weight. With the exception of hydrogen, with an atomic weight of one, the number of electrons in all other atoms had to be approximately equal to half the atomic weight. It was now possible to nail down the number of electrons in a helium atom, for example, as two, where previously it could have been as many as four. However, this reduction in the number of electrons implied that Rutherford's atom radiated energy even more strongly than had previously been suspected.

As Rutherford recounted tales from the first Solvay conference for Bohr's benefit, he failed to mention that in Brussels neither he nor anyone else discussed his nuclear atom.

Back in Cambridge, the intellectual rapport that Bohr sought with Thomson never happened. Years later, Bohr identified one possible reason for the failure: 'I had no great knowledge of English and therefore I did not know how to express myself. And I could say only that this is incorrect. And he was not interested in the accusation that it was not correct.'[57] Infamous for neglecting papers and letters from students and colleagues alike, Thomson was also no longer actively engaged in electron physics.

Increasingly disenchanted, Bohr met Rutherford again at the Cavendish research students' annual dinner. Held in early December, it was a rowdy, informal affair with toasts, songs and limericks following a ten-course meal. Once again struck by the personality of the man, Bohr seriously began thinking about swapping Cambridge and Thomson for Manchester and Rutherford. Later that month he went to Manchester and discussed the possibility with Rutherford. A young man separated from his fiancée, Bohr desperately wanted something tangible to show for their year apart. Telling Thomson that he wanted 'to know something about radioactivity', Bohr was granted permission to leave at the end of the new term.[58] 'The whole thing was very interesting in Cambridge,' he admitted many years later, 'but it was absolutely useless.'[59]

With only four months left in England, Bohr arrived in Manchester in the middle of March 1912 to begin a seven-week course in the experimental techniques of radioactive research. With no time to lose, Bohr spent his evenings working on the application of electron physics to provide a better understanding of the physical properties of metals. With Geiger and Marsden among the instructors, he successfully completed the course and was given a small research project by Rutherford.

'Rutherford is a man whom one cannot be mistaken about,' Bohr wrote to Harald, 'he comes regularly to hear how things are going and talk about every little thing.'[60] Unlike Thomson, who seemed to him unconcerned about the progress of his students, Rutherford was 'really interested in the work of all people who are around him'. He had an uncanny ability to recognise scientific promise. Eleven of his students, along with several close collaborators, would win the Nobel Prize. As soon as Bohr arrived in Manchester, Rutherford wrote to a friend: 'Bohr, a Dane, has pulled out of Cambridge and turned up here to get some experience in radioactive work.'[61] Yet there was nothing in what Bohr had done to date to suggest that he was any different from the other eager young men in his laboratory, except the fact that he was a theorist.

Rutherford held a generally low opinion of theorists and never lost an opportunity to air it. 'They play games with their symbols,' he once told a colleague, 'but we turn out the real solid facts of Nature.'[62] On another occasion when invited to deliver a lecture on the trends of modern physics, he replied: 'I can't give a paper on that. It would only take two minutes. All I could say would be that the theoretical physicists have got their tails up and it is time that we experimentalists pulled them down again!'[63] Yet he had immediately liked the 26-year-old Dane. 'Bohr's different', he would say. 'He's a football player!'[64]

Late every afternoon, work in the laboratory stopped as the research students and staff gathered to chat over tea, cakes and slices of bread and butter. Rutherford would be there, sitting on a stool with plenty to say, whatever the subject. But most of the time the talk was simply of physics, particularly of the atom and radioactivity. Rutherford had succeeded in creating a culture where there was an almost tangible sense of discovery in the air, where ideas were openly exchanged and discussed in the spirit of co-operation, with no one afraid to speak – even a newcomer. At its centre was Rutherford, who Bohr knew was always prepared 'to listen to every young man, when he felt he had any idea, however modest, on his mind'.[65] The only thing Rutherford could not stand was 'pompous talk'. Bohr loved to talk.

Unlike Einstein who spoke and wrote fluently, Bohr frequently paused as he struggled to find the right words to express himself, whether in

Danish, English or German. When Bohr spoke, he was often only thinking aloud in search of clarity. It was during the tea breaks that he got to know the Hungarian Georg von Hevesy, who would win the 1943 Nobel Prize for chemistry for developing the technique of radioactive tracing that was to become a powerful diagnostic tool in medicine, with widespread applications in chemical and biological research.

Strangers in a strange country, speaking a language that both had yet to master, the pair formed an easy friendship that lasted a lifetime. 'He knew how to be helpful to a foreigner', Bohr said as he recalled how Hevesy, only a few months older, helped him ease into the life of the laboratory.[66] It was during their conversations that Bohr first began to focus on the atom, as Hevesy explained that so many radioactive elements had been discovered that there was not enough room to accommodate them all in the periodic table. The very names given to these 'radioelements', spawned in the process of radioactive disintegration of one atom into another, captured the sense of uncertainty and confusion surrounding their true place within the atomic realm: uranium-X, actinium-B, thorium-C. But there was, Hevesy told Bohr, a possible solution proposed by Rutherford's former Montreal collaborator, Frederick Soddy.

In 1907 it was discovered that two elements produced during radioactive decay, thorium and radiothorium, were physically different but chemically identical. Every chemical test they were subjected to failed to tell them apart. During the next few years, other such sets of chemically inseparable elements were discovered. Soddy, now based at Glasgow University, suggested that the only difference between these new radioelements and those with which they shared 'complete chemical identity' was their atomic weight.[67] They were like identical twins whose only distinguishing feature was a slight difference in weight.

Soddy proposed in 1910 that chemically inseparable radioelements, 'isotopes' as he later called them, were just different forms of the same element and should therefore share its slot in the periodic table.[68] It was an idea at odds with the existing organisation of elements within the periodic table, which listed them in order of increasing atomic weight, with

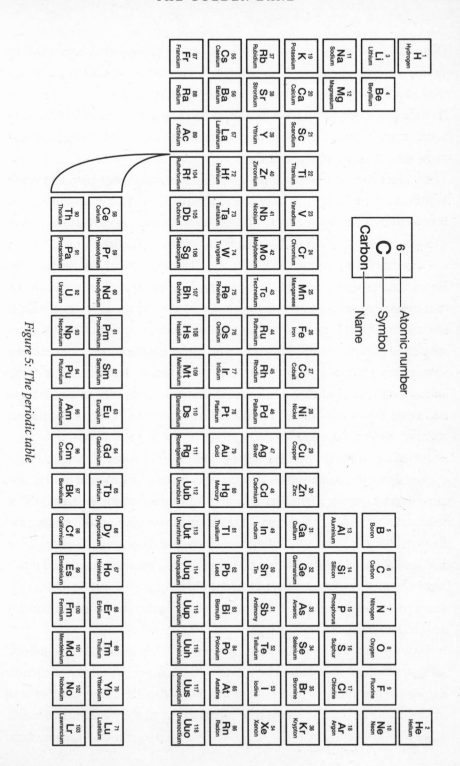

Figure 5: The periodic table

hydrogen first and uranium last. Yet the fact that radiothorium, radioactinium, ionium, and uranium-X were all chemically identical to thorium was strong evidence in favour of Soddy's isotopes.[69]

Until his chats with Hevesy, Bohr had shown no interest in Rutherford's atomic model. But he now had an idea: it was not enough to distinguish between the physical and chemical properties of an atom; one had to differentiate between nuclear and atomic phenomena. Ignoring the problem of its inevitable collapse, Bohr took Rutherford's nuclear atom seriously as he tried to reconcile isotopes with the use of atomic weights to order the periodic table. 'Everything,' he said later, 'then fell into line.'[70]

Bohr understood that it was the charge of the nucleus in Rutherford's atom that fixed the number of electrons it contained. Since an atom was neutral, possessing no overall charge, he knew that the positive charge of the nucleus had to be balanced by the combined negative charge of all its electrons. Therefore the Rutherford model of the hydrogen atom must consist of a nuclear charge of plus one and a single electron with a charge of minus one. Helium with a nuclear charge of plus two must have two electrons. This increase in nuclear charge coupled to a corresponding number of electrons led all the way up to the then heaviest-known element, uranium, with a nuclear charge of 92.

For Bohr the conclusion was unmistakable: it was nuclear charge and not atomic weight that determined the position of an element within the periodic table. From here he took the short step to the concept of isotopes. It was Bohr, not Soddy, who recognised nuclear charge as being the fundamental property that tied together different radioelements that were chemically identical but physically different. The periodic table could accommodate all the radioelements; they just had to be housed according to nuclear charge.

At a stroke, Bohr was able to explain why Hevesy had been unable to separate lead and radium-D. If the electrons determined the chemical properties of an element, then any two with the same number and arrangement of electrons would be identical twins, chemically inseparable. Lead and radium-D had the same nuclear charge, 82, and therefore the same number of electrons, 82, resulting in 'complete chemical identity'. Physically they were distinct because of their different nuclear masses: approximately 207 for lead and 210 for radium-D. Bohr had worked out that radium-D was an isotope of lead and as a result it was impossible to separate the two by any chemical means. Later, all isotopes were labelled with the name of the element of which they were an isotope and their atomic weight. Radium-D was lead-210.

Bohr had grasped the essential fact that radioactivity was a nuclear and not an atomic phenomenon. It allowed him to explain the process of radioactive disintegration in which one radioelement decayed into another with the emission of alpha, beta or gamma radiation as a nuclear event. Bohr realised that if radioactivity originated in the nucleus, then a uranium nucleus with a charge of plus 92 transmuting into uranium-X by emitting an alpha particle lost two units of positive charge, leaving behind a nucleus with a charge of plus 90. This new nucleus could not hold on to all of the original 92 atomic electrons, quickly losing two to form a new neutral atom. Every new atom formed as the product of radioactive decay immediately either acquires or loses electrons so as regain its neutrality. Uranium-X with a positive nuclear charge of 90 is an isotope of thorium. They both 'possessed the same nuclear charge and differed only in the mass and intrinsic structure of the nucleus', explained Bohr.[71] It was the reason why those who tried, failed to separate thorium, with an atomic weight of 232, and 'uranium-X', thorium-234.

His theory of what was happening at the nuclear level in radioactive disintegration implied, Bohr said later, 'that by radioactive decay the element, quite independently of any change in its atomic weight, would shift its place in the periodic table by two steps down or one step up, corresponding to the decrease or increase in the nuclear charge accompanying the emission of alpha or beta rays, respectively'.[72] Uranium decaying with the emission of an alpha particle into thorium-234 ended up two places further back in the periodic table.

Beta particles, being fast-moving electrons, have a negative charge of minus one. If a nucleus emits a beta particle, its positive charge increases by one – as if two particles, one positive and the other negative, that existed in harmony as a neutral pair had been ripped apart with the ejection of the electron, leaving behind its positive partner. The new atom produced by beta decay has a nuclear charge that is one greater than the disintegrating atom, moving it one place to the right in the periodic table.

When Bohr took his ideas to Rutherford he was warned about the danger of 'extrapolating from comparatively meagre experimental evidence'.[73] Surprised by this muted reception, he attempted to convince Rutherford 'that it would be the final proof of his atom'.[74] He failed. Part of the problem lay in Bohr's inability to express his ideas clearly. Rutherford, preoccupied with writing a book, did not make the time to fully grasp the significance of what Bohr had done. Rutherford believed that although alpha particles were emitted from the nucleus, beta particles were just atomic electrons somehow ejected from a radioactive atom. Despite Bohr's trying on five separate occasions to persuade him, Rutherford hesitated in following his logic all the way to its conclusion.[75] Sensing that Rutherford was by now becoming 'a bit impatient' with him and his ideas, Bohr decided to let the matter rest.[76] Others did not.

Frederick Soddy soon spotted the same 'displacement laws' as Bohr, but unlike the young Dane, he was able to publish his research without first having to seek approval of a superior. Nobody was surprised that Soddy was at the forefront of these breakthroughs. But no one could have guessed that an eccentric 42-year-old Dutch lawyer would introduce an idea of fundamental importance. In July 1911, in a short letter to the journal *Nature*, Antonius Johannes van den Broek speculated that the nuclear charge of a particular element is determined by its place in the periodic table, its atomic number, not its atomic weight. Inspired by Rutherford's atomic model, van den Broek's idea was based upon various assumptions that turned out to be wrong, such as nuclear charge being equal to half the atomic weight of the element. Rutherford was suitably annoyed that a lawyer should publish 'a lot of guesses for fun without sufficient foundation'.[77]

Having failed to gain any support, on 27 November 1913 in another letter to *Nature*, van den Broek dropped the assumption that the nuclear charge was equal to half the atomic weight. He did so after the publication of the extensive study by Geiger and Marsden into alpha particle scattering. A week later, Soddy wrote to *Nature* explaining that van den Broek's idea made clear the meaning of the displacement laws. Then came an endorsement from Rutherford: 'The original suggestion of van den Broek that the charge on the nucleus is equal to the atomic number and not to half the atomic weight seems to me very promising.' He was writing in praise of van den Broek's proposal a little more than eighteen months after advising Bohr against pursuing similar ideas.

Bohr never complained that he had missed out on being the first to publish the concept of atomic number, or those ideas that won Soddy the Nobel Prize for chemistry in 1921, due to Rutherford's lack of enthusiasm.[78] 'The confidence in his judgement,' Bohr fondly remembered, 'and our admiration for his powerful personality was the basis for the inspiration felt by all in his laboratory, and made us all try our best to deserve the kind and untiring interest he took in the work of everyone.'[79] In fact, Bohr continued to regard an approving word from Rutherford as 'the greatest encouragement for which any of us could wish'.[80] The reason why he could afford to be so generous, when others would have been left feeling disappointed and bitter, was what happened next.

After Rutherford dissuaded him from publishing his innovative ideas, by chance Bohr came across a recently published paper that grabbed his attention.[81] It was the work of the only theoretical physicist on Rutherford's staff, Charles Galton Darwin, the grandson of the great naturalist. The paper concerned the energy lost by alpha particles as they passed through matter rather than being scattered by atomic nuclei. It was a problem that J.J. Thomson had originally investigated using his own atomic model, but which Darwin now re-examined on the basis of Rutherford's atom.

Rutherford had developed his atomic model using the large-angle alpha particle scattering data gathered by Geiger and Marsden. He knew that atomic electrons could not be responsible for such large-angle scattering and so ignored them. In formulating his scattering law that predicted the fraction of scattered alpha particles to be found at any angle of deflection, Rutherford had treated the atom as a naked nucleus. Afterwards he simply placed the nucleus at the centre of the atom and surrounded it with electrons without saying anything about their possible arrangement. In his paper, Darwin adopted a similar approach when he ignored any influence that the nucleus may have exerted on the passing alpha particles and concentrated solely on the atomic electrons. He pointed out that the energy lost by an alpha particle as it passed through matter was due almost entirely to collisions between it and atomic electrons.

Darwin was unsure how electrons were arranged inside Rutherford's atom. His best guess was that they were evenly distributed either throughout the atom's volume or over its surface. His results depended only on the size of the nuclear charge and the atom's radius. Darwin found that his values for various atomic radii were in disagreement with existing estimates. As he read this paper, Bohr quickly identified where Darwin had gone wrong. He had mistakenly treated the negatively-charged electrons as if they were free, instead of being bound to the positively-charged nucleus.

Bohr's greatest asset was his ability to identify and exploit failures in existing theory. It was a skill that served him well throughout his career, as he started much of his own work from spotting errors and inconsistencies in that of others. On this occasion, Darwin's mistake was Bohr's point of departure. While Rutherford and Darwin had considered the nucleus and the atomic electrons separately, each ignoring the other component of the atom, Bohr realised that a theory that succeeded in explaining how alpha particles interacted with atomic electrons might reveal the true structure of the atom.[82] Any lingering disappointment over Rutherford's reaction to his earlier ideas was forgotten as he set about trying to rectify Darwin's mistake.

Bohr abandoned his usual practice of drafting letters even to his brother. 'I am not getting along badly at the moment,' Bohr reassured Harald, 'a couple of days ago I had a little idea with regard to understanding the absorption of alpha-rays (it happened in this way: a young mathematician here, C.G. Darwin (grandson of the real Darwin), has just published a theory about this problem, and I felt that it not only wasn't quite right mathematically (however, only slightly wrong) but very unsatisfactory in the basic conception, and I have worked out a little theory about it, which, even if it isn't much, perhaps may throw some light on certain things connected with the structure of atoms). I am planning to publish a little paper about it very soon.'[83] Not having to go to the laboratory 'has been wonderfully convenient for working out my little theory', he admitted.[84]

Until he had put flesh onto the bare bones of his emerging ideas, the only person in Manchester whom Bohr was willing to confide in was Rutherford. Though surprised by the direction the Dane had taken, Rutherford listened and this time encouraged him to continue. With his approval, Bohr stopped going to the laboratory. He was under pressure, since his time in Manchester was almost up. 'I believe I have found out a few things; but it is certainly taking more time to work them out than I was foolish enough to believe at first', he wrote to Harald on 17 July, a month after first sharing his secret. 'I hope to have a little paper ready to show Rutherford before I leave, so I'm busy, so busy; but the unbelievable heat here in Manchester doesn't exactly help my diligence. How I look forward to talking to you!'[85] He wanted to tell his brother that he hoped to fix Rutherford's flawed nuclear atom by turning it into the quantum atom.

Chapter 4

THE QUANTUM ATOM

Slagelse, Denmark, Thursday, 1 August 1912. The cobbled streets of the small, picturesque town some 50 miles south-west of Copenhagen were decked out in flags. Yet it was not in the beautiful medieval church, but in the civic hall that Niels Bohr and Margrethe Nørland were married in a two-minute ceremony conducted by the chief of police. The mayor was away on holiday, Harald was best man, and only close family were present. Like his parents before him, Bohr did not want a religious ceremony. He had stopped believing in God as a teenager, when he had confessed to his father: 'I cannot understand how I could be so taken in by all this; it means nothing whatsoever to me.'[1] Had he lived, Christian Bohr would have approved when, a few months before the wedding, his son formally resigned from the Lutheran Church.

Originally intending to spend their honeymoon in Norway, the couple were forced to change their plans as Bohr failed to finish a paper on alpha particles in time. Instead the newlyweds travelled to Cambridge for a two-week stay during their month-long honeymoon.[2] In between visits to old friends and showing Margrethe around Cambridge, Bohr completed his paper. It was a joint effort. Niels dictated, always struggling for the right word to make his meaning clear, while Margrethe corrected and improved his English. They worked so well together that for the next few years Margrethe effectively became his secretary.

Bohr disliked writing and avoided doing so whenever he could. He was able to complete his doctoral thesis only by dictating it to his mother. 'You mustn't help Niels so much, you must let him learn to write himself', his father had urged, to no avail.[3] When he did put pen to paper, Bohr wrote slowly and in an almost indecipherable scrawl. 'First and foremost,' recalled a colleague, 'he found it difficult to think and write at the same time.'[4] He needed to talk, to think aloud as he developed his ideas. He thought best while on the move, usually circling a table. Later, an assistant, or anyone he could find for the task, would sit with pen poised as he paced about dictating in one language or other. Rarely satisfied with the composition of a paper or lecture, Bohr would 'rewrite' it up to a dozen times. The end result of this excessive search for precision and clarity was often to lead the reader into a forest where it was difficult to see the wood for the trees.

With the manuscript finally completed and safely packed away, Niels and Margrethe boarded the train to Manchester. On meeting his bride, Ernest and Mary Rutherford knew that the young Dane had been lucky enough to find the right woman. The marriage indeed proved to be a long and happy one that was strong enough to endure the death of two of their six sons. Rutherford was so taken with Margrethe that for once there was little talk of physics. But he made time to read Bohr's paper and promised to send it to the *Philosophical Magazine* with his endorsement.[5] Relieved and happy, a few days later the Bohrs travelled to Scotland to enjoy the remainder of their honeymoon.

Returning to Copenhagen at the beginning of September, they moved into a small house in the prosperous coastal suburb of Hellerup. In a country with only one university, physics posts rarely became vacant.[6] Just before his wedding day, Bohr had accepted a job as a teaching assistant at the Læreanstalt, the Technical College. Each morning, Bohr would cycle to his new office. 'He would come into the yard, pushing his bicycle, faster than anybody else', recalled a colleague later.[7] 'He was an incessant worker and seemed always to be in a hurry.' The relaxed, pipe-smoking elder statesman of physics lay in the future.

Bohr also began teaching thermodynamics as a *privatdozent* at the university. Like Einstein, he found preparing a lecture course arduous. Nevertheless, at least one student appreciated the effort and thanked Bohr for 'the clarity and conciseness' with which he had 'arranged the difficult material' and 'the good style' with which it had been delivered.[8] But teaching combined with his duties as an assistant left him precious little time to tackle the problems besetting Rutherford's atom. Progress was painfully slow for a young man in a hurry. He had hoped that a report written for Rutherford while still in Manchester on his nascent ideas about atomic structure, later dubbed the 'Rutherford Memorandum', would serve as the basis of a paper ready for publication soon after his honeymoon.[9] It was not to be.

'You see,' Bohr said 50 years later in one of the last interviews he gave, 'I'm sorry because most of that was wrong.'[10] However, he had identified the key problem: the instability of Rutherford's atom. Maxwell's theory of electromagnetism predicted that an electron circling the nucleus should continuously emit radiation. This incessant leaking of energy sends the electron spiralling into the nucleus as its orbit rapidly decays. Radiative instability was such a well known failing that Bohr did not even mention it in his Memorandum. What really concerned him was the mechanical instability that plagued Rutherford's atom.

Beyond assuming that electrons revolved around the nucleus in the manner of planets around the sun, Rutherford had said nothing about their possible arrangement. A ring of negatively-charged electrons circling the nucleus was known to be unstable due to the repulsive forces the electrons exert on each other because they have the same charge. Nor could the electrons be stationary; since opposite charges attract, the electrons would be dragged towards the positively-charged core. It was a fact that Bohr recognised in the opening sentence of his memo: 'In such an atom there can be no equilibrium [con]figuration without the motion of electrons.'[11] The problems that the young Dane had to overcome were mounting up. The electrons could not form a ring, they could not be stationary, and they could not orbit the nucleus. Lastly, with a tiny, point-like nucleus at its heart, there was no way in Rutherford's model to fix the radius of an atom.

Whereas others had interpreted these problems of instability as damning evidence against Rutherford's nuclear atom, for Bohr they signalled the limitations of the underlying physics that predicted its demise. His identification of radioactivity as a 'nuclear' and not an 'atomic' phenomenon, his pioneering work on radioelements, what Soddy later called isotopes, and on nuclear charge convinced Bohr that Rutherford's atom was indeed stable. Although it could not bear the weight of established physics, it did not suffer the predicted collapse. The question that Bohr had to answer was: why not?

Since the physics of Newton and Maxwell had been impeccably applied and forecast electrons crashing into the nucleus, Bohr accepted that the 'question of stability must therefore be treated from a different point of view'.[12] He understood that to save Rutherford's atom would require a 'radical change', and he turned to the quantum discovered by a reluctant Planck and championed by Einstein.[13] The fact that in the interaction between radiation and matter, energy was absorbed and emitted in packets of varying size rather than continuously, was something beyond the realm of time-honoured 'classical' physics. Even though like almost everyone else he did not believe in Einstein's light-quanta, it was clear to Bohr that the atom 'was in some way regulated by the quantum'.[14] But in September 1912 he had no idea how.

All his life, Bohr loved to read detective stories. Like any good private eye, he looked for clues at the crime scene. The first were the predictions of instability. Certain that Rutherford's atom was stable, Bohr came up with an idea that proved crucial to his ongoing investigation: the concept of stationary states. Planck had constructed his blackbody formula to explain the available experimental data. Only then did he attempt to derive his equation and in the process stumbled across the quantum. Bohr adopted a similar strategy. He would begin by rebuilding Rutherford's atomic model so that electrons did not radiate energy as they orbited the nucleus. Only later would he try to justify what he had done.

Classical physics placed no restrictions on an electron's orbit inside an atom. But Bohr did. Like an architect designing a building to the strict requirements of a client, he restricted electrons to certain 'special' orbits in which they could not continuously emit radiation and spiral into the nucleus. It was a stroke of genius. Bohr believed that certain laws of physics were not valid in the atomic world and so he 'quantised' electron orbits. Just as Planck had quantised the absorption and emission of energy by his imaginary oscillators so as to derive his blackbody equation, Bohr abandoned the accepted notion that an electron could orbit an atomic nucleus at any given distance. An electron, he argued, could occupy only a few select orbits, the 'stationary states', out of all the possible orbits allowed by classical physics.

It was a condition that Bohr was perfectly entitled to impose as a theorist trying to piece together a viable working atomic model. It was a radical proposal, and for the moment all he had was an unconvincing circular argument that contradicted established physics – electrons occupied special orbits in which they did not radiate energy; electrons did not radiate energy because they occupied special orbits. Unless he could offer a real physical explanation for his stationary states, the permissible electron orbits, they would be dismissed as nothing more than theoretical scaffolding erected to hold up a discredited atomic structure.

'I hope to be able to finish the paper in a few weeks,' Bohr wrote to Rutherford at the beginning of November.[15] Reading the letter and sensing Bohr's mounting anxiety, Rutherford replied that there was no reason 'to feel pressed to publish in a hurry' since it was unlikely anyone else was working along the same lines.[16] Bohr was unconvinced as the weeks passed without success. If others were not already actively engaged in trying to solve the mystery of the atom, then it was only a matter of time. Struggling to make headway, in December he asked for and was granted a few months' sabbatical by Knudsen. Together with Margrethe, Bohr found a secluded cottage in the countryside as he set about searching for more atomic clues. Just before Christmas he found one in the work of John Nicholson. At first he feared the worst, but he soon realised that the Englishman was not the competitor he dreaded.

Bohr had met Nicholson during his abortive stay in Cambridge, and had not been overly impressed. Only a few years older at 31, Nicholson had since been appointed professor of mathematics at King's College, University of London. He had also been busy building an atomic model of his own. He believed that the different elements were actually made up of various combinations of four 'primary atoms'. Each of these 'primary atoms' consisted of a nucleus surrounded by a different number of electrons that formed a rotating ring. Though, as Rutherford said, Nicholson had made an 'awful hash' of the atom, Bohr had found his second clue. It was the physical explanation of the stationary states, the reason why electrons could occupy only certain orbits around the nucleus.

An object moving in a straight line has momentum. It is nothing more than the object's mass times its velocity. An object moving in a circle possesses a property called 'angular momentum'. An electron moving in a circular orbit has an angular momentum, labelled L, that is just the mass of the electron multiplied by its velocity multiplied by the radius of its orbit, or simply $L=mvr$. There were no limits in classical physics on the angular momentum of an electron or any other object moving in a circle.

When Bohr read Nicholson's paper, he found his former Cambridge colleague arguing that the angular momentum of a ring of electrons could change only by multiples of $h/2\pi$, where h is Planck's constant and π (pi) is the well-known numerical constant from mathematics, 3.14....[17] Nicholson showed that the angular momentum of a rotating electron ring could only be $h/2\pi$ or $2(h/2\pi)$ or $3(h/2\pi)$ or $4(h/2\pi)$ ... all the way to $n(h/2\pi)$ where n is an integer, a whole number. For Bohr it was the missing clue that underpinned his stationary states. Only those orbits were permitted in which the angular momentum of the electron was an integer n multiplied by h and then divided by 2π. Letting n=1, 2, 3 and so on generated the stationary states of the atom in which an electron did not emit radiation and could therefore orbit the nucleus indefinitely. All other orbits, the non-stationary states, were forbidden. Inside an atom, angular momentum was quantised. It could only have the values $L=nh/2\pi$ and no others.

Just as a person on a ladder can stand only on its steps and nowhere in between, because electron orbits are quantised, so are the energies that an electron can possess inside an atom. For hydrogen, Bohr was able to use classical physics to calculate its single electron's energy in each orbit. The set of allowed orbits and their associated electron energies are the quantum states of the atom, its energy levels E_n. The bottom rung of this atomic energy ladder is n=1, when the electron is in the first orbit, the lowest-energy quantum state. Bohr's model predicted that the lowest energy level, E_1, called the 'ground state', for the hydrogen atom would be –13.6eV, where an electron volt (eV) is the unit of measurement adopted for energy on the atomic scale and the minus sign indicates that the electron is bound to the nucleus.[18] If the electron occupies any other orbit but n=1, then the atom is said to be in an 'excited state'. Later called the principal quantum number, n is always an integer, a whole number, which designates the series of stationary states that an electron can occupy and the corresponding set of energy levels, E_n, of the atom.

Bohr calculated the values of the energy levels of the hydrogen atom and found that the energy of each level was equal to the energy of the ground state divided by n^2, (E_1/n^2)eV. Thus, the energy value for n=2, the first excited state, is –13.6/4 = –3.40eV. The radius of the first electron orbit, n=1, determines the size of the hydrogen atom in the ground state. From his model, Bohr calculated it as 5.3 nanometres (nm), where a nanometre is a billionth of a metre – in close agreement with the best experimental estimates of the day. He discovered that the radius of the other allowed orbits increased by a factor of n^2: when n=1, the radius is r; when n=2, then the radius is 4r; when n=3, the radius is 9r and so on.

'I hope very soon to be able to send you my paper on the atoms,' Bohr wrote to Rutherford on 31 January 1913, 'it has taken far more time than I had thought; I think, however, that I have made some progress in it in the latest time.'[19] He had stabilised the nuclear atom by quantising the angular momentum of the orbiting electrons, and thereby explained why

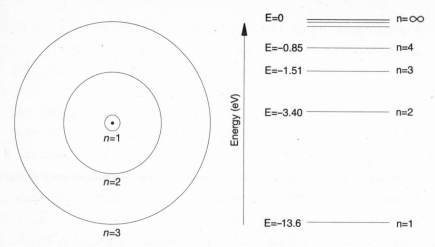

Figure 6: Some of the stationary states and the corresponding energy levels of the hydrogen atom (not drawn to scale)

they could occupy only a certain number, the stationary states, of all possible orbits. Within days of writing to Rutherford, Bohr came across the third and final clue that allowed him to complete the construction of his quantum atomic model.

Hans Hansen, a year younger and a friend of Bohr's from their student days in Copenhagen, had just returned to the Danish capital after completing his studies in Göttingen. When they met, Bohr told him about his latest ideas on atomic structure. Having conducted research in Germany in spectroscopy, the study of the absorption and emission of radiation by atoms and molecules, Hansen asked Bohr if his work shed any light on the production of spectral lines. It had long been known that the appearance of a naked flame changed colour depending upon which metal was being vaporised: bright yellow with sodium, deep red with lithium, and violet with potassium. In the nineteenth century it had been discovered that each element produced a unique set of spectral lines, spikes in the spectrum of light. The number, spacing and wavelengths of the spectral lines produced by the atoms of any given element are unique, a fingerprint of light that can be used to identify it.

Spectra appeared far too complicated, given the enormous variety of patterns displayed by the spectral lines of the different elements, for anyone to seriously believe that they could be the key to unlocking the inner workings of the atom. The beautiful array of colours on a butterfly's wing were all very interesting, Bohr said later, 'but nobody thought that one could get the basis of biology from the colouring of the wing of a butterfly'.[20] There was obviously a link between an atom and its spectral lines, but at the beginning of February 1913 Bohr had no inkling what it could be. Hansen suggested that he take a look at Balmer's formula for the spectral lines of hydrogen. As far as Bohr could remember, he had never heard of it. More likely he had simply forgotten it. Hanson outlined the formula and pointed out that no one knew why it worked.

Johann Balmer was a Swiss mathematics teacher at a girls' school in Basel and a part-time lecturer at the local university. Knowing that he was interested in numerology, a colleague told Balmer about the four spectral lines of hydrogen after he had complained about having nothing interesting to do. Intrigued, he set out to find a mathematical relationship between the lines where none appeared to exist. The Swedish physicist, Anders Ångström, had in the 1850s measured the wavelengths of the four lines in the red, green, blue and violet regions of the visible spectrum of hydrogen with remarkable accuracy. Labelling them alpha, beta, gamma and delta respectively, he found their wavelengths to be: 656.210, 486.074, 434.01 and 410.12nm.[21] In June 1884, as he approached 60, Balmer found a formula that reproduced the wavelengths (λ) of the four spectral lines: $\lambda = b[m^2/(m^2-n^2)]$ in which m and n are integers and b is a constant, a number determined by experiment as 364.56nm.

Balmer discovered that if n was fixed as 2 but m set equal to 3, 4, 5 or 6, then his formula gave an almost exact match for each of the four wavelengths in turn. For example, when n=2 and m=3 is plugged into the formula, it gives the wavelength of the red alpha line. However, Balmer did more than just generate the four known spectral lines of hydrogen, later named the Balmer series in his honour. He predicted the existence of a fifth line when n=2 but m=7, unaware that Ångström, whose work was published in Swedish, had already discovered and measured its wavelength. The two values, experimental and theoretical, were in near-perfect agreement.

Had Ångström lived (he died in 1874 aged 59), he would have been astounded by Balmer's use of his formula to predict the existence of other series of spectral lines for the hydrogen atom in the infrared and ultraviolet regions by simply setting n to 1, 3, 4 and 5 while letting m cycle through different numbers, as he had done with n set at 2 to generate the four original lines. For example, with n=3 and m=4 or 5 or 6 or 7..., Balmer predicted the series of lines in the infrared that were discovered in 1908 by Friedrich Paschen. Each of the series forecast by Balmer was later discovered, but no one had been able to explain what lay behind the success of his formula. What physical mechanism did the formula, arrived at through a process of trial and error, symbolise?

'As soon as I saw Balmer's formula,' Bohr said later, 'the whole thing was immediately clear to me.'[22] It was electrons jumping between different allowed orbits that produced the spectral lines emitted by an atom. If a hydrogen atom in the ground state, n=1, absorbs enough energy, then the electron 'jumps' to a higher-energy orbit such as n=2. The atom is then in an unstable, excited state and quickly returns to the stable ground state when the electron jumps down from n=2 to n=1. It can do so only by emitting a quantum of energy that is equivalent to the difference in energy of the two levels, 10.2eV. The wavelength of the resulting spectral line can be calculated using the Planck-Einstein formula, $E=h\nu$, where ν is the frequency of the emitted electromagnetic radiation.

An electron jumping from a range of higher energy levels to the same lower energy level produced the four spectral lines of the Balmer series. The size of the quanta emitted depended only on the initial and final energy levels involved. This was why Balmer's formula generated the correct wavelengths when n was set equal to 2 but m was 3, 4, 5 or 6 in turn. Bohr was able to derive the other spectral series predicted by Balmer by fixing the lowest energy level that the electron could jump to. For example, transitions ending with the electron jumping to n=3 produced the Paschen series in the infrared, while those that ended at n=1 generated the so-called Lyman series in the ultraviolet region of the spectrum.[23]

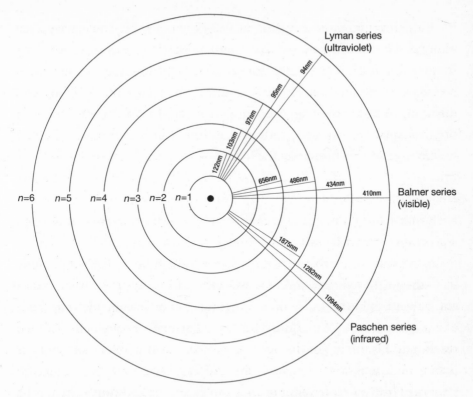

Figure 7: Energy levels, line spectra and quantum jumps (not drawn to scale)

There is, as Bohr discovered, a very strange feature associated with an electron's quantum leap. It is impossible to say where an electron actually is during a jump. The transition between orbits, energy levels, has to occur instantaneously. Otherwise as the electron travelled from one orbit to another it would radiate energy continuously. In Bohr's atom, an electron could not occupy the space between orbits. As if by magic, it disappeared while in one orbit and instantly reappeared in another.

'I'm fully convinced that the problem of spectral lines is intimately tied to the question of the nature of the quantum.' Remarkably, it was Planck, in February 1908, who wrote these words in a notebook.[24] But in his ongoing struggle to minimise the impact of the quantum, and before the Rutherford atom, it was as far as Planck could go. Bohr embraced the idea that electromagnetic radiation was emitted and absorbed by atoms in quanta, but in 1913 he did not accept that electromagnetic radiation itself was quantised. Even six years later, in 1919, few believed in Einstein's quantum of light when Planck declared in his Nobel Prize lecture that Bohr's quantum atom was 'the long-sought key to the entrance-gate into the wonderland' of spectroscopy.[25]

On 6 March 1913, Bohr sent Rutherford the first of a trilogy of papers and asked him to send it on to the *Philosophical Magazine*. At the time, and for many years to come, every junior scientist like Bohr needed someone of Rutherford's seniority to 'communicate' a paper to a British journal to ensure swift publication. 'I am very anxious to know what you may think of it all', he wrote to Rutherford.[26] He was particularly concerned about the reaction to his mixing of the quantum and classical physics. Bohr did not have to wait long for the answer: 'Your ideas as to the mode of origin of spectra in hydrogen are very ingenious and seem to work out well; but the mixture of Planck's ideas with the old mechanics make it very difficult to form a physical idea of what is the basis of it all.'[27]

Rutherford, as others would, was having trouble picturing how the electron in the hydrogen atom 'jumped' between energy levels. The difficulty lay in the fact that Bohr had violated a cardinal rule of classical physics. An electron moving in a circle is an oscillating system, with one complete orbit being an oscillation and the number of orbits per second being the frequency of oscillation. An oscillating system radiates energy at the frequency of its oscillation, but since two energy levels are involved in an electron making a 'quantum jump', there are two frequencies of oscillation. Rutherford was complaining that there was no link between these frequencies, between the 'old' mechanics and the frequency of the radiation emitted as the electron jumps between energy levels.

He also identified another more serious problem: 'There appears to me one grave difficulty in your hypothesis, which I have no doubt you fully realize, namely, how does an electron decide what frequency it is going to vibrate at when it passes from one stationary state to the other? It seems to me that you would have to assume that the electron knows beforehand where it is going to stop.'[28] An electron in the n=3 energy level can jump down to either the n=2 or the n=1 levels. In order to make the jump, the electron appears to 'know' to which energy level it is heading so that it can emit radiation of the correct frequency. These were weakness of the quantum atom to which Bohr had no answer.

There was another, more minor criticism that concerned Bohr far more deeply. Rutherford thought the paper 'really ought to be cut down', since 'long papers have a way of frightening readers, who feel that they have not time to dip into them'.[29] After offering to correct Bohr's English where necessary, Rutherford added a postscript: 'I suppose you have no objection to my using my judgement to cut out any matter I may consider unnecessary in your paper? Please reply.'[30]

When Bohr received the letter he was horrified. For a man who agonised over the choice of every word and went through endless drafts and revisions, the idea that someone else, even Rutherford, would make changes was appalling. Two weeks after posting the original paper, Bohr sent a longer revised manuscript containing alterations and additions. Rutherford agreed that the changes were 'excellent and appear quite reasonable', but he once again urged Bohr to cut the length. Even before he received this latest letter, he wrote to Rutherford telling him that he was coming to Manchester on holiday.[31]

When Bohr knocked on the front door, Rutherford was busy entertaining his friend Arthur Eve. He later recalled that Rutherford immediately took the 'slight-looking boy' into his study, leaving Mrs Rutherford to explain that the visitor was a young Dane and her husband thought 'very highly indeed of his work'.[32] Through hour after hour of discussions over several long evenings during the days that followed, Bohr admitted that Rutherford 'showed an almost angelic patience' as he tried to defend every word in his paper.[33]

An exhausted Rutherford finally gave in and afterwards began regaling his friends and colleagues with tales of the encounter: 'I could see that he had weighed up every word in it, and it impressed me how determinedly he held on to every sentence, every expression, every quotation; everything had a definite reason, and although I first thought that many sentences could be omitted, it was clear, when he explained to me how closely knit the whole was, that it was impossible to change anything.'[34] Ironically, Bohr admitted years later that Rutherford had been right 'in objecting to the rather complicated presentation'.[35]

Bohr's trilogy was published virtually unchanged in the *Philosophical Magazine* as 'On the Constitution of Atoms and Molecules'. The first, dated 5 April 1913, appeared in July. The second and third parts, published in September and November, were a presentation of ideas concerning the possible arrangements of electrons inside atoms that would preoccupy Bohr for the next decade as he used the quantum atom to explain the periodic table and the chemical properties of the elements.

Bohr had built his atom using a heady cocktail of classical and quantum physics. In the process he had violated tenets of accepted physics by proposing that: electrons inside atoms can occupy only certain orbits, the stationary states; electrons cannot radiate energy while in those orbits; an atom can be in only one of a series of discrete energy states, the lowest being the 'ground state'; electrons can 'somehow' jump from a stationary state of high energy to a stationary state of low energy and the difference in energy between the two is emitted in a quantum of energy. Yet his model correctly predicted various properties of the hydrogen atom such as its radius, and it provided a physical explanation for the production of spectral lines. The quantum atom, Rutherford said later, was 'a triumph of mind over matter' and until Bohr unveiled it, he believed that 'it would require centuries' to solve the mystery of the spectral lines.[36]

A true measure of Bohr's achievement was the initial reactions to the quantum atom. It was discussed publicly for the first time on

12 September 1913 at the 83rd annual meeting of the British Association for the Advancement of Science (BAAS), held that year in Birmingham. With Bohr in the audience, it received a muted and mixed reception. J.J. Thomson, Rutherford, Rayleigh and Jeans were all there, while the distinguished foreign contingent included Lorentz and Curie. 'Men over seventy should not be hasty in expressing opinions on new theories', was Rayleigh's diplomatic response when pressed for his opinion about Bohr's atom. In private, however, Rayleigh did not believe 'that Nature behaved in that way' and admitted that he had 'difficulty in accepting it as a picture of what actually takes place'.[37] Thomson objected to Bohr's quantisation of the atom as totally unnecessary. James Jeans begged to differ. He pointed out in a report to the packed hall that the only justification that Bohr's model required was 'the very weighty one of success'.[38]

In Europe, the quantum atom was greeted with disbelief. 'This is all nonsense! Maxwell's equations are valid under all circumstances', said Max von Laue during one heated discussion. 'An electron in a circular orbit must emit radiation.'[39] While Paul Ehrenfest confessed to Lorentz that Bohr's atom 'has driven me to despair'.[40] 'If this is the way to reach the goal,' he continued, 'I must give up doing physics.'[41] In Göttingen, Bohr's brother Harald reported that there was great interest in his work, but that his assumptions were deemed too 'bold' and 'fantastic'.[42]

One early triumph for Bohr's theory clinched the support of some, including Einstein. Bohr predicted that a series of lines found in the spectrum of light from the sun that had been attributed to hydrogen actually belonged to ionised helium, helium with one of its two electrons removed. This interpretation of the so-called 'Pickering-Fowler lines' was at odds with that of its discoverers. Who was right? The issue was settled by one of Rutherford's team in Manchester after a detailed study of the spectral lines instigated at the behest of Bohr. Just in time for the BAAS meeting in Birmingham, it was found that the Dane had been correct in his assignment of the Pickering-Fowler lines to helium. Einstein heard the news during a conference in Vienna at the end of September from Bohr's friend Georg von Hevesy. 'The big eyes of Einstein,' reported Hevesy in a letter to Rutherford, 'looked still bigger and he told me: "Then it is one of the greatest discoveries".'[43]

By the time Part III of the trilogy was published in November 1913, another member of Rutherford's group, Henry Moseley, had confirmed the idea that the nuclear charge of an atom, its atomic number, was a unique whole number for a given element and the key parameter that decided its position within the periodical table. It was only after Bohr visited Manchester in July that year and spoke to Moseley about the atom that the young Englishman began shooting beams of electrons at different elements and examined the resulting X-ray spectra.

By then it was known that X-rays were a form of electromagnetic radiation with wavelengths thousands of times shorter than those of visible light, and that they were produced when electrons with sufficient energy struck a given metal. Bohr believed that X-rays were emitted because one of the innermost electrons was knocked out of an atom and an electron moved down from a higher energy level to fill the vacancy. The difference in the two energy levels was such that the quantum of energy emitted in the transition was an X-ray. Bohr realised that, using his atomic model, it was possible to determine the charge of the nucleus using the frequencies of the emitted X-rays. It was this intriguing fact that he had discussed with Moseley.

With a prodigious capacity for work matched only by his stamina, while others slept Moseley stayed in the laboratory working through the night. Within a couple of months he had measured the frequencies of X-rays emitted by every element between calcium and zinc. He discovered that as the elements he bombarded got heavier, there was a corresponding increase of frequencies of the emitted X-rays. Moseley predicted the existence of missing elements with atomic numbers 42, 43, 72 and 75 on the basis that each element produced a characteristic set of X-ray spectral lines and those adjacent to each other in the periodic table had very similar ones.[44] All four were later discovered, but by then Moseley was dead. When the First World War began he enlisted in the Royal Engineers and served as a signals officer. He died, shot through the head, in Gallipoli on 10 August

1915. His tragic death at the age of 27 robbed him of a certain Nobel Prize. Rutherford personally gave him the highest possible accolade: he hailed Moseley as 'a born experimenter'.

Bohr's correct assignment of the 'Pickering-Fowler lines' and Moseley's ground-breaking work on nuclear charge were beginning to win support for the quantum atom. A more significant turning point in its acceptance came in April 1914, when the young German physicists James Franck and Gustav Hertz bombarded mercury atoms with electrons and found that the electrons lost 4.9eV of energy during these collisions. Franck and Hertz believed they had succeeded in measuring the amount of energy required to rip an electron from a mercury atom. Not having read his paper, due to the initial widespread scepticism that greeted it in Germany, it was left to Bohr to provide the correct interpretation of their data.

When the electrons fired at the mercury atoms had energies of less than 4.9eV, nothing happened. But when a bombarding electron with energy above 4.9eV scored a direct hit, it lost that amount of energy and the mercury atom emitted an ultraviolet light. Bohr pointed out that 4.9eV was the energy difference between the ground state of the mercury atom and its first excited state. It corresponded to an electron jumping between the first two energy levels in the mercury atom, and the energy difference between these levels was exactly as predicted by his atomic model. When the mercury atom returns to its ground state, as the electron jumps down to the first energy level, it emits a quantum of energy that produces an ultraviolet light of wavelength 253.7nm in the mercury line spectra. The Franck-Hertz results provided direct experimental evidence for Bohr's quantised atom and the existence of atomic energy levels. Despite initially having misinterpreted their data, Franck and Hertz were awarded the 1925 Nobel Prize in physics.

Just as Part I of the trilogy was published in July 1913, Bohr had finally been appointed to a lectureship at Copenhagen University. Before long he was unhappy, as his major responsibility was to teach elementary physics to medical students. At the beginning of 1914, with his reputation on the rise, Bohr set about trying to establish a new professorship in theoretical physics for himself. It would be difficult, as theoretical physics as a distinct discipline was still poorly recognised as such outside Germany. 'In my opinion Dr Bohr is one of the most promising and able of the young Mathematical Physicists in Europe today', wrote Rutherford in the testimonial to the Department of Religious and Educational Affairs in support of Bohr and his proposal.[45] The immense interest that his work had attracted internationally ensured that Bohr received the backing of the faculty, but once again the university hierarchy chose to postpone any decision. It was then that a dejected Bohr received a letter from Rutherford offering an escape route.

'I daresay you know Darwin's tenure of readership has expired, and we are now advertising for a successor at £200', Rutherford wrote.[46] 'Preliminary inquiries show that not many men of promise are available. I should like to get a young fellow with some originality in him.' Having already told the Dane that his work showed 'great originality and merit', Rutherford wanted Bohr without explicitly saying so.[47]

In September 1914, having been granted a year's leave of absence, as any decision on the professorship he wanted was unlikely before then, Niels and Margrethe Bohr arrived in Manchester to a warm welcome at their safe arrival after a stormy voyage around Scotland. The First World War had begun and much had changed. The wave of patriotism that swept the country had virtually emptied the laboratories as those eligible to fight signed up. The hope that the war would be short and sharp receded by the day as the Germans smashed through Belgium and into France. Men who had only recently been colleagues were now fighting on opposing sides. Marsden was soon at the western front. Geiger and Hevesy had joined the armies of the Central Powers.

Rutherford was not in Manchester when Bohr arrived. He had left in June to attend the annual meeting of the British Association for the Advancement of Science, being held that year in Melbourne, Australia. Recently knighted, he visited his family in New Zealand before travelling on to America and Canada as planned. Once back in Manchester, Rutherford devoted much of his time to anti-submarine warfare. Since Denmark was neutral, Bohr was not allowed to take part in any war-related activities. He concentrated largely on teaching, and what research was possible was impeded by the lack of journals and the censorship of letters from and to Europe.

Originally planning to spend just a year in Manchester, Bohr was still there when in May 1916 he was formally appointed to the newly created post of professor of theoretical physics in Copenhagen. The growing recognition of his work had secured the post, but despite its successes there were problems that the quantum atom could not solve. The answers it gave for atoms with more than one electron failed to tally with experiments. It could not even account for helium with just two electrons. Worse, Bohr's atomic model predicted spectral lines that could not be found. In spite of the introduction of ad hoc 'selection rules' to explain why some lines were observed and others were not, all the central elements of Bohr's atom were accepted by the end of 1914: the existence of discrete energy levels, the quantisation of angular momentum of the orbiting electrons, and the origin of spectral lines. However, if there existed a single spectral line that could not be explained, even with the imposition of some new rule, then the quantum atom was in trouble.

In 1892, improved equipment appeared to show that the red alpha and blue gamma Balmer lines of the hydrogen spectrum were not single lines at all, but were each split in two. For more than twenty years, it remained an open question whether these lines were 'true doublets' or not. Bohr thought not. It was at the beginning of 1915 that he changed his mind as new experiments revealed that the red, blue and violet Balmer lines were all doublets. Using his atomic model, Bohr could not explain this 'fine structure', as the splitting of the lines was called. As he settled into his new role as a professor in Copenhagen, Bohr found a batch of papers waiting for him from a German who had solved the problem by modifying his atom.

Arnold Sommerfeld was a 48-year-old distinguished professor of theoretical physics at Munich University. Over the years, some of the most brilliant young physicists and students would work under his watchful eye as he turned Munich into a thriving centre of theoretical physics. Like Bohr, he loved skiing and would invite students and colleagues to his house in the Bavarian Alps to ski and talk physics. 'But let me assure you that if I were in Munich and had the time, I would sit in on your lectures in order to perfect my knowledge of mathematical physics', Einstein had written to Sommerfeld in 1908 while still at the Patent Office.[48] It was some compliment coming from a man described as a 'lazy dog' by his maths professor in Zurich.

To simplify his model, Bohr had confined electrons to move only in circular orbits around the nucleus. Sommerfeld decided to lift this restriction, allowing electrons to move in elliptical orbits, like the planets in their journey around the sun. He knew that, mathematically speaking, circles were just a special class of ellipse, therefore circular electron orbits were only a subset of all possible quantised elliptical orbits. The quantum number n in the Bohr model specified a stationary state, a permitted circular electron orbit, and the corresponding energy level. The value of n also determined the radius of a given circular orbit. However, two numbers are required to encode the shape of an ellipse. Sommerfeld therefore introduced k, the 'orbital' quantum number, to quantise the shape of an elliptical orbit. Of all the possible shapes of an elliptical orbit, k determined those that were allowed for a given value of n.

In Sommerfeld's modified model, the principal quantum number n determined the values that k could have.[49] If n=1, then k=1; when n=2, k=1 and 2; when n=3, k=1, 2 and 3. For a given n, k is equal to every whole number from 1 up to and including the value of n. When n=k, the orbit is always circular. However, if k is less than n, then the orbit is elliptical. For example, when n=1 and k=1, the orbit is circular with a radius r, called the Bohr radius. When n=2 and k=1, the orbit is elliptical; but n=2 and k=2 is a circular orbit with a radius 4r. Thus, when the hydrogen atom is in the n=2 quantum state, its single electron can be in either the k=1 or k=2 orbits. In the n=3 state, the electron can occupy any one of three orbits: n=3 and k=1, elliptical; n=3, k=2, elliptical; n=3 and k=3, circular. Whereas in Bohr's model n=3 was just one circular orbit, in Sommerfeld's modified quantum atom there were three permitted orbits. These extra stationary states could explain the splitting of the spectral lines of the Balmer series.

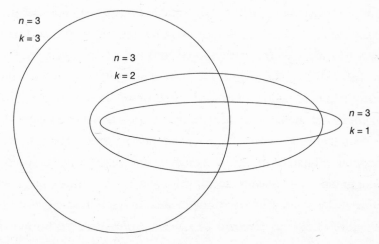

Figure 8: Electron orbits for n=3 and k=1, 2, 3 in the Bohr-Sommerfeld model of the hydrogen atom

To account for the splitting of the spectral lines, Sommerfeld turned to Einstein's theory of relativity. Like a comet in orbit about the sun, as an electron in an elliptical orbit heads towards the nucleus its speed increases. Unlike a comet, the speed of the electron is great enough for its mass to increase as predicted by relativity. This relativistic mass increase gives rise to a very small energy change. The n=2 states, the two orbits, k=1 and k=2, have different energies because k=1 is elliptical and k=2 circular. This minor energy difference leads to two energy levels that yield two spectral lines where only one was predicted by Bohr's model. However, the Bohr-Sommerfeld quantum atom was still unable to explain two other phenomena.

In 1897 the Dutch physicist Pieter Zeeman discovered that in a magnetic field, a single spectral line split into a number of separate lines or components. This was called the Zeeman effect, and once the magnetic field was switched off, the splitting disappeared. Then in 1913 the German physicist Johannes Stark found that a single spectral line splits up into several lines when atoms are placed in an electric field.[50] Rutherford contacted Bohr as Stark published his findings: 'I think it is rather up to you at the present time to write something on the Zeeman and electric effects, if it is possible to reconcile them with your theory.'[51]

Rutherford was not the first to ask. Soon after the publication of Part I of his trilogy, Bohr had received a letter of congratulation from Sommerfeld. 'Will you also apply your atomic model to the Zeeman effect?' he asked. 'I want to tackle this.'[52] Bohr was unable to explain it, but Sommerfeld did. His solution was ingenious. Earlier he had opted for elliptical orbits and thereby increased the number of possible quantised orbits that an electron could occupy when an atom was in a given energy state, such as n=2. Bohr and Sommerfeld had both pictured orbits, whether circular or elliptical, as lying in a plane. As he tried to account for the Zeeman effect, Sommerfeld realised that the orientation of an orbit was the vital missing component. In a magnetic field, an electron can select from more permitted orbits pointing in various directions with respect to the field. Sommerfeld introduced what he called the 'magnetic' quantum number m to quantise the orientation of those orbits. For a given principal quantum number n, m can only have values that range from –n to n.[53] If n=2, then m has the values: –2, –1, 0, 1, 2.

'I do not believe ever to have read anything with more joy than your beautiful work', Bohr wrote to Sommerfeld in March 1916. The orientation of electron orbits, or 'space quantisation' as it became known, was experimentally confirmed five years later in 1921. It made available extra energy states, now labelled by the three quantum numbers n, k and m, which an electron could occupy in the presence of an external magnetic field, leading to the Zeeman effect.

Necessity being the mother of invention, Sommerfeld had been forced to introduce his two new quantum numbers k and m to explain facts revealed by experiments. Leaning heavily on the work of Sommerfeld, others explained the Stark effect as resulting from the changes in the spacing between energy levels due to the presence of an electric field. Although there were still weaknesses, such as the inability to reproduce the relative intensity of the spectral lines, the successes of the Bohr-Sommerfeld atom further enhanced Bohr's reputation and earned him an institute of his own in Copenhagen. He was on his way to becoming, as Sommerfeld called him later, 'the director of atomic physics' through his work and the inspiration he gave others.[54]

It was a compliment that would have pleased Bohr, who had always wanted to replicate the way in which Rutherford had run his laboratory, and the spirit he had succeeded in creating among all those who worked there. Bohr had learnt more than just physics from his mentor. He saw how Rutherford was able to galvanise a group of young physicists into producing their best. In 1917 Bohr set out to replicate what he had been fortunate enough to experience in Manchester. He approached the authorities in Copenhagen about establishing an institute for theoretical physics at the university. The institute was approved, as friends raised the money necessary for buildings and land. Construction began the following year, soon after the end of the war, at a site on the edge of a beautiful park not far from the city centre.

Work had only just begun when a letter arrived that unsettled Bohr. It was from Rutherford, who was offering him a permanent professorship in theoretical physics back in Manchester. 'I think the two of us could try and make physics boom', wrote Rutherford.[55] It was tempting, but Bohr could not leave Denmark just as he was about to be given everything that he wanted. Maybe if he had gone, Rutherford would not have left Manchester in 1919 to take over from J.J. Thomson as the director of the Cavendish Laboratory at Cambridge.

Always known as the Bohr Institute, the Universitetets Institut for Teoretisk Fysik was formally opened on 3 March 1921.[56] The Bohrs had already moved into the seven-room flat on the first floor with their growing family. Following the upheavals of war and the hardship of the years that followed in its wake, the institute was soon the creative haven Bohr hoped it would be. It quickly became a magnet for many of the world's brightest physicists, but the most talented of them all would always remain an outsider.

Chapter 5

WHEN EINSTEIN MET BOHR

'Those are the madmen who do not occupy themselves with quantum theory', Einstein told a colleague as they looked out of the window of his office in the Institute of Theoretical Physics at the German University in Prague.[1] After his arrival from Zurich in April 1911, he had been puzzled as to why only women used the grounds in the mornings and only men in the afternoons. As he struggled with his own demon he discovered that the beautiful garden next door belonged to a lunatic asylum. Einstein was finding it difficult to live with the quantum and the dual nature of light. 'I wish to assure you in advance that I am not the orthodox light-quantizer for whom you take me', he told Hendrik Lorentz.[2] It was a faulty impression that arose, he claimed, 'from my imprecise way of expressing myself in my papers'.[3] Soon he gave up even asking if 'quanta really exist'.[4] By the time he returned from the first Solvay conference in November 1911 on 'The Theory of Radiation and the Quanta', Einstein had decided that enough was enough and pushed the lunacy of the quantum to one side. Over the next four years, as Bohr and his atom took centre stage, Einstein effectively abandoned the quantum to concentrate on extending his theory of relativity to encompass gravity.

Founded in the mid-fourteenth century, Prague University was divided in 1882 along lines of nationality and language into two separate universities, one Czech and the other German. It was a division that reflected a society where Czechs and Germans harboured a deep-seated suspicion and mistrust of each other. After the easy-going, tolerant atmosphere of Switzerland and the cosmopolitan mix of Zurich, Einstein was ill at ease in spite of the full professorship and the salary that enabled him to live in some comfort. It provided just a quantum of solace against the creeping sense of isolation.

By the end of 1911, as Bohr contemplated his move from Cambridge to Manchester, Einstein desperately wanted to return to Switzerland, and it was then that an old friend came to his rescue. Recently appointed as the dean of the mathematics and physics section of the Swiss Federal Technical University (ETH), Marcel Grossmann offered Einstein a professorship in Zurich at the renamed former Polytechnic. Although the job was his, there were formalities that Grossmann had to observe. High on the list was seeking the advice of eminent physicists about Einstein's possible appointment. One of those asked was France's premier theorist, Henri Poincaré, who described Einstein as 'one of the most original minds' he knew.[5] The Frenchman admired the ease with which he adapted to new concepts, his ability to see beyond classical principles, and when 'faced with a physics problem, [he] promptly envisages all possibilities'.[6] Where Einstein had once failed to get a job as an assistant, in July 1912 he returned as a master physicist.

It was inevitable that sooner rather than later Einstein would become a prime target for the men in Berlin. In July 1913 Max Planck and Walther Nernst boarded the train to Zurich. They knew that it would not be easy to persuade Einstein to return to a country he had left almost twenty years ago, but they were prepared to make him an offer he simply could not refuse.

As Einstein met them off the train, he knew why Planck and Nernst had come, but not the details of the proposal they were about to make. Having just been elected a member of the prestigious Prussian Academy of Sciences, he was being offered one of its two salaried positions. This alone was a great honour, but the two emissaries of German science also offered a unique research professorship without any teaching duties and the directorship of the Kaiser Wilhelm Institute of Theoretical Physics once it was established.

He needed time to mull over the unprecedented package of three jobs. Planck and Nernst went on a short sightseeing train ride as he considered whether or not to accept. Einstein told them they would have his answer when they returned by the colour of the rose he carried. If red, he would go to Berlin; if white, he would stay in Zurich. As they got off the train, Planck and Nernst knew they had got their man when they saw Einstein clutching a red rose.

Part of the lure of Berlin for Einstein was the freedom to 'give myself over completely to rumination' with no obligations to teach.[7] But with it came the pressure of having to deliver the sort of physics that made him the hottest property in science. 'The Berliners are speculating with me as with a prize-winning laying hen,' he told a colleague after his farewell dinner, 'but I don't know if I can still lay eggs.'[8] After celebrating his 35th birthday in Zurich, Einstein moved to Berlin at the end of March 1914. Whatever reservations he might have had about returning to Germany, he was soon enthusing: 'Intellectual stimulation abounds here, there is just too much of it.'[9] The likes of Planck, Nernst and Rubens were all within easy reach, but there was another reason why he found 'odious' Berlin exciting – his cousin Elsa Löwenthal.[10]

Two years earlier, in March 1912, Einstein had begun an affair with the 36-year-old divorcee with two young daughters – Ilse, aged thirteen, and Margot, eleven. 'I treat my wife like an employee whom I cannot fire', he told Elsa.[11] Once in Berlin, Einstein would often disappear for days without a word of explanation. Soon he moved out of the family home altogether and drew up a remarkable list of conditions under which he was willing to return. If Mileva accepted his terms she would indeed become an employee, and one her husband was determined to fire.

Einstein demanded: '1. that my clothes and laundry are kept in good order and repair; 2. that I receive my three meals regularly *in my room*; 3. that my bedroom and my office are always kept neat, in particular, that the desk is available *to me alone*.' Further, she was to 'renounce all personal relations' and refrain from criticising him 'either in word or deed in front of my children'. Finally he insisted that Mileva adhere to 'the following points: 1. You are neither to expect intimacy from me nor reproach me in any way. 2. You must desist immediately from addressing me if I request it. 3. You must leave my bedroom or office immediately without protest if I so request.'[12]

Mileva agreed to his demands and Einstein returned. But it could not last. At the end of July, after just three months in Berlin, Mileva and the boys went back to Zurich. As he stood on the platform waving goodbye, Einstein wept, if not for Mileva and the memories of what had been, then for his two departing sons. But within a matter of weeks he was happily enjoying living alone 'in my large apartment in undiminished tranquillity'.[13] It was a tranquillity that few would enjoy as Europe descended into war.

———

'One day the great European war will come out of some damned foolish thing in the Balkans', Bismarck was once reported as saying.[14] That day was Sunday, 28 June 1914, and it was the assassination in Sarajevo of Archduke Franz Ferdinand, the heir to the crowns of Austria and Hungary. Austria, supported by Germany, declared war on Serbia. Germany declared war on Serbia's ally Russia on 1 August and on France two days later. Britain, who guaranteed Belgian independence, declared war on Germany on 4 August after it had violated Belgium's neutrality.[15] 'Europe in its madness has now embarked on something incredibly preposterous', Einstein wrote on 14 August to his friend Paul Ehrenfest.[16]

While Einstein felt 'only a mixture of pity and disgust', Nernst at 50 volunteered as an ambulance driver.[17] Planck, unable to contain his patriotism, declared: 'It is a great feeling to be able to call oneself a

German.'[18] Believing that it was a glorious time to be alive, as rector of Berlin University, Planck sent his students to the trenches in the name of a 'just war'. Einstein could hardly believe it when he discovered that Planck, Nernst, Röntgen and Wien were among the 93 luminaries who signed the *Appeal to the Cultured World*.

This manifesto was published on 4 October 1914 in leading German newspapers and in others abroad, its signatories protesting against 'the lies and defamations with which our enemies are trying to besmirch Germany's pure cause in the hard life-and-death struggle forced upon it'.[19] They asserted that Germany bore no responsibility for the war, had not violated Belgian neutrality, and had committed no atrocities. Germany was 'a cultured nation to whom the legacy of Goethe, Beethoven and Kant is fully as sacred as its hearths and plots of land'.[20]

Planck quickly regretted having signed, and in private began apologising to his friends among foreign scientists. Of all those that lent their names to the falsehoods and half-truths of the *Manifesto of the Ninety-Three*, as it became known, Einstein had expected better from Planck. Even the German chancellor had publicly admitted that Belgium's neutral status had been violated: 'The wrong that we are committing, we will endeavour to make good as soon as our military goal is reached.'[21]

As a Swiss citizen, Einstein was not asked to add his signature. However, he was deeply concerned at the long-term effect of the unbridled national chauvinism of the manifesto and was involved in producing a counter-manifesto entitled an *Appeal to Europeans*. It called on 'educated men of all states' to ensure that 'the conditions of peace did not become the source of future wars'.[22] It challenged the attitude expressed by the *Manifesto of the Ninety-Three* as 'unworthy of what until now the whole world has understood by the term culture, and it would be a disaster if it were to become the common property of educated people'.[23] It castigated German intellectuals for behaving 'almost to a man, as though they had relinquished any further desire for the continuance of international relations'.[24] However, including Einstein, there were only four signatories.

By the spring of 1915 the attitudes of his colleagues at home and abroad had left Einstein feeling deeply disheartened: 'Even scholars of the various nations behave as if their cerebrums had been amputated eight months ago.'[25] Soon all hope that the war would be short-lived evaporated, leaving him by 1917 'constantly very depressed about the endless tragedy we must witness'.[26] 'Even the habitual flight into physics does not always help', he confessed to Lorentz.[27] Yet the four years of war proved to be among his most productive and creative, as Einstein published a book and some 50 scientific papers, and in 1915 completed his masterpiece – general relativity.

Even before Newton, it had been assumed that time and space were fixed and distinct, the stage on which the never-ending drama of the cosmos was played out. It was an arena where mass, length and time were absolute and unvarying. It was a theatre in which spatial distances and time intervals between events were identical for all observers. Einstein, however, discovered that mass, length and time were not absolute and unchanging. Spatial distances and time intervals depended on the relative motion of observers. Compared to his earth-bound twin, for an astronaut travelling at near light-speed, time slows down (the hands on a moving clock are slower), space contracts (the length of a moving object shrinks), and a moving object gains mass. These were the consequences of 'special' relativity, and each would be confirmed by experiments in the twentieth century, but the theory did not incorporate acceleration. 'General' relativity did. In the midst of his struggle to construct it, Einstein said that it made special relativity look like 'child's play'.[28] Just as the quantum was challenging the accepted view of reality in the atomic realm, Einstein took humanity closer to understanding the true nature of space and time. General relativity was his theory of gravity, and it would lead others to the big bang origin of the universe.

In Newton's theory of gravity, the force of attraction between two objects, such as the sun and the earth, is proportional to the product of their respective masses and inversely proportional to the square of the distance separating their centres of mass. With no contact between the masses, in Newtonian physics gravity is a mysterious 'action-at-a-distance' force. In general relativity, however, gravity is due to the warping of space caused by the presence of a large mass. The earth moves around the sun not because some mysterious invisible force pulls it, but because of the warping of space due to the sun's enormous mass. In short, matter warps space and warped space tells matter how to move.

In November 1915, Einstein tested general relativity by applying it to a feature of Mercury's orbit that could not be explained using Newton's gravitational theory. In its journey around the sun, Mercury does not trace out exactly the same path every orbit. Astronomers had precise measurements that revealed that the planet's orbit rotated slightly. Einstein used general relativity to calculate this orbital shift. When he saw that the number matched the data within the margins of error, he had palpitations of the heart and felt as if something had snapped inside. 'The theory is beautiful beyond comparison', he wrote.[29] With his boldest dreams fulfilled, Einstein was content but the Herculean effort left him worn out. When he recovered he turned to the quantum.

Even as he worked on the general theory, in May 1914, Einstein was among the first to grasp that the Franck-Hertz experiment was a confirmation of the existence of energy levels in atoms and 'a striking verification of the quantum hypothesis'.[30] By the summer of 1916, Einstein had 'a brilliant idea' of his own about an atom's emission and absorption of light.[31] It led him to an 'astonishingly simple derivation, I should say, *the* derivation of Planck's formula'.[32] Soon Einstein was convinced that 'light-quanta are as good as established'.[33] However, it came at a price. He had to abandon the strict causality of classical physics and introduce probability into the atomic domain.

Einstein had offered alternatives before, but this time he could derive Planck's law from Bohr's quantum atom. Starting with a simplified Bohr atom with only two energy levels, he identified three ways in which an electron could jump from one level to the other. When an electron jumps from a higher to a lower energy level and emits a quantum of light, Einstein called this 'spontaneous emission'. It occurs only when an atom is in an excited state. The second type of quantum leap happens when an atom becomes excited as an electron absorbs a light-quantum and jumps from a lower to a higher energy level. Bohr had invoked both types of quantum leap to explain the origin of atomic emission and absorption spectra, but Einstein now revealed a third: 'stimulated emission'. It occurs when a light-quantum strikes an electron in an atom that is already in an excited state. Instead of absorbing the incoming light-quantum, the electron is 'stimulated', nudged, to leap to a lower energy, emitting a light-quantum. Four decades later, stimulated emission formed the basis of the laser, an acronym for 'light amplification by stimulated emission of radiation'.

Einstein also discovered that light-quanta had momentum, which, unlike energy, is a vector quantity; it has direction as well as magnitude. However, his equations clearly showed that the exact time of spontaneous transition from one energy level to another and the direction in which an atom emits a light-quantum was entirely random. Spontaneous emission was like the half-life of a radioactive sample. Half the atoms will decay in a certain amount of time, the half-life, but there was no way of knowing when any given atom would decay. Likewise, the probability that a spontaneous transition will take place could be calculated but the exact details were entirely left to chance, with no connection between cause and effect. This concept of a transition probability that left the time and direction of the emission of a light-quantum down to pure 'chance' was for Einstein a 'weakness' of his theory. It was something he was prepared to tolerate for the moment in the hope that it would be removed with the further development of quantum physics.[34]

Einstein was uneasy with this discovery of chance and probability at work in the heart of the quantum atom. Causality appeared to be at risk even though he no longer doubted the reality of quanta.[35] 'That business about causality causes me a lot of trouble, too', he wrote to Max Born three years later in January 1920.[36] 'Can the quantum absorption and emission of light ever be understood in the sense of the complete causality requirement, or would a statistical residue remain? I must admit that there I lack the courage of my convictions. But I would be very unhappy to renounce *complete* causality.'

What troubled Einstein was a situation akin to an apple being held above the ground, that when let go did not fall. Once the apple is let go, it is in an unstable state with respect to the state of lying on the ground, so gravity acts immediately on the apple, *causing* it to fall. If the apple behaved like an electron in an excited atom, then instead of falling back as soon as it was let go, it would hover above the ground, falling at some unpredictable time that can be calculated only in terms of probability. There may be a high probability that the apple will fall within a very short time, but there is a small probability that the apple will just hover above the ground for hours. An electron in an excited atom will fall to a lower energy level, resulting in the more stable ground state of the atom, but the exact moment of the transition is left to chance.[37] In 1924, Einstein was still struggling to accept what he had unearthed: 'I find the idea quite intolerable that an electron exposed to radiation should choose *of its own free will*, not only its moment to jump off, but also its direction. In that case, I would rather be a cobbler, or even an employee in a gaming-house, than a physicist.'[38]

It was inevitable that the years of intense intellectual effort coupled with his bachelor lifestyle would take their toll. In February 1917, aged only 38, Einstein collapsed with intense stomach pains and the diagnosis was a liver complaint. Within two months he lost 56 pounds as his health deteriorated. It was the beginning of a series of illnesses, including gallstones, a duodenal ulcer and jaundice, that dogged him over the next few years. Plenty of rest and a strict diet were the prescribed cure. It was easier said than done, as life was transformed beyond recognition by the trials and tribulations of war. Even potatoes were a rarity by then in Berlin, and most Germans went hungry. Few actually starved to death, but malnutrition claimed lives – an estimated 88,000 in 1915. The following year it rose to more than 120,000 as riots erupted in more than 30 German cities. It was hardly surprising, as people were forced to eat bread made from ground straw instead of wheat.

There was an ever-growing list of such *ersatz* foods. Plant husks mixed with animal hides replaced meat, and dried turnips were used to make 'coffee'. Ash masqueraded as pepper, and people spread a mixture of soda and starch on their bread, pretending it was butter. Constant hunger made cats, rats and horses appear tasty alternatives for Berliners. If a horse dropped dead in the street it was swiftly butchered. 'They fought each other for the best pieces, their faces and clothing covered in blood', reported an eyewitness to one such incident.[39]

Real food was scarce, but still available to those who could afford to pay. Einstein was luckier than most, as he received food parcels from relatives in the south and from friends in Switzerland. Amid all the suffering, Einstein felt 'like a drop of oil on water, isolated by mentality and outlook on life'.[40] Yet he could not look after himself and reluctantly moved into a vacant apartment next door to Elsa's. With Mileva still unwilling to grant a divorce, Elsa finally had Einstein as near to her as propriety would allow. Nursing Albert slowly back to health gave Elsa the perfect opportunity to pressurise him into doing whatever it took to get a divorce. Einstein initially had no intention of rushing into marriage a second time, as the first felt like 'ten years of prison', but eventually he relented.[41] Mileva agreed after Einstein proposed to increase his existing payments, make her the recipient of his widow's pension, and offer her the money when he won the Nobel Prize. By 1918, having been nominated in six of the previous eight years, he was a dead certainty to be awarded the prize some time soon.

Einstein and Elsa married in June 1919. He was 40, she three years older. What happened next was beyond anything that Elsa could have imagined. Before the end of the year, the lives of the newlyweds were transformed as Einstein became world-famous. He was hailed as the 'new Copernicus' by some, derided by others.

In February 1919, just as Einstein and Mileva were finally divorced, two expeditions set off from Britain. One headed to the island of Principe off the coast of West Africa, the other to Sobral in the north-west of Brazil.

Each destination had been carefully chosen by astronomers as a perfect site from which to observe the solar eclipse on 29 May. Their aim was to test a central prediction of Einstein's general theory of relativity, the bending of light by gravity. The plan was to photograph stars in close proximity to the sun that would be visible only during the few minutes of blackout of a total solar eclipse. In reality, of course, these stars were nowhere near the sun, but their light passed very close to it before reaching the earth.

The photographs would be compared with those taken at night six months earlier when the earth's position in relation to the sun ensured that the light from these same stars passed nowhere near the neighbourhood of the sun. The bending of light due to the presence of the sun warping the space-time in its vicinity would be revealed by small changes in the position of the stars in the two sets of photographs. Einstein's theory predicted the exact amount of displacement due to the bending or deflection of light that should be observed. At a rare joint meeting of the Royal Society and the Royal Astronomical Society on 6 November in London, the cream of British science gathered to hear whether Einstein was right or not.[42]

REVOLUTION IN SCIENCE
NEW THEORY OF THE UNIVERSE
Newtonian Ideas Overthrown

… were the headlines on page twelve of the London *Times* the following morning. Three days later, on 10 November, the *New York Times* carried an article with six headings: 'Lights all askew in the heavens/Men of science more or less agog over results of eclipse observation/Einstein theory triumphs/Stars not where they seem or were calculated to be, but nobody need worry/A book for 12 wise men/ No more in all world could comprehend it, said Einstein, when his daring publishers accepted it.'[43] Einstein had never said any such thing, but it made good copy as the press latched onto the mathematical sophistication of the theory and the idea of warped space.

One of those who unwittingly contributed to the mystique surrounding general relativity was Sir J.J. Thomson, the president of the Royal Society. 'Perhaps Einstein has made the greatest achievement in human thought,' he told a journalist afterwards, 'but no one has yet succeeded in stating in clear language what the theory of Einstein's really is.'[44] In fact, by the end of 1916 Einstein had already published the first popular book on both the special and general theories.[45]

'The general theory of relativity is being received with downright enthusiasm among my colleagues', Einstein reported to his friend Heinrich Zangger in December 1917.[46] However, in the days and weeks that followed the first press reports, there were many who came forth to pour scorn on 'the suddenly famous Dr Einstein' and his theory.[47] One critic described relativity as 'voodoo nonsense' and 'the moronic brainchild of mental colic'.[48] With supporters like Planck and Lorentz, Einstein did the only sensible thing; he ignored his detractors.

In Germany, Einstein was already a well-known public figure when the *Berliner Illustrirte Zeitung* gave over its entire front page to a photograph of him. 'A new figure in world history whose investigations signify a complete revision of nature, and are on a par with insights of Copernicus, Kepler, and Newton', read the accompanying caption. Just as he refused to be riled by his critics, Einstein kept a sense of perspective about being anointed the successor of three of history's great scientists. 'Since the light deflection result became public, such a cult has been made out of me that I feel like a pagan idol', he wrote after the *Berliner Illustrirte Zeitung* hit the news-stands. 'But this, too, God willing, will pass.'[49] It never did.

Part of the widespread public fascination with Einstein and his work lay in a world still coming to terms with the upheavals in the aftermath of the First World War, which ended at 11am on 11 November 1918. Two days earlier, on 9 November, Einstein had cancelled his relativity course lecture 'because of revolution'.[50] Later that day, Kaiser Wilhelm II abdicated and fled to Holland as a republic was proclaimed from a balcony of the Reichstag. Germany's economic problems were among the most difficult challenges facing the new Weimar Republic. Inflation was quickly on the rise, as Germans lost confidence in the mark and were busy either selling it or buying anything they could before it fell further.

It was a vicious circle that war reparations sent spiralling out of control, and the economy went into meltdown as Germany defaulted on its payments of wood and coal towards the end of 1922, and 7,000 marks bought one US dollar. However, that was nothing to the hyperinflation that occurred throughout 1923. In November that year, one dollar was worth 4,210,500,000,000 marks, a glass of beer cost 150 billion marks and a loaf of bread 80 billion. With the country in danger of imploding, the situation was brought under control only with the help of American loans and a reduction in reparation payments.

Amid the suffering, talk of warped space, bending light beams, and shifting stars that only '12 wise men' could comprehend fired the public imagination. However, everyone thought they had an intuitive grasp of concepts like space and time. As a result, the world appeared to Einstein to be a 'curious madhouse' as 'every coachman and every waiter argues about whether or not relativity theory is correct'.[51]

Einstein's international celebrity and his well-known anti-war stance made him an easy target for a campaign of hate. 'Anti-semitism is strong here and political reaction is violent', Einstein wrote to Ehrenfest in December 1919.[52] Soon he began receiving threatening mail and on occasions suffered verbal abuse as he left his apartment or office. In February 1920, a group of students disrupted his lecture at the university, one of them shouting, 'I'm going to cut the throat of that dirty Jew.'[53] But the political leaders of the Weimar Republic knew what an asset Einstein was, as its scientists faced exclusion from international conferences after the war. The minister of culture wrote to reassure him that Germany, 'was, and will forever be, proud to count you, highly honoured *Herr Professor*, among the finest ornaments of our science'.[54]

Niels Bohr did as much as anyone to ensure that personal relations between scientists on opposing sides were restored as quickly as possible after the war. As a citizen of a neutral country, Bohr felt no resentment towards his German colleagues. He was among the first to extend an invitation to a German scientist when he asked Arnold Sommerfeld to lecture in Copenhagen. 'We had long discussions on the general principle of the quantum theory and the application of all kinds of detailed atomic problems', Bohr said after Sommerfeld's visit.[55] Excluded for the foreseeable future from international meetings, German scientists and their hosts knew the value of these personal invitations. So when he received one from Max Planck to give a lecture on the quantum atom and the theory of atomic spectra in Berlin, Bohr gladly accepted. When the date was fixed as Tuesday, 27 April 1920, he was excited at the prospect of meeting Planck and Einstein for the first time.

'His must be a first-rate mind, extremely critical and far-seeing, which never loses track of the grand design', was Einstein's assessment of the young Dane, six years his junior.[56] It was October 1919 and such an appraisal was a spur for Planck to get Bohr to Berlin. Einstein had long been an admirer. In the summer of 1905 as the creative storm that had broken loose in his mind began to subside, Einstein found nothing that was 'really exciting' to tackle next.[57] 'There would of course be the topic of spectral lines,' he told his friend Conrad Habicht, 'but I believe that a simple relationship between these phenomena and those already investigated does not exist at all, so that for the moment, the thing looks rather unpromising to me.'[58]

Einstein's nose for a physics problem ripe for attack was second to none. Having passed on the mystery of spectral lines, he came up with $E=mc^2$, which said that mass and energy were interconvertible. But for all he knew, God Almighty was having a laugh at his expense by leading him 'around by the nose'.[59] So when in 1913 Bohr showed how his quantised atom solved the mystery of atomic spectra, it appeared to Einstein 'like a miracle'.[60]

The uneasy mixture of excitement and apprehension that had taken hold of his stomach as Bohr made his way from the station to the university vanished as soon as he met Planck and Einstein. They put him at his ease by moving quickly from pleasantries to talk of physics. The two men could not have been more dissimilar. Planck was the epitome of Prussian formality and rectitude, while Einstein with his big eyes, unruly hair and trousers that were just a little too short gave the impression of a man at ease with himself, if not the troubled world in which he lived. Bohr accepted Planck's invitation to stay at his home during the visit.

His days in Berlin, Bohr said later, were spent 'discussing theoretical physics from morning to night'.[61] It was the perfect break for the man who just loved to talk physics. He particularly enjoyed the lunch that the younger university physicists had thrown for him, from which they excluded all the 'bigwigs'. It was a chance for them to quiz Bohr after his lecture had left them 'somewhat depressed because we had the feeling that we had understood very little'.[62] Einstein, however, understood perfectly well what Bohr was arguing and he did not like it.

Like virtually everyone else, Bohr did not believe in the existence of Einstein's light-quanta. He accepted, like Planck, that radiation was emitted and absorbed in quanta, but not that radiation itself was quantised. For him there was just too much evidence in favour of the wave theory of light, but with Einstein in the audience, Bohr told the assembled physicists: 'I shall not consider the problem of the nature of radiation.'[63] However, he had been deeply impressed by Einstein's work of 1916 on spontaneous and stimulated emission of radiation and electron transitions between energy levels. Einstein had succeeded where he had failed by showing that it was all a matter of chance and probability.

Einstein continued to be troubled by the fact that his theory could not predict either the time or the direction in which the light-quantum emitted as an electron jumps from one energy level to a lower one. 'Nevertheless,' he had written in 1916, 'I fully trust in the reliability of the road taken.'[64] He believed it was a road that would eventually lead to a restoration of causality. In his lecture, Bohr argued that no exact determination of time and direction was ever possible. The two men found themselves on opposite sides. In the days that followed, each tried to convert the other to his point of view as they walked the streets of Berlin together or dined at Einstein's home.

'Seldom in my life has a person given me such pleasure by his mere presence as you have', Einstein wrote to Bohr soon after he returned to Copenhagen. 'I am now studying your great publications and – unless I happen to get stuck somewhere – have the pleasure of seeing before me your cheerful boyish face, smiling and explaining.'[65] The Dane had made a deep and lasting impression. 'Bohr was here, and I am just as enamoured of him as you are', Einstein told Paul Ehrenfest a few days later. 'He is like a sensitive child and walks about this world in a kind of hypnosis.'[66] Bohr was equally intent in trying to convey, in his less than polished German, what it meant to him to have met Einstein: 'It was to me one of my greatest experiences to have met you and to talk to you. You cannot imagine what a great inspiration it was for me to hear your views from you in person.'[67] Bohr did so again quite soon, as Einstein made a fleeting visit as he stopped off in Copenhagen in August on his way back from a trip to Norway.

'He is a highly gifted and excellent man', Einstein wrote to Lorentz after seeing Bohr.[68] 'It is a good omen for physics that prominent physicists are mostly also splendid people.' Einstein had become the target of two men who were not. Philipp Lenard, whose experimental work on the photo-electric effect Einstein had used in 1905 in support of his light-quanta, and Johannes Stark, the discoverer of the splitting of spectral lines by an electric field, had become rabid anti-Semites. The two Nobel laureates were behind an organisation calling itself the Working Group of German Scientists for the Preservation of Pure Science, whose prime aim was to denounce Einstein and relativity.[69] On 24 August 1920 the group held a meeting at Berlin's Philharmonic Hall and attacked relativity as 'Jewish physics' and its creator as both a plagiarist and a charlatan. Not to be intimidated, Einstein went along with Walther Nernst and watched the proceedings from a private box as he was vilified. Refusing to rise to the bait, he said nothing.

Nernst, Heinrich Rubens and Max von Laue wrote to the newspapers defending Einstein against the outrageous charges levelled at him. Many of his friends and colleagues were therefore dismayed when Einstein wrote an article for the *Berliner Tageblatt* entitled 'My Reply'. He pointed out that had he not been Jewish and an internationalist he would not have been denounced, nor his work attacked. Einstein immediately regretted having been riled into writing the article. 'Everyone has to sacrifice at the altar of stupidity from time to time, to please the Deity and the human race', he wrote to the physicist Max Born and his wife.[70] He was well aware that his celebrity status meant that 'like a man in the fairy tale who turned everything into gold – so with me everything turns into a fuss in the newspapers'.[71] Soon there were rumours that Einstein might leave the country, but he chose to stay in Berlin, 'the place to which I am most closely tied by human and scientific connections'.[72]

In the two years after their meetings in Berlin and Copenhagen, Einstein and Bohr continued their individual struggles with the quantum. Both were beginning to feel the strain. 'I suppose it's a good thing that I have so much to distract me,' Einstein wrote to Ehrenfest in March 1922, 'else the quantum problem would have got me into a lunatic asylum.'[73] A month later, Bohr confessed to Sommerfeld: 'In the last few years, I have often felt myself scientifically very lonesome, under the impression that my effort to develop the principles of the quantum theory systematically to the best of my ability has been received with very little understanding.'[74] His feelings of isolation were about to end. In June 1922, he travelled to Germany and gave a celebrated series of seven lectures spread over eleven days at Göttingen University that became known as the 'Bohr Festspiele'.

More than a hundred physicists, old and young, came from all over the country to hear Bohr explain his electron shell model of the atom. It was his new theory about the arrangement of electrons inside atoms that explained the placing and grouping of elements within the periodic table. He proposed that orbital shells, like layers of an onion, surrounded an atomic nucleus. Each such shell was actually made up of a set or subset of electron orbits and was able to accommodate only a certain maximum number of electrons.[75] Elements that shared the same chemical properties, Bohr argued, did so because they had the same numbers of electrons in their outermost shell.

According to Bohr's model, sodium's eleven electrons are arranged 2, 8 and 1. Caesium's 55 electrons are arranged in a 2, 8, 18, 18, 8, 1 configuration. It is because the outer shell of each element has a single electron that sodium and caesium share similar chemical properties. During the lectures Bohr used his theory to make a prediction. The unknown element with atomic number 72 would be chemically similar to zirconium, atomic number 40, and titanium, atomic number 22, the two elements in the same column of the periodic table. It would not, Bohr said, belong to the 'rare earth' group of elements that were on either side of it in the table, as predicted by others.

Einstein did not attend Bohr's Göttingen lectures, as he feared for his life following the murder of Germany's Jewish foreign minister. Walther Rathenau, a leading industrialist, had been in office only a few short months when he was gunned down in broad daylight on 24 June 1922 to become the 354th political assassination by the right since the end of the war. Einstein was one of those who had urged Rathenau not to take such a high-profile post within government. When he did, it was deemed 'an absolutely unheard of provocation of the people!' by the right-wing press.[76]

'Here our daily lives have been nerve-racking since the shameful assassination of Rathenau', Einstein wrote to Maurice Solovine.[77] 'I am always on the alert; I have stopped my lectures and am officially absent, though I am actually here all the time.' Warned by reliable sources that he was a prime target for assassination, Einstein confided to Marie Curie that he was thinking about giving up his post at the Prussian Academy to find a quiet place to settle down as a private citizen.[78] For the man who in his youth had hated authority had now become a figure of authority. He was no longer simply a physicist, but was a symbol of German science and of Jewish identity.

Despite the turmoil, Einstein read Bohr's published papers, including 'The Structure of the Atoms and the Physical and Chemical Properties of the Elements', which appeared in the *Zeitschrift für Physik* in March 1922. He recalled nearly half a century later how Bohr's 'electron-shells of the atoms together with their significance for chemistry appeared to me like a miracle – and appears to me as a miracle even today'.[79] It was, Einstein said, 'the highest form of musicality in the sphere of thought'. What Bohr had done was indeed as much art as science. Using evidence gathered from a variety of different sources such as atomic spectra and chemistry, Bohr had built up a particular atom, one electron shell at a time, layer by onion layer, until he had reconstructed every element in the entire periodic table.

At the heart of his approach lay Bohr's belief that quantum rules apply on the atomic scale, but any conclusion drawn from them must not conflict with observations made on the macroscopic scale where classical physics rules. Calling it the 'correspondence principle' allowed him to eliminate ideas on the atomic scale that when extrapolated did not correspond to results that were known to be correct in classical physics. Since 1913 the correspondence principle had helped Bohr bridge the divide between quantum and classical. Some viewed it as a 'magic wand, which did not act outside Copenhagen', recalled Bohr's assistant Hendrik Kramers.[80] Others might have struggled to wave it, but Einstein recognised a fellow sorcerer at work.

Whatever reservations there might have been at the lack of hard mathematics to underpin Bohr's theory of the periodic table, everyone had been impressed by the Dane's latest ideas and gained a greater appreciation of the problems that remained. 'My entire stay in Göttingen was a wonderful and instructive experience for me,' Bohr wrote on his return to Copenhagen, 'and I cannot say how happy I was for all the friendship shown me by everybody.'[81] He was no longer feeling under-appreciated and isolated. Later that year there was further confirmation, if he needed it.

———

As the telegrams of congratulation landed on Bohr's desk in Copenhagen, none meant more to him than the one from Cambridge. 'We are delighted that you have been awarded the Nobel Prize', Rutherford wrote. 'I knew it was merely a question of time, but there is nothing like the accomplished fact. It is well merited recognition of your great work and everybody here is delighted in the news.'[82] In the days that followed the announcement, Rutherford had never been far from Bohr's thoughts. 'I have felt so strongly how much I owe you,' he told his old mentor, 'not only for your direct influence on my work and your inspiration, but also for your friendship in these twelve years since I had the great fortune of meeting you for the first time in Manchester.'[83]

The other person Bohr could not help thinking about was Einstein. He was delighted and relieved that the day he received the 1922 prize, Einstein had been awarded the Nobel Prize for 1921 that had been deferred for a year. 'I know how little I have deserved it,' he wrote to Einstein, 'but I should like to say that I consider it a good fortune that your fundamental contribution in the special area in which I work as well as contributions by Rutherford and Planck should be recognized before I was considered for such an honour.'[84]

Einstein was on a ship bound for the other side of the world when the Nobel Prize winners were announced. On 8 October, still fearing for his safety, Einstein and Elsa had left for a lecture tour of Japan. He 'welcomed the opportunity of a long absence from Germany, which took me away from temporarily increased danger'.[85] He did not return to Berlin until February 1923. The original six-week itinerary turned into a grand tour lasting five months, during which he had received Bohr's letter. He replied during the voyage home: 'I can say without exaggeration that [your letter] pleased me as much as the Nobel Prize. I find especially charming your fear that you might have received the award before me – that is typically Bohr-like.'[86]

A blanket of snow covered the Swedish capital on 10 December 1922 as the invited guests assembled in the Great Hall of the Academy of Music in Stockholm to watch the presentation of the Nobel Prizes. The ceremony began at five o'clock in the presence of King Gustav V. The German ambassador to Sweden received the prize on behalf of the absent Einstein, but only after winning a diplomatic argument with the Swiss over the physicist's nationality. The Swiss were claiming Einstein as one of their own, until the Germans discovered that by accepting the appointment at the Prussian Academy in 1914 Einstein had automatically become a German citizen, even though he had not given up his Swiss nationality.

Having renounced his German citizenship in 1896 and taken Swiss citizenship five years later, Einstein was surprised to learn that he was a German after all. Whether he liked it or not, the needs of the Weimar Republic meant that Einstein officially had dual nationality. 'By an application of the theory of relativity to the taste of readers,' Einstein had written in November 1919 in an article for the London *Times*, 'today in Germany I am called a German man of science and in England I am represented as a Swiss Jew. If I come to be regarded as a *bête noire*, the descriptions will be reversed and I shall become a Swiss Jew for the Germans and a German man of science for the English!'[87] Einstein might have recalled these words had he been at the Nobel banquet and heard the German ambassador propose a toast that expressed the 'joy of my people that once again one of them has been able to achieve something for all of mankind'.[88]

Bohr rose after the German ambassador and gave a short speech as tradition demanded. After paying tribute to J.J. Thomson, Rutherford, Planck and Einstein, Bohr proposed a toast to the international cooperation for the advancement of science, 'which is, I may say, in these so manifoldly depressing times, one of the bright spots visible in human existence'.[89] Given the occasion, it is understandable that he chose to forget the continuing exclusion of German scientists from international conferences. The next day Bohr was on firmer ground as he gave his Nobel lecture on 'The structure of the atom'. 'The present state of atomic theory is characterized by the fact that we not only believe the existence of atoms to be proved beyond a doubt,' he began, 'but we even believe that we have an intimate knowledge of the constituents of the individual atoms.'[90] Having given a survey of the developments in atomic physics of which he had been such a central figure in the past decade, Bohr conclude his lecture with a dramatic announcement.

In his Göttingen lectures, Bohr had predicted the properties that the missing element with an atomic number of 72 should possess, based upon his theory of the arrangement of electrons in atoms. At exactly that time a paper was published outlining an experiment performed in Paris that confirmed a long-standing rival French claim that element 72 was a member of the 'rare earth' family of elements that occupied slots 57 to 71 in the periodic table. After the initial shock, Bohr began having serious doubts about the validity of the French results. Fortunately his old friend Georg von Hevesy, who was now in Copenhagen, and Dirk Coster devised an experiment to settle the dispute about element 72.

Bohr had already left for Stockholm by the time Hevesy and Coster completed their investigation. Coster telephoned Bohr shortly before his lecture and he was able to announce that 'appreciable quantities' of element 72 had been isolated, 'the chemical properties of which show a great similarity to those of zirconium and a decided difference from those of the rare earths'.[91] Later called hafnium after the ancient name for Copenhagen, it was a fitting conclusion to Bohr's work on the configuration of electrons within atoms that he had begun in Manchester a decade earlier.[92]

In July 1923, Einstein gave his Nobel lecture on the theory of relativity as part of the 300th anniversary celebrations of the founding of the Swedish city of Göteborg. He broke with tradition by choosing relativity, when he had been awarded the prize 'for his attainments in mathematical physics and especially for his discovery of the law of the photoelectric effect'.[93] By limiting the award of the prize for the 'law', the mathematical formula that accounted for the photoelectric effect, the committee deftly sidestepped endorsing Einstein's controversial underlying physical explanation – the light-quantum. 'In spite of its heuristic value, however, the hypothesis of light-quanta, which is quite irreconcilable with so-called interference phenomena, is not able to throw light on the nature of radiation', Bohr had said during his own Nobel lecture.[94] It was a familiar refrain echoed by every self-respecting physicist. But as Einstein went to meet Bohr for the first time in nearly three years, he knew that an experiment performed by a young American meant that he no longer stood alone in defence of the quantum of light. Bohr had heard the dreaded news before Einstein.

————

In February 1923 Bohr received a letter dated 21 January, from Arnold Sommerfeld, alerting him to the 'most interesting thing that I have experienced scientifically in America'.[95] He had swapped Munich, Bavaria for Madison, Wisconsin for a year and managed to escape the worst of the hyperinflation about to engulf Germany. It had been a shrewd financial move for Sommerfeld. To get an early glimpse of the work of Arthur Holly Compton before his European colleagues was an unexpected bonus.

Compton had made a discovery that challenged the validity of the wave theory of X-rays. Since X-rays were electromagnetic waves, a form of short-wavelength invisible light, Sommerfeld was saying that the wave nature of light, contrary to all the evidence in its favour, was in serious trouble. 'I do not know if I should mention his results', wrote Sommerfeld somewhat coyly, since Compton's paper had not yet been published. 'I want to call your attention to the fact that eventually we may expect a completely fundamental and new lesson.'[96] It was a lesson that Einstein had been trying to teach with varying degrees of enthusiasm since 1905. Light was quantised.

Compton was one of America's leading young experimenters. He had been appointed professor and head of physics at the University of Washington in St Louis, Missouri in 1920 at just 27. His investigations into the scattering of X-rays conducted two years later would be described as 'the turning point in twentieth-century physics'.[97] What Compton did was fire a beam of X-rays at a variety of elements such as carbon (in the form of graphite) and measure the 'secondary radiation'. When the X-rays slammed into the target most of them passed straight through, but some were scattered at a variety of angles. It was these 'secondary' or scattered X-rays that interested Compton. He wanted to find out if there was any change in their wavelength compared to the X-rays that had struck the target.

He found that the wavelengths of the scattered X-rays were always slightly longer than those of the 'primary' or incident X-rays. According to the wave theory they should have been exactly the same. Compton understood that the difference in wavelength (and therefore frequency) meant the secondary X-rays were not the same as the ones that had been fired at the target. It was as strange as shining a beam of red light at a metal surface and finding blue light being reflected.[98] Unable to make his scattering data tally with the predictions of a wavelike theory of X-rays, Compton turned to Einstein's light-quanta. Almost at once he found 'that the wavelength and the intensity of the scattered rays are what they should be if a quantum of radiation bounced from an electron, just as one billiard ball bounces from another'.[99]

If X-rays came in quanta, then a beam of X-rays would be similar to a collection of microscopic billiard balls slamming into the target. Although some would pass through without hitting anything, others would collide with electrons inside atoms of the target. During such a collision an X-ray quantum would lose energy as it was scattered and the electron sent recoiling from the impact. Since the energy of an X-ray quantum is given by $E=h\nu$, where h is Planck's constant and ν its frequency, then any loss of energy must result in a drop in frequency. Given that frequency is inversely proportional to wavelength, the wavelength associated with a scattered X-ray quantum increases. Compton constructed a detailed mathematical analysis of how the energy lost by the incoming X-ray and the resulting change in the wavelength (frequency) of the scattered X-ray was dependent upon the angle of scattering.

No one had ever observed the recoiling electrons that Compton believed should accompany the scattered X-rays. But then no one had been looking for them. When he did, Compton soon found them. 'The obvious conclusion,' he said, 'would be that X-rays, and so also light, consist of discrete units, proceeding in definite directions, each unit possessing the energy $h\nu$ and the corresponding momentum $h\lambda$.'[100] The 'Compton effect', the increase in wavelength of X-rays when they are scattered by electrons, was irrefutable evidence for the existence of light-quanta, which until then many had dismissed at best as science fiction. It was by assuming that energy and momentum are conserved in the collision between an X-ray quantum and an electron that Compton was able to explain his data. It was Einstein, in 1916, who had been the first to suggest that light-quanta possessed momentum, a particle-like property.

In November 1922 Compton announced his discovery at a conference in Chicago.[101] However, although he sent his paper to the *Physical Review* just before Christmas, it was not published until May 1923 as the editors failed to understand the significance of its content. The avoidable delay meant that the Dutch physicist Pieter Debye beat Compton into print with the first complete analysis of the discovery. A former Sommerfeld assistant, Debye had submitted his paper to a German journal in March. Unlike their American counterparts, the German editors recognised the importance of the work and published it the following month. However, Debye and everyone else gave the talented young American the credit and recognition he deserved. It was sealed when Compton was awarded the Nobel Prize in 1927. By then, Einstein's light-quantum had been rechristened the photon.[102]

There had been 2,000 at his Nobel lecture in July 1923, but Einstein knew that most of them had come to see rather than to listen to him. Sitting on the train as he made his way from Göteborg to Copenhagen, Einstein was looking forward to meeting a man who would listen to his every word and probably disagree. When he got off the train, Bohr was there to greet him. 'We took the streetcar and talked so animatedly that we went much too far', Bohr recalled almost 40 years later.[103] Speaking in German, they were oblivious to the curious stares of fellow passengers. Whatever was discussed, as they rode back and forth missing their stop, it was sure to include the Compton effect, soon to be described by Sommerfeld as 'probably the most important discovery that could have been made in the current state of physics'.[104] Bohr was unconvinced and refused to accept that light was made up of quanta. It was he, not Einstein, who was now in the minority. Sommerfeld was in no doubt that 'the death-knell of the wave theory of radiation' had been sounded by Compton.[105]

Like the doomed hero in the westerns that he later liked to watch, Bohr was outnumbered as he made one last stand against the quantum of light. In collaboration with his assistant Hendrik Kramers and a visiting young American theorist, John Slater, Bohr proposed sacrificing the law of conservation of energy. It was a vital component in the analysis leading to the Compton effect. If the law was not strictly enforced on the atomic scale as it was in the everyday world of classical physics, then Compton's effect was no longer incontrovertible evidence for Einstein's light-quanta. The BKS proposal, as it became known (after Bohr, Kramers and Slater), appeared to be a radical suggestion but was in truth an act of desperation that showed how much Bohr abhorred the quantum theory of light.

The law had never been experimentally tested at the atomic level and Bohr believed that the extent of its validity remained an open question in processes such as the spontaneous emission of light-quanta. Einstein believed that energy and momentum were conserved in every single collision between a photon and an electron, while Bohr believed they were valid only as a statistical average. It was 1925 before experiments by Compton, then at Chicago University, and by Hans Geiger and Walther Bothe at the Physikalische-Technische Reichsanstalt, confirmed that energy and momentum were conserved in collisions between a photon and an electron. Einstein had been right and Bohr wrong.

Confident as ever, on 20 April 1924, more than a year before experiments silenced the doubters, Einstein eloquently summed up the situation for the readers of the *Berliner Tageblatt*: 'There are therefore now two theories of light, both indispensable and – as one must admit today despite twenty years of tremendous effort on the part of theoretical physicists – without any logical connection.'[106] Einstein meant that both the wave theory of light and quantum theory of light were in some way valid. Light-quanta could not be invoked to explain the wave phenomena associated with light, such as interference and diffraction. Conversely, a full explanation of Compton's experiment and the photoelectric effect could not be provided without recourse to the quantum theory of light. Light had a dual, wave-particle character, which physicists just had to accept.

One morning, not long after the article appeared, Einstein received a parcel with a Paris postmark. Opening it, he discovered a note from an old friend seeking his opinion of the accompanying doctoral thesis written by a French prince on the nature of matter.

Chapter 6

THE PRINCE OF DUALITY

'Science is an old lady who does not fear mature men', his father had once said.[1] But he, like his elder brother, had been seduced by science. Prince Louis Victor Pierre Raymond de Broglie, a member of one of France's leading aristocratic families, had been expected to follow in the footsteps of his illustrious forebears. The de Broglie family, originally from Piedmont, had served French kings as soldiers, statesmen and diplomats with high distinction since the middle of the seventeenth century. In recognition of the service he had rendered, an ancestor was given the hereditary title of Duc in 1742 by Louis XV. The Duke's son, Victor-François, inflicted a crushing defeat on an enemy of the Holy Roman Empire and a grateful Emperor rewarded him with the title of Prinz. Henceforth, all of his descendants would be either a prince or a princess. So it was that the young scientist would one day be both a German prince and a French duke.[2] It is an unlikely family history for the man who made a fundamental contribution to quantum physics, which Einstein described as 'the first feeble ray of light on this worst of our physics enigmas'.[3]

———

The youngest of the four surviving children, Louis was born in Dieppe on 15 August 1892. In keeping with their elevated position in society, the de Broglies were educated at the ancestral home by private tutors. While other boys might have been able to recite the names of the great steam engines of the day, Louis could recite the names of all the ministers of the Third Republic. To the amusement of the family, he began giving speeches based upon the political coverage in the newspapers. With a grandfather who had been prime minister, before long 'a great future as a statesman was predicted for Louis', recalled his sister Pauline.[4] It might have been the case had his father not died, in 1906, when he was fourteen.

His elder brother, Maurice, at 31, was now the head of the family. As tradition demanded, Maurice had pursued a military career but had chosen the navy rather than the army. At naval college he excelled at science. As a promising young officer he found a navy in a period of transition as it prepared for the twentieth century. Given his scientific interests, it was only a matter of time before Maurice became involved in attempts at establishing a reliable ship-to-ship wireless communication system. In 1902 he wrote his first paper on 'radioelectric waves' and, despite the opposition of his father, it strengthened his determination to leave the navy and devote himself to scientific research. In 1904, after nine years in the service, he quit the navy. Two years later his father was dead and he had to shoulder new responsibilities as the sixth Duc.

On Maurice's advice, Louis was sent to school. 'Having experienced myself the inconvenience of a pressure exercised on the studies of a young man I refrained from imparting a rigid direction to the studies of my brother, although at times his vacillations gave me some concerns', he wrote almost half a century later.[5] Louis did well in French, history, physics and philosophy. In mathematics and chemistry he was indifferent. After three years Louis graduated in 1909 at the age of seventeen, with both the *baccalauréat* of philosophy and that of mathematics. A year earlier Maurice had acquired his PhD under Paul Langevin at the Collège de France and set up a laboratory in his Parisian mansion on the rue Châteaubriand. Rather than seek employment in a university, the creation of a private laboratory in which to pursue his new vocation helped soften the disappointment of some of the family at a de Broglie abandoning military service for science.

Unlike Maurice, Louis at the time was set for a more traditional career as he studied medieval history at the University of Paris. However, the twenty-year-old prince soon discovered that the critical study of texts, sources and documents of the past held little interest for him. Maurice said later that his brother was 'not far from losing faith in himself'.[6] Part of the problem was a burgeoning interest in physics fostered by time spent with Maurice in the laboratory. The enthusiasm of his elder brother about his research on X-rays had proved contagious. However, Louis was consumed by doubts about his abilities that were aggravated by failing a physics exam. Was he, Louis wondered, destined to be a failure? 'Gone the gaiety and high spirits of his adolescence! The brilliant chatter of his childhood has been muted by the depth of his reflections', was how Maurice remembered the introvert he hardly recognised.[7] Louis would become, according to his brother, 'an austere and fairly untamed scholar', who did not like leaving his own home.[8]

The first time Louis travelled abroad it was to Brussels in October 1911.[9] He was nineteen. In the years since he left the navy, Maurice had become a much-respected scientist specialising in X-ray physics. When the invitation arrived to be one of the two scientific secretaries entrusted with the smooth running of the first Solvay conference, he readily accepted. Even though it was an administrator's role, the chance to discuss the quantum with the likes of Planck, Einstein and Lorentz was just too tempting to forgo. The French would be well represented. Curie, Poincaré, Perrin, and his former supervisor Langevin would all be there.

Staying at the Hotel Metropole with all the delegates, Louis kept his distance. It was only after they returned and Maurice recounted the discussions about the quantum that took place in the small room on the first floor that Louis began taking an even greater interest in the new physics. When the proceedings of the conference were published, Louis read them and resolved to become a physicist. By then he had already swapped history books for those of physics, and in 1913 he obtained his Licence és Science, the equivalent of a degree. His plans had to wait as a year of military service beckoned. Despite the three Marshals of France that the de Broglies could boast, Louis entered the army as a lowly private in a company of engineers stationed just outside Paris.[10] With Maurice's help, he was soon transferred to the Service of Wireless Communication. Any hopes of a quick return to his study of physics evaporated with the outbreak of the First World War. He spent the next four years as a radio engineer stationed underneath the Eiffel Tower.

Discharged in August 1919, he deeply resented having spent six years, from the age of 21 to 27, in uniform. Louis was more determined than ever to continue down his chosen path. He was helped and encouraged by Maurice and spent time in his well-equipped laboratory following the research being done on X-rays and the photoelectric effect. The brothers had long discussions on the interpretation of the experiments being conducted. Maurice reminded Louis of 'the educational value of the experimental sciences' and 'that theoretical constructions of science have no value unless they are supported by facts'.[11] He wrote a series of papers on the absorption of X-rays while thinking about the nature of electromagnetic radiation. The brothers accepted that both the wave and particle theories of light were in some sense correct, since neither on its own could explain diffraction and interference and also the photoelectric effect.

In 1922, the year Einstein lectured in Paris at the invitation of Langevin and received a hostile reception for having remained in Berlin throughout the war, de Broglie wrote a paper in which he explicitly adopted 'the hypothesis of quanta of light'. He had already accepted the existence of 'atoms of light' at a time when Compton had yet to make any sort of announcement concerning his experiments. By the time the American published his data and analysis of the scattering of X-rays by electrons and thereby confirmed the reality of Einstein's light-quanta, de Broglie had already learned to live with the strange duality of light. Others, however, were only half-joking when they complained about having to teach the wave theory of light on Mondays, Wednesdays and Fridays, and the particle theory on Tuesdays, Thursdays and Saturdays.

'After long reflection in solitude and meditation,' de Broglie wrote later, 'I suddenly had the idea, during the year 1923, that the discovery made by Einstein in 1905 should be generalized by extending it to all material particles and notably to electrons.'[12] De Broglie had dared to ask the simple question: if light waves can behave like particles, can particles such as electrons behave like waves? His answer was yes, as de Broglie discovered that if he assigned to an electron a 'fictitious associated wave' with a frequency ν and wavelength λ, he could explain the exact location of the orbits in Bohr's quantum atom. An electron could occupy only those orbits that could accommodate a whole number of wavelengths of its 'fictitious associated wave'.

In 1913, to prevent Rutherford's model of the hydrogen atom from collapsing as its orbiting electron radiated energy and spiralled into the nucleus, Bohr had been forced to impose a condition for which he could offer no other justification: an electron in a stationary orbit around the nucleus did not emit radiation. De Broglie's idea of treating electrons as standing waves was a radical departure from thinking about electrons as particles orbiting an atomic nucleus.

Standing waves can easily be generated in strings tethered at both ends, such as those used in violins and guitars. Plucking such a string produces a variety of standing waves with the defining characteristic that they are made up of a whole number of half-wavelengths. The longest standing wave possible is one with a wavelength twice as long as the string. The next standing wave is made up of two such half-wavelength units, giving a wavelength equal to the physical length of the string. The next is a standing wave consisting of three half-wavelengths, and so on up the scale. This whole number sequence of standing waves is the only one that is physically possible, and each has its own energy. Given the relationship between frequency and wavelength, this is equivalent to the fact that a plucked guitar string can vibrate only at certain frequencies beginning with the fundamental tone, the lowest frequency.

De Broglie realised that this 'whole number' condition restricted the possible electron orbits in the Bohr atom to those with a circumference that permitted the formation of standing waves. These electron standing waves were not bound at either end like those on a musical instrument, but

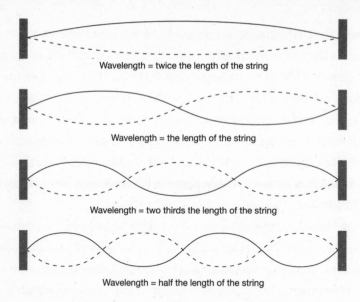

Wavelength = twice the length of the string

Wavelength = the length of the string

Wavelength = two thirds the length of the string

Wavelength = half the length of the string

Figure 9: Standing waves of a string tethered at both ends

were formed because a whole number of half-wavelengths could be fitted into the circumference of the orbit. Where there was no exact fit, there could be no standing wave and therefore no stationary orbit.

Figure 10: Standing electron waves in the quantum atom

If viewed as a standing wave around the nucleus instead of a particle in orbit, an electron would experience no acceleration and therefore no continual loss of radiation sending it crashing into the nucleus as the atom collapsed. What Bohr had introduced simply to save his quantum atom, found its justification in de Broglie's wave-particle duality. When he did the calculations, de Broglie found that Bohr's principal quantum number, n, labelled only those orbits in which electron standing waves could exist around the nucleus of the hydrogen atom. It was the reason why all other electron orbits were forbidden in the Bohr model.

When de Broglie outlined why all particles should be viewed as having a dual wave-particle character in three short papers in the autumn of 1923, it was not immediately clear what was the nature of the relationship between the billiard ball-like particles and the 'fictitious associated wave'. Was de Broglie suggesting that it was akin to a surfer riding a wave? It was later established that such an interpretation would not work and that electrons, and all other particles, behaved exactly like photons: they are both wave and particle.

De Broglie wrote up his ideas in an expanded form and presented them as his PhD thesis in the spring of 1924. The necessary formalities of acceptance and its reading by the examiners meant that de Broglie did not defend his doctoral dissertation until 25 November. Three of the four examiners were professors at the Sorbonne: Jean Perrin, who had been instrumental in testing Einstein's theory of Brownian motion; Charles Mauguin, a distinguished physicist working on the properties of crystals; and Elie Cartan, a renowned mathematician. The last member of the quartet was the external examiner, Paul Langevin. He alone was well versed in quantum physics and relativity. Before officially submitting his dissertation, de Broglie approached Langevin and asked him to look at his conclusions. Langevin agreed and afterwards told a colleague: 'I am taking with me the little brother's thesis. Looks far-fetched to me.'[13]

Louis de Broglie's ideas may have been fanciful, but Langevin did not quickly dismiss them. He needed to consult another. Langevin knew that Einstein had publicly stated in 1909 that future research into radiation would reveal a kind of fusion of the particle and wave. Compton's experiments had convinced almost everyone that Einstein had been right about light. It did after all appear to be a particle in collisions with electrons. Now, de Broglie was suggesting the same kind of fusion, wave-particle duality, for all of matter. He even had a formula that linked the wavelength λ of the 'particle' to its momentum p, $\lambda=h/p$ where h is Planck's constant. Langevin asked the physicist prince for a second copy of the dissertation and sent it to Einstein. 'He has lifted a corner of the great veil', Einstein wrote back to Langevin.[14]

The judgement of Einstein was enough for Langevin and the other examiners. They congratulated de Broglie for 'having pursued with a remarkable mastery an effort that had to be attempted in order to overcome the difficulties in the midst of which the physicists found themselves'.[15] Mauguin later admitted that he 'did not believe at the time in the physical reality of the waves associated with grains of matter'.[16] All Perrin knew for sure was that de Broglie was 'very intelligent'.[17] As for the rest he had no idea. With Einstein's support, aged 32, he was no longer just Prince Louis Victor Pierre Raymond de Broglie, but had earned the right to call himself plain Dr Louis de Broglie.

Having an idea was one thing, but could it be tested? De Broglie had quickly realised in September 1923 that if matter has wave properties, then a beam of electrons should spread out like a beam of light – they should be diffracted. In one of his short papers written that year, de Broglie had predicted that a 'group of electrons that passes through a small aperture should show diffraction effects'.[18] He tried, but failed to convince any of the skilled experimentalists working in his brother's private laboratory to put his idea to the test. Busy with other projects, they simply thought the experiments too difficult to perform. Already indebted to his brother Maurice for continually directing his 'attention to the importance and the undeniable accuracy of the dual particulate and wave properties of radiation', Louis did not pursue the matter.[19]

However Walter Elsasser, a young physicist at Göttingen University, soon pointed out that if de Broglie was right, a simple crystal would diffract a beam of electrons hitting it: since the spacing between adjacent atoms in a crystal would be small enough for an object the size of an electron to reveal its wave character. 'Young man, you are sitting on a gold mine', Einstein told Elsasser when he heard of his proposed experiment.[20] It was not a gold mine, but something a bit more precious: a Nobel Prize. But as in any gold rush, one cannot wait too long before getting started. Elsasser did, and two others staked their claims first and grabbed the prize.

Thirty-four-year-old Clinton Davisson of the Western Electric Company in New York, later better known as the Bell Telephone Laboratories, had been investigating the consequences of smashing a beam of electrons into various metal targets when, one day in April 1925, a strange thing happened. A bottle of liquefied air exploded in his laboratory and broke the evacuated tube containing the nickel target that he was using. The air caused the nickel to rust. As a result of cleaning the nickel by heating it, Davisson had accidentally turned the array of tiny nickel crystals into just a few large ones, which caused electron diffraction. When he continued his experiments he soon realised that his results were different. Unaware that he had diffracted electrons, he simply wrote up the data and published it.

'It seems impossible that we will be in Oxford a month from today – doesn't it? We should have a lovely time – Lottie darling – It will be a second honeymoon – and should be sweeter even than the first', Davisson wrote to his wife in July 1926.[21] With the children being looked after by relatives back home, the Davissons could enjoy a much-needed break touring England before heading to Oxford and the British Association for the Advancement of Science conference. It was there that Davisson was astonished to learn that some physicists believed that the data from his experiment supported the idea of a French prince. He had not heard of de Broglie or his suggestion that wave-particle duality be extended to encompass all matter. Davisson was not alone.

Few people had read de Broglie's three short papers because they had been published in the French journal *Compte Rendu*. Fewer still knew of the existence of the doctoral dissertation. On returning to New York, Davisson and a colleague, Lester Germer, immediately set about checking whether electrons really were diffracted. It was January 1927 before they had conclusive evidence that matter was diffracted, it did behave like waves, when Davisson calculated the wavelengths of the diffracted electrons from the new results and found they matched those predicted by de Broglie's theory of wave-particle duality. Davisson later admitted that the original experiments were really 'undertaken as a sort of sideline' in the wake of others that he had been conducting on behalf of his employers, who were defending a lawsuit from a rival company.

Max Knoll and Ernst Ruska quickly utilised the wave nature of the electron with the invention in 1931 of the electron microscope. No particle smaller than approximately half the wavelength of white light can absorb or reflect light waves so as to make the particle visible through an ordinary microscope. However, with wavelengths more than 100,000 times smaller than that of light, electron waves could. The construction of the first commercial electron microscope began in England in 1935.

Meanwhile in Aberdeen, Scotland, the English physicist George Paget Thomson was carrying out his own experiments with electron beams as Davisson and Germer were busy conducting theirs. He too had attended the BAAS conference in Oxford where de Broglie's work had been widely discussed. Thomson, who had a very personal interest in the nature of the electron, immediately began experiments to detect electron diffraction. But instead of crystals, he used specially prepared thin films that gave a diffraction pattern whose features were exactly as de Broglie predicted. Sometimes matter behaves like a wave, smeared over an extended region of space, and at others like a particle, located at a single position in space.

In a remarkable twist of fate, the dual nature of matter was embodied in the Thomson family. George Thomson was awarded the Nobel Prize for physics in 1937, together with Davisson, for discovering that the electron was a wave. His father, Sir J.J. Thomson, had been awarded the Nobel Prize for physics in 1906 for discovering that the electron was a particle.

Over a quarter of a century, the developments in quantum physics – from Planck's blackbody radiation law to Einstein's quantum of light, from Bohr's quantum atom to de Broglie's wave-particle duality of matter – were the product of an unhappy marriage of quantum concepts and classical physics. It was a union that by 1925 was increasingly under strain. 'The more successes the quantum theory enjoys, the more stupid it looks', Einstein had written as early as May 1912.[22] What was needed was a new theory, a new mechanics of the quantum world.

'The discovery of quantum mechanics in the mid-1920s,' said the American Nobel laureate Steven Weinberg, 'was the most profound revolution in physical theory since the birth of modern physics in the seventeenth century.'[23] Given the pivotal role of young physicists in making the revolution that shaped the modern world, these were the years of *knabenphysik* – 'boy physics'.

PART II

BOY PHYSICS

'Physics at the moment is again very muddled; in any case, for me it is too complicated, and I wish I were a film comedian or something of that sort and had never heard anything about physics.'
—WOLFGANG PAULI

'The more I think about the physical portion of the Schrödinger theory, the more repulsive I find it. What Schrödinger writes about the visualizability of his theory "is probably not quite right", in other words it's crap.'
—WERNER HEISENBERG

'If all this damned quantum jumping were really here to stay, I should be sorry I ever got involved with quantum theory.'
—ERWIN SCHRÖDINGER

Chapter 7

SPIN DOCTORS

'One wonders what to admire most, the psychological understanding for the development of ideas, the sureness of mathematical deduction, the profound physical insight, the capacity for lucid, systematic presentation, the knowledge of the literature, the complete treatment of the subject matter, or the sureness of critical appraisal.'[1] Einstein was certainly impressed by the 'mature, grandly conceived work' he had just reviewed. It was difficult for him to believe that the 237-page article, with 394 footnotes, on relativity was the work of a 21-year-old physicist who had been a student, and just nineteen, when asked to write it. Wolfgang Pauli, later nicknamed 'The Wrath of God', was acerbic and regarded as 'a genius comparable only with Einstein'.[2] 'Indeed from the point of view of pure science,' said Max Born, his one-time boss, 'he was possibly even greater than Einstein.'[3]

Wolfgang Pauli was born on 25 April 1900 in Vienna, a city still in the grip of *fin de siècle* anxiety while enjoying the good times. His father, also called Wolfgang, had been a physician, but abandoned medicine for science and in the process changed his family name from Pascheles to Pauli. The transformation was complete as he converted to Catholicism amid fears that the rising tide of anti-Semitism threatened his academic ambitions. His son grew up knowing nothing of the family's Jewish ancestry. At university, when another student said that he must be Jewish, Wolfgang junior was astonished: 'I? No. Nobody has ever told me that and I don't believe that I am.'[4] He learnt the truth from his parents during his next visit home. His father felt vindicated by the decision to assimilate when, in 1922, he was appointed to a coveted professorship and became the director of a new institute for medical chemistry at Vienna University.

Pauli's mother, Bertha, was a well-known Viennese journalist and writer. Her circle of friends and acquaintances meant that Wolfgang and his sister Hertha, six years his junior, grew up accustomed to seeing leading figures from the arts as well as science and medicine at the family home. His mother, a pacifist and a socialist, exerted a strong influence on Pauli. The longer the First World War dragged on through his formative teenage years, 'the keener became his opposition against it and, generally, against the whole "establishment"', recalled a friend.[5] When she died two weeks before her 49th birthday in November 1927, an obituary in the *Neue Freie Presse* described Bertha as 'one of the few truly strong personalities among Austrian women'.[6]

Pauli was academically gifted but far from a model pupil, finding school unchallenging. He began having private tuition in physics to compensate. Before long, when bored by a particularly tedious lesson at school, he began reading Einstein's papers on general relativity hidden under his desk. Physics had always loomed large in his young life in the form of the influential Austrian physicist and philosopher of science Ernst Mach, his godfather. For one who would later enjoy the company and friendship of the likes of Einstein and Bohr, Pauli said that contact with Mach, whom he last saw in the summer of 1914, was 'the most important event in my intellectual life'.[7]

In September 1918 Pauli left what he called the 'spiritual desert' that was Vienna.[8] With the Austria-Hungarian empire on the verge of extinction and Vienna's past glories faded, it was the lack of top-flight physicists at the city's university that he was lamenting. He could have gone almost anywhere, but went to Munich to study with Arnold Sommerfeld. Having recently turned down a professorship in Vienna, Sommerfeld had already been in charge of theoretical physics at Munich University for a dozen years when Pauli arrived. From the beginning, in 1906, Sommerfeld set out to create an institute that would be 'a nursery of theoretical physics'.[9] It was not as grand as the institute Bohr would soon create in Copenhagen, consisting as it did of only four rooms: Sommerfeld's office, a lecture theatre, a seminar room, and a small library. There was also a large laboratory in the basement where in 1912 Max von Laue's theory that X-rays were short-wavelength electromagnetic waves was tested and confirmed, bringing quick recognition to the 'nursery'.

Sommerfeld was an exceptional teacher with the uncanny knack of setting his students problems that tested, but did not exceed, their abilities. Having already supervised more than his fair share of talented young physicists, Sommerfeld soon recognised Pauli as someone of rare and exceptional promise. He was a man not easily impressed, but in January 1919 a paper on general relativity written by Pauli before leaving Vienna had just been published. In his 'nursery' he had a first-year student, not yet nineteen years old, who was already regarded by others as an expert in relativity.

Pauli quickly became known, and feared, for his sharp and incisive criticism of new and speculative ideas. Some would later call him the 'conscience of physics' for his uncompromising principles. Stout with bulging eyes, he was every inch the Buddha of physics, albeit one with a biting tongue. Whenever he was lost deep in thought, Pauli unconsciously rocked back and forth. It was acknowledged far and wide that his intuitive grasp of physics was unmatched among his contemporaries and probably not even surpassed by Einstein. He judged his own work even more harshly than that of others. At times Pauli understood physics and its problems too well, and that hampered the free exercise of his creative powers. Discoveries that he might have made if his imagination and intuition had roamed a little more freely went instead to colleagues less talented and unconstrained.

The only person towards whom he was, and remained, diffident was Sommerfeld. Even as a celebrated physicist, whenever Pauli found himself in the presence of his former professor, those who had been on the receiving end of his sharp judgements were always amazed to see the 'Wrath of God' responding with 'Ja, Herr Professor', 'Nein, Herr Professor'. They hardly recognised the man who had once ticked off a colleague: 'I do not mind if you think slowly, but I do object when you publish more quickly than you think.'[10] Or on another occasion saying of a paper he had just read: 'It is not even wrong.'[11] He spared no one. 'You know, what Mr Einstein said is not so stupid', he told a packed lecture theatre while still a student.[12] Sommerfeld, sitting in the front row, would not have tolerated such a remark coming from any of his other students. But then he knew none of them would have uttered it. When it came to evaluating physics, Pauli was self-confident and uninhibited even in the presence of Einstein.

In a clear sign of the high regard in which he held Pauli, Sommerfeld asked him to help write a major article on relativity for the *Encyklopädie der Mathematischen Wissenschaffen*. Sommerfeld had accepted the task of editing the fifth volume of the *Encyklopädie* that dealt with physics. After Einstein declined, Sommerfeld decided to write on relativity himself but found he had little time to do so. He needed help and turned to Pauli. When Sommerfeld saw the first draft, 'it proved to be so masterly that I renounced all collaboration'.[13] It was not only a brilliant exposition of the special and general theories of relativity, but an unrivalled review of the existing literature. It remained for decades the definitive work in the field and drew Einstein's wholehearted praise. The article appeared in 1921, two months after Pauli received his doctorate.

As a student, Pauli preferred to spend his evenings enjoying the Munich nightlife in some café or other, returning to his lodgings to work through much of the night. He rarely attended lectures the following morning, turning up only around noon. But he attended enough to be drawn to the mysteries of quantum physics by Sommerfeld. 'I was not spared the shock which every physicist accustomed to the classical way of thinking experienced when he came to know Bohr's basic postulate of quantum theory for the first time', Pauli said more than 30 years later.[14] But he quickly got over it as he set about tackling his doctoral thesis.

Sommerfeld had set Pauli the task of applying the quantum rules of Bohr and his own modifications to the ionised hydrogen molecule, in which one of the two hydrogen atoms that make up the molecule has had its electron ripped off. As expected, Pauli produced a theoretically impeccable analysis. The only problem was that his results did not agree with the experimental data. Used to one success after another, Pauli was despondent at this lack of agreement between theory and experiment. However, his thesis was regarded as the first strong evidence that the outer limits of the Bohr-Sommerfeld quantum atom had been reached. The ad hoc way in which quantum physics was bolted onto classical physics had always been unsatisfactory, and now Pauli had shown that the Bohr-Sommerfeld model could not even deal with the ionised hydrogen molecule, let alone more complex atoms. In October 1921, armed with his doctorate, Pauli left Munich for Göttingen to take up the post of assistant to the professor of theoretical physics.

Max Born, 38, a key figure in the future development of quantum physics, had arrived in the small university town from Frankfurt just six months before Pauli. Growing up in Breslau, capital of the then Prussian province of Silesia, it was mathematics, not physics, that attracted Born. His father, like Pauli's, was a highly cultured medical man and academic. A professor of embryology, Gustav Born advised his son not to specialise too early once he enrolled at Breslau University. Dutifully, Born settled on astronomy and mathematics only after having attended courses in physics, chemistry, zoology, philosophy and logic. His studies, including time at the universities of Heidelberg and Zurich, ended in 1906 with a doctorate in mathematics from Göttingen.

Immediately afterwards he began a year of compulsory military service that was cut short because of asthma. After spending six months in Cambridge as an advanced student, where he attended the lectures of J.J. Thomson, Born returned to Breslau to begin experimental work. But quickly discovering that he possessed neither the patience nor the skills required to be even a competent experimenter, Born turned to theoretical physics. By 1912 he had done enough to become a *privatdozent* in the world-renowned mathematics department at Göttingen, where they believed that 'physics is much too hard for physicists'.[15]

Born's success in tackling a string of problems by harnessing the power of mathematical techniques unknown to most physicists led in 1914 to an extraordinary professorship in Berlin. Just before war broke out, another newcomer arrived at the epicentre of German science: Einstein. Before long the two men, who shared a passion for music, became firm friends. When war came, Born was called up for military service. After a spell as a radio operator with the air force, he spent the rest of the war conducting artillery research for the army. Fortunately stationed near Berlin, Born was able to attend seminars at the university, meetings of the German Physical Society, and musical evenings at Einstein's home.

After the war, in the spring of 1919, Max von Laue, an ordinary professor at Frankfurt, suggested to Born that they swap posts. Laue had won the 1914 Nobel Prize for the theory behind the diffraction of X-rays by crystals, and wanted to work with Planck, his former supervisor and a scientist he idolised. Born, encouraged by Einstein to 'definitely accept', agreed, as the exchange meant promotion to a full professorship and independence.[16] Less than two years later, he moved to Göttingen to head the university's institute of theoretical physics. It consisted of one small room, one assistant, and a part-time secretary, but Born was determined to build on these humble beginnings an institute to rival Sommerfeld's in Munich. High on his list of priorities was getting Wolfgang Pauli, whom he described as 'the greatest talent in the physics area that has emerged in the last years'.[17] Born had already tried once before and failed, as Pauli opted to stay in Munich to finish his doctorate. This time he got his man.

'W. Pauli is now my assistant; he is amazingly intelligent, and very able', Born wrote to Einstein.[18] Soon he discovered that the hired help had his own way of doing things. Pauli might have been brilliant, but he put in long hours of hard thinking as he continued his practice of working into the middle of the night and sleeping late. Whenever Born was unable to give his eleven o'clock lecture, the only way he could ensure Pauli would be there to teach in his place was by sending the maid to wake him up at 10.30am.

It was clear from the beginning that Pauli was an 'assistant' in name only. Born admitted later that he learnt more from Pauli, despite his bohemian ways and poor time-keeping, than he was able to teach the 'infant prodigy'. He was sad to see him go when in April 1922 Pauli left to become an assistant at Hamburg University. Swapping the quiet life of the small university town that he could hardly bear for the bustling nightlife of the big city was not the only reason he left so quickly. Pauli trusted his sense of physical intuition in pursuit of a logically flawless argument when tackling any physics problem. Born, however, turned much more readily to mathematics and allowed it to lead his search for a solution.

Two months later, in June 1922, Pauli was back in Göttingen to hear Bohr's celebrated lecture series and met the great Dane for the first time. Impressed, Bohr asked Pauli if he would come to Copenhagen for a year as his assistant to help edit work in progress for publication in German. Pauli was taken aback by the offer. 'I answered with that certainty of which only a young man is capable: "I hardly think that the scientific demands which you will make on me will cause me any difficulty, but the learning of a foreign tongue like Danish far exceeds my abilities." I went to Copenhagen in the fall of 1922, where both my contentions were shown to be wrong.'[19] It was also, he recognised later, the beginning of 'a new phase' in his life.[20]

Aside from helping Bohr, Pauli made a serious effort in Copenhagen to explain the 'anomalous' Zeeman effect – a feature of atomic spectra that the Bohr-Sommerfeld model could not explain. If atoms were exposed to a strong magnetic field, then the resulting atomic spectra contained lines that were split. It was quickly shown by Lorentz that classical physics predicted a splitting of a line into a doublet or a triplet: a phenomenon known as the 'normal' Zeeman effect which Bohr's atom could not accommodate.[21] Fortunately, Sommerfeld came to the rescue with two new quantum numbers and the modified quantum atom resolved the problem. It involved a series of new rules governing electrons jumping from one orbit (or energy level) to another based on three 'quantum numbers', n, k, and m, that described the size of the orbit, the shape of the orbit, and the direction in which the orbit was pointing. But the celebrations were short-lived when it was discovered that the splitting of the red alpha line in the spectrum of hydrogen was smaller than expected. The situation grew worse with the confirmation that some spectral lines actually split up into a quartet or more instead of just two or three lines.

Although called the 'anomalous' Zeeman effect because the extra lines could not be explained using either existing quantum physics or classical theory, it was in fact far more common than the 'normal' effect. For Pauli it signalled nothing less than the 'deep seated failure of the theoretical principles known till now'.[22] Having set himself the task of rectifying this miserable state of affairs, Pauli could not come up with an explanation. 'Up till now I have thoroughly gone wrong', he wrote to Sommerfeld in June 1923.[23] Consumed by the problem, Pauli later admitted that he was in complete despair for some time.

One day another physicist from the institute met him while strolling around the streets of Copenhagen. 'You look very unhappy', said his colleague. Pauli turned on him: 'How can one look happy when he is thinking about the anomalous Zeeman effect?'[24] The use of ad hoc rules to describe the complex structure of atomic spectra was just too much for Pauli. He wanted a deeper, more fundamental explanation of the phenomena. Part of the problem, he believed, was the guesswork involved in Bohr's theory of the periodic table. Did it really describe the correct arrangement of electrons inside atoms?

By 1922 the electrons in the Bohr-Sommerfeld model were believed to move in three-dimensional 'shells'. These were not physical shells, but energy levels within atoms around which electrons seemed to cluster. A vital clue in helping Bohr construct this new electron shell model was the stability of the so-called noble gases: helium, neon, argon, krypton, xenon and radon.[25] With atomic numbers of 2, 10, 18, 36, 54 and 86, the relatively high energies required to ionise any noble gas atom – to rip away an electron and turn it into a positive ion – together with their reluctance to chemically bond with other atoms to form compounds, suggested that the electron configurations in these atoms were extremely stable and consisted of 'closed shells'.

The chemical properties of the noble gases were in stark contrast to the elements that preceded them in the periodic table – hydrogen and the halogens: fluorine, chlorine, bromine, iodine, and astatine. With atomic numbers 1, 9, 17, 35, 53 and 85, all of these elements easily formed compounds. Unlike the chemically inert noble gases, hydrogen and the halogens united with other atoms because in the process they picked up another electron and thereby filled the single vacancy in the outermost electron shell. By doing so, the resulting negative ion had a completely full or 'closed' set of electron shells and acquired the highly stable electronic configuration of a noble gas atom. Mirroring the halogens, the alkalis group – lithium, sodium, potassium, rubidium, caesium and francium – were quick to lose an electron as they formed compounds and became positive ions with the electron distribution of a noble gas.

The chemical properties of these three groups of elements formed part of the evidence that led Bohr to propose that the atom of each element in a row of the periodic table is built up from the previous element by the addition of another electron to the outer electron shell. Each row would end with a noble gas in which the outer shell was full. Since only electrons outside the closed shells, called valence electrons, took part in chemical reactions, atoms with the same number of valence electrons shared similar chemical properties and occupied the same column in the periodic table. The halogens all have seven electrons in the outermost shell, requiring just one more electron to close it and acquire an electron configuration of a noble gas. The alkalis, on the other hand, all have one valence electron.

It was these ideas that Pauli heard Bohr outline during the Göttingen lectures in June 1922. Sommerfeld had greeted the shell model as 'the greatest advance in atomic structure since 1913'.[26] If he could mathematically reconstruct the numbers 2, 8, 18 … of the elements in the rows of the periodic table, then it would be, Sommerfeld told Bohr, 'the fulfilment of the boldest hopes of physics'.[27] In truth, there was no hard mathematical reasoning to back up the new electron shell model. Even Rutherford told Bohr that he was struggling 'to form an idea of how you arrive at your conclusions'.[28] Nevertheless, Bohr's ideas had to be taken seriously, especially after the announcement in his Nobel lecture in December 1922 that the unknown element with atomic number 72, later called hafnium, did not belong to the 'rare earth' group of elements was later confirmed to be correct. However, there was no organising principle or criteria behind Bohr's shell model. It was an ingenious improvisation based on an array of chemical and physical data that could in large part explain the chemical properties of the various groupings of elements in the periodic table. Its crowning glory was hafnium.

As he continued to fret over the anomalous Zeeman effect and the shortcomings of the electron shell model, Pauli's time in Copenhagen came to an end. In September 1923 he returned to Hamburg, where the following year he was promoted from assistant to *privatdozent*. But with Copenhagen a short train journey and a ferry across the Baltic Sea, Pauli was still a regular visitor to the institute. He concluded that Bohr's model could work only if there was a restriction on how many electrons could occupy any given shell. Otherwise, in contradiction of the results of atomic spectra, there seemed nothing to prevent all the electrons in any atom from occupying the same stationary state, the same energy level. At the end of 1924 Pauli discovered the fundamental organising rule, the 'exclusion principle', that provided the theoretical justification that had been missing in Bohr's empirically devised electron shell atomic model.

Pauli was inspired by the work of a Cambridge postgraduate student. Edmund Stoner, 35, was still working on his doctorate under Rutherford when in October 1924 his paper 'The Distribution of Electrons Among Atomic Levels' was published in the *Philosophical Magazine*. Stoner argued that the outermost or valence electron of an alkali atom has as many energy states to choose from as there are electrons in the last closed shell of the first inert noble gas that follows it in the periodic table. For example, lithium's valence electron could occupy any one of eight possible energy states, exactly the number of electrons in the corresponding closed shell of the gas neon. Stoner's idea implied that a given principal quantum number n corresponds to a Bohr electron shell which would be completely full or 'closed' when the number of electrons it contains reaches twice its number of possible energy states.

If each electron in an atom is assigned the quantum numbers n, k, m, and each unique set of numbers labels a distinct electron orbit or energy level, then according to Stoner, the number of possible energy states for, say, n=1, 2 and 3 would be 2, 8 and 18. For the first shell n=1, k=1 and m=0. These are the only possible values the three quantum numbers can have and they label the energy state (1,1,0). But according to Stoner, the first shell is closed when it contains 2 electrons, double the number of available energy states. For n=2, either k=1 and m=0 or k=2 and m=−1,0,1. Thus in this second shell there are four possible sets of quantum numbers that can be assigned to the valence electron and the energy states it can occupy: (2,1,0), (2,2,−1), (2,2,0), (2,2,1). Therefore, the shell n=2 can accommodate 8 electrons when it is full. The third shell, n=3, has 9 possible electron energy states: (3,1,0), (3,2,−1), (3,2,0), (3,2,1), (3,3,−2), (3,3,−1), (3,3,0), (3,3,1), (3,3,2).[29] Using Stoner's rule, the n=3 shell can contain a maximum of 18 electrons.

Pauli had seen the October issue of the *Philosophical Magazine*, but ignored Stoner's paper. Not known for his athleticism, Pauli ran to the library to read it after Sommerfeld mentioned Stoner's work in the preface to the fourth edition of his textbook *Atomic Structure and Spectral Lines*.[30] Pauli realised that for a given value of n, the number of available energy states, N, in an atom that an electron could occupy was equivalent to all the possible values that the quantum numbers k and m could take, and was equal to $2n^2$. Stoner's rule yielded the correct series of numbers 2, 8, 18, 32 … for the elements in the rows of the periodic table. But why was the number of electrons in a closed shell twice the value of N or n^2? Pauli came up with the answer – a fourth quantum number had to be assigned to electrons in atoms.

Unlike the other numbers n, k, and m, Pauli's new number could have only two values, so he called it *Zweideutigkeit*. It was this 'two-valuedness' that doubled the number of electron states. Where there had previously been a single energy state with a unique set of three quantum numbers n, k, and m, there were now two energy states: n, k, m, A and n, k, m, B. These extra states explained the enigmatic splitting of spectral lines of the anomalous Zeeman effect. Then the 'two-valued' fourth quantum number led Pauli to the exclusion principle, one of the great commandments of nature: no two electrons in an atom can have the same set of four quantum numbers.

The chemical properties of an element are not determined by the total number of electrons in its atom but only by the distribution of its valence electrons. If all the electrons in an atom occupied the lowest energy level, then all the elements would have the same chemistry.

It was Pauli's exclusion principle that managed the occupancy of the electron shells in Bohr's new atomic model and prevented all of them from gathering in the lowest energy level. The exclusion principle provided the underlying explanation for the arrangement of the elements in the periodic table and the closing of shells with chemically inert rare gases. Yet despite these successes, Pauli admitted in his paper, 'On the Connection between the Closing of Electron Groups in Atoms and the Complex Structure of Spectra', published on 21 March 1925 in *Zeitschrift für Physik*: 'We cannot give a more precise reason for this rule.'[31]

Why four quantum numbers, and not three, were needed to specify the position of electrons in an atom was a mystery. It had been accepted since the seminal work of Bohr and Sommerfeld that an atomic electron in orbital motion around a nucleus moves in three dimensions and therefore requires three quantum numbers for its description. What was the physical basis of Pauli's fourth quantum number?

In the late summer of 1925 two Dutch postgraduate students, Samuel Goudsmit and George Uhlenbeck, realised that the property of 'two-valuedness' that Pauli had proposed was not just another quantum number. Unlike the three existing quantum numbers n, k, and m that specified the angular momentum of the electron in its orbit, the shape of that orbit, and its spatial orientation respectively, 'two-valuedness' was an intrinsic property of an electron that Goudsmit and Uhlenbeck called 'spin'.[32] It was an unfortunate choice of name that conjured up images of spinning objects, but electron 'spin' was a purely quantum concept that solved some of the problems still besetting the theory of atomic structure while neatly providing the physical justification of the exclusion principle.

George Uhlenbeck, 24, had enjoyed his time in Rome as a private tutor to the son of the Dutch ambassador. He had secured the position in September 1922 after having gained the equivalent of a bachelor's degree in physics from Leiden University. No longer wishing to be a financial burden to his parents, it was the perfect opportunity for Uhlenbeck to be self-sufficient as he worked towards his master's degree. With no formal lectures to attend, he learned most of what he needed from books, with only the summer back at the university. Unsure whether to pursue a doctorate when he returned to Leiden in June 1925, Uhlenbeck went to see Paul Ehrenfest, who had succeeded Hendrik Lorentz as professor of physics, in 1912, after Einstein chose Zurich.

Ehrenfest, born in Vienna in 1880, had been a student of the great Boltzmann. Together with his Russian wife, Tatiana, who was a mathematician, Ehrenfest had produced a series of important papers in statistical mechanics as he eked out a living as a physicist in Vienna, Göttingen and St Petersburg. Over the twenty years as Lorentz's successor, Ehrenfest established Leiden as a centre of theoretical physics and in the process became one of the most respected figures in the field. He was renowned for his ability to clarify difficult areas of physics, rather than for any original theories of his own. His friend Einstein later described Ehrenfest as 'the best teacher in our profession' and one 'passionately preoccupied with the development and destiny of men, especially his students'.[33] It was this concern for his students that led Ehrenfest to offer the wavering Uhlenbeck a two-year post as an assistant while he set about getting a doctorate. The offer proved irresistible. Ehrenfest, who ensured whenever possible that his trainee physicists worked together in pairs, introduced him to another graduate student, Samuel Goudsmit.

A year and a half younger than Uhlenbeck, Goudsmit had already published well-received papers on atomic spectra. He had arrived in Leiden in 1919 not long after Uhlenbeck, who called Goudsmit's first paper at only eighteen 'a most presumptuous display of self-confidence' but 'highly creditable'.[34] Given his doubts, a clearly talented younger collaborator might have intimidated others, but not Uhlenbeck. 'Physics,' Goudsmit said towards the end of his life, 'was not a profession but a calling, like creative poetry, music composition or painting.'[35] However, he had chosen physics simply because he had enjoyed science and mathematics at school. It was Ehrenfest who kindled a real passion for physics in the teenager as he set him tasks related to analysing and finding order in the fine structure of atomic spectra. While he was not the most studious, Goudsmit possessed an uncanny skill at making sense out of empirical data.

By the time Uhlenbeck returned to Leiden from his time in Rome, Goudsmit was spending three days a week in Amsterdam working in Pieter Zeeman's spectroscopy laboratory. 'The trouble with you is I don't know what to ask, all you know is spectral lines', Ehrenfest complained as he fretted about setting Goudsmit a much-delayed exam.[36] Despite concerns that his flair for spectroscopy was having a detrimental impact on his all-round development as a physicist, Ehrenfest asked Goudsmit to teach Uhlenbeck the theory of atomic spectra. After Uhlenbeck was brought up to date on the latest developments, Ehrenfest wanted the pair to work on the alkali doublet lines – the splitting of spectral lines due to an external magnetic field. 'He knew nothing; he asked all those questions which I never asked', said Goudsmit.[37] Whatever his shortcomings, Uhlenbeck had a thorough knowledge of classical physics that led him to pose intelligent questions that challenged Goudsmit's understanding. It was an inspired piece of pairing by Ehrenfest that ensured that each would learn from the other.

Throughout the summer of 1925 Goudsmit taught Uhlenbeck everything he knew about spectral lines. Then one day they discussed the exclusion principle, which Goudsmit thought was no more than another ad hoc rule that brought a little more order to the unholy mess of atomic spectra.

However, Uhlenbeck immediately hit upon an idea that Pauli had already dismissed.

An electron could move up and down, back and forth, and side to side. Each of these different ways of moving physicists called a 'degree of freedom'. Since each quantum number corresponds to a degree of freedom of the electron, Uhlenbeck believed that Pauli's new quantum number must mean that the electron had an additional degree of freedom. To Uhlenbeck, a fourth quantum number implied that the electron must be rotating. However, spin in classical physics is a rotational motion in three dimensions. So if electrons spin in the same way, like the earth about its axis, there was no need for a fourth number. Pauli argued that his new quantum number referred to something 'which cannot be described from the classical point of view'.[38]

In classical physics, angular momentum, everyday spin, can point in any direction. What Uhlenbeck was proposing was quantum spin – 'two-valued' spin, spin 'up' or spin 'down'. He pictured these two possible spin states as an electron spinning either clockwise or anti-clockwise about a vertical axis as it orbits the atomic nucleus. As it did so, the electron would generate its own magnetic field and act like a subatomic bar magnet. The electron can line up either in the same or in the opposite direction as an external magnetic field. Initially it was believed that any allowed electron orbit could accommodate a pair of electrons provided that one had spin 'up' and the other spin 'down'. However, these two spin directions have very similar but not identical energies, resulting in the two slightly different energy levels that gave rise to the alkali doublet lines – two closely spaced lines in the spectra instead of one.

Uhlenbeck and Goudsmit showed that electron spin could be either plus or minus half, values that satisfied Pauli's restriction for the fourth quantum number to be 'two-valued'.[39]

By the middle of October, Uhlenbeck and Goudsmit had written a one-page paper and showed it to Ehrenfest. He suggested that the normal alphabetical order of names be reversed. Since Goudsmit had already published several well-received papers on atomic spectra, Ehrenfest was concerned that readers would think that Uhlenbeck was the junior partner. Goudsmit agreed, as 'it was Uhlenbeck who had thought of spin'.[40] But as to the soundness of the concept itself, Ehrenfest was unsure. He wrote to Lorentz asking for 'his judgement and advice on a very witty idea'.[41]

Although 72, retired and living in Haarlem, Lorentz still travelled to Leiden once a week to teach. Uhlenbeck and Goudsmit met him one Monday morning after his lecture. 'Lorentz was not discouraging', said Uhlenbeck. [42] 'He was a little bit reticent, said that it was interesting and that he would think about it.' A week or two later, Uhlenbeck went back to receive Lorentz's verdict and was given a stack of papers full of calculations in support of an objection to the very notion of spin. A point on the surface of a spinning electron, Lorentz pointed out, would move faster than the speed of light – something forbidden by Einstein's special theory of relativity. Then another problem was discovered. The separation of the alkali doublet spectral lines, predicted using electron spin, was twice the measured value. Uhlenbeck asked Ehrenfest not to submit the paper. It was too late. He had already sent it to a journal. 'You are both young enough to be able to afford a stupidity', Ehrenfest reassured him.[43]

When the paper was published on 20 November, Bohr was deeply sceptical. The following month he travelled to Leiden to participate in the celebrations to mark the 50th anniversary of Lorentz receiving his doctorate. As his train pulled into Hamburg, Pauli was waiting on the platform to ask Bohr what he thought about electron spin. The concept was 'very interesting', said Bohr. His well-worn put-down meant he believed that electron spin was flawed. How, he asked, could an electron moving in the electric field of the positively-charged nucleus experience the magnetic field necessary for producing the fine structure? When he arrived at Leiden, two men impatient to know his views on spin met Bohr at the station: Einstein and Ehrenfest.

Bohr outlined his objection about the magnetic field and was amazed when Ehrenfest said that Einstein had already resolved the problem by invoking relativity. Einstein's explanation, Bohr admitted later, was a 'complete revelation'. He now felt confident that any remaining problems surrounding electron spin would all sooner rather than later be overcome. Lorentz's objection was based on classical physics, of which he was a master. However, electron spin was a quantum concept. So this particular problem was not as serious as it first appeared. The British physicist Llewellyn Thomas solved the second. He showed that an error in the calculation of the relative motion of the electron in its orbit around the nucleus was responsible for the extra factor of two in the separation of the doublet lines. 'I have never since faltered in my conviction that we are at the end of our sorrows', Bohr wrote in March 1926.[44]

On the return leg of his trip, Bohr met more physicists eager to hear what he had to say about quantum spin. When his train stopped at Göttingen, Werner Heisenberg, who just a few months earlier had finished his stint as Bohr's assistant, and Pascual Jordan were waiting at the station. Electron spin, he told them, was a great advance. He then travelled to Berlin to attend the 25th anniversary celebrations of Planck's famous lecture to the German Physical Society in December 1900 that was the official birthday of the quantum. Pauli lay in wait at the station, having travelled from Hamburg to quiz the Dane once again. As he feared, Bohr had changed his mind and was now the prophet of electron spin. Unmoved by initial attempts to convert him, Pauli called quantum spin 'a new Copenhagen heresy'.[45]

A year earlier he had dismissed the idea of electron spin when a 21-year-old German-American, Ralph Kronig, had first proposed it. On a two-year odyssey around some of Europe's leading centres of physics after gaining his PhD at Columbia University, Kronig arrived in Tübingen on 9 January 1925, prior to spending the next ten months at Bohr's institute. Interested in the anomalous Zeeman effect, Kronig was excited when his host, Alfred Landé, told him that Pauli was expected the following day. He was coming to talk to Landé about the exclusion principle before submitting his paper for publication. Having studied under Sommerfeld and later served as Born's assistant in Frankfurt, Landé was highly regarded by Pauli. Landé showed Kronig a letter Pauli had written to him the previous November.

In the course of his life, Pauli wrote thousands of letters. As his reputation grew and the number of correspondents increased, his letters were highly prized and passed around and studied. For Bohr, who saw past the sarcastic wit, a letter from Pauli was an event. He would slip it into his jacket pocket and carry it around for days, showing it to anyone remotely interested in whatever problem or idea Pauli was dissecting. Under the cover of drafting a reply, Bohr would conduct an imaginary dialogue as though Pauli were seated in front of him smoking his pipe. 'Probably all of us are afraid of Pauli; but then again we are not so afraid of him that we dare not admit it', he once playfully declared.[46]

Kronig later recalled that as he read Pauli's letter to Landé his 'curiosity was aroused'.[47] Pauli had outlined the need to label every electron inside an atom with a unique set of four quantum numbers and its consequences. Immediately Kronig began thinking about the possible physical interpretation of the fourth quantum number, and came up with the idea of an electron rotating about its axis. He was quick to appreciate the difficulties attached to such a spinning electron. However, finding it 'a fascinating idea', Kronig spent the rest of the day developing the theory and doing the mathematics.[48] He had worked out much of what Uhlenbeck and Goudsmit would announce in November. When he explained his findings to Landé, both men were impatient for Pauli to arrive and give his seal of approval. Kronig was taken aback when Pauli ridiculed the notion of electron spin: 'That is surely quite a clever idea, but nature is not like that.'[49] So fervent had Pauli been in rejecting the proposal, Landé tried to soften the blow: 'Yes, if Pauli says so, then it is not like that.'[50] Dejected, Kronig abandoned the idea.

Unable to contain his anger when electron spin was quickly embraced, in March 1926 Kronig wrote to Bohr's assistant Hendrik Kramers. He reminded Kramers that he had been the first to suggest electron spin and had not published because of Pauli's derisive reaction. 'In future I shall trust my own judgement more and that of others less', he lamented, having learnt the lesson too late.[51] Disturbed by Kronig's letter, Kramers showed it to Bohr. No doubt remembering his own dismissal of electron spin when Kronig had discussed it with him and others during his stay in Copenhagen, Bohr wrote to express his 'consternation and deep regret'.[52] 'I should not have mentioned the matter at all if it were not to take a fling at the physicists of the preaching variety, who are always so damned sure of, and inflated with, the correctness of their own opinion', replied Kronig.[53]

Despite feeling robbed, Kronig was sensitive enough to ask Bohr not to mention the whole sorry affair in public, since 'Goudsmit and Uhlenbeck would hardly be very happy about it'.[54] He knew they were entirely blameless. However, both Goudsmit and Uhlenbeck became aware of what had happened. Uhlenbeck later openly acknowledged that he and Goudsmit 'were clearly not the first to propose a quantized rotation of the electron, and there is no doubt that Ralph Kronig anticipated what certainly was the main part of our ideas in the spring of 1925, and that he was discouraged mainly by Pauli from publishing his results'.[55] It was proof, a physicist told Goudsmit, 'that the infallibility of the Deity does not extend to his self-styled vicar on earth'.[56]

In private, Bohr believed that Kronig 'was a fool'.[57] If he was convinced of the correctness of his idea, then he should have published no matter what others thought. 'Publish or perish' is a rule not to be forgotten in science. In his heart, Kronig must have reached a similar conclusion. His initial outburst of bitterness towards Pauli amid the disappointment of missing out on electron spin had dissipated by the end of 1927. At only 28, Pauli was appointed professor of theoretical physics at the ETH in Zurich. He asked Kronig, who was once again spending time in Copenhagen, to become his assistant. 'Every time I say something, contradict me with detailed arguments', Pauli wrote to Kronig after he accepted the offer.[58]

By March 1926 the problems that had led Pauli to reject electron spin had all been resolved. 'Now there is nothing else I can do than to *capitulate completely*', he wrote to Bohr.[59] Years later, most physicists assumed that Goudsmit and Uhlenbeck had received the Nobel Prize – after all, electron spin was one of the seminal ideas of twentieth-century physics, an entirely new quantum concept. But the Pauli-Kronig affair meant that the Nobel committee shied away from giving them the prestigious award.

Pauli always felt guilty for discouraging Kronig. Just as he did for receiving the Nobel Prize in 1945 for the discovery of the exclusion principle while the Dutchmen were denied. 'I was so stupid when I was young!' he said later.[60]

On 7 July 1927, Uhlenbeck and Goudsmit received their doctorates within an hour of each other. Flouting convention, the ever-thoughtful Ehrenfest had arranged it that way. He had also secured both of them jobs at the University of Michigan. With few positions then available, Goudsmit said towards the end of his life, the post in America 'was for me a far more significant award than a Nobel Prize'.[61]

Goudsmit and Uhlenbeck provided the first concrete evidence that existing quantum theory had reached the limits of its applicability. Theorists could no longer use classical physics to gain a foothold before 'quantising' a piece of existing physics, because there was no classical counterpart to the quantum concept of electron spin. The discoveries of Pauli and the Dutch spin doctors brought to a close the achievements of the 'old quantum theory'. There was a sense of crisis. The state of physics 'was from a methodological point of view, a lamentable hodgepodge of hypothesis, principles, theorems, and computational recipes rather than a logical, consistent theory.'[62] Progress was often based on artful guessing and intuition rather than scientific reasoning.

'Physics at the moment is again very muddled; in any case, for me it is too complicated, and I wish I were a film comedian or something of that sort and had never heard anything about physics', wrote Pauli in May 1925, some six months after discovering the exclusion principle.[63] 'Now I do hope nevertheless that Bohr will save us with a new idea. I beg him to do so urgently, and convey to him my greetings and many thanks for all his kindness and patience towards me.' However, Bohr had no answers to 'our present theoretical troubles'.[64] That spring, it seemed that only a quantum magician could conjure up the yearned-for 'new' quantum theory – quantum mechanics.

Chapter 8

THE QUANTUM MAGICIAN

‘On a Quantum-Theoretical Reinterpretation of Kinematics and Mechanical Relations’ was the paper that everyone had been waiting for and some had hoped to write. The editor of the *Zeitschrift für Physik* received it on 29 July 1925. In the preamble that scientists call an ‘abstract’, the author boldly stated his ambitious plan: ‘to establish a basis for theoretical quantum mechanics, founded exclusively on relationships between quantities which, in principle, are observable.’ Some fifteen pages later, his goal achieved, Werner Heisenberg had laid the foundations for the physics of the future. Who was this young German wunderkind and how he had succeeded where all others had failed?

Werner Karl Heisenberg was born on 5 December 1901 in Würzburg, Germany. He was eight when his father was appointed to the country's only professorship of Byzantine philology at Munich University and the family moved to the Bavarian capital. For Heisenberg and his brother Erwin, almost two years older, home became a spacious apartment in the fashionable suburb of Schwabing on the northern outskirts of the city. They attended the prestigious Maximilians Gymnasium, where Max Planck had been a student 40 years earlier. It was also the school where their grandfather was now in charge. If the staff were tempted to treat the headmaster's grandsons more leniently than other pupils, then they quickly discovered there was no need. 'He has an eye for what is essential, and never gets lost in details', Werner's first-year teacher reported.[1] 'His thought processes in grammar and mathematics operate rapidly and usually without mistakes.'

August Heisenberg's father, forever the teacher, devised all manner of intellectual games for Werner and Erwin. In particular he always encouraged mathematical games and problem-solving. Pitting one brother against the other as they raced to solve them, it was evident that Werner was the more mathematically talented. Around the age of twelve he started learning calculus and asked his father to get him maths books from the university library. Seeing this as an opportunity to improve his son's grasp of languages, he started supplying him with books written in Greek and Latin. It was the beginning of Werner's fascination with the work of the Greek philosophers. Then came the First World War and the end of Heisenberg's comfortable and secure world.

The end of the war brought in its wake political and economic chaos throughout Germany, but few places experienced this more intensely than Munich and Bavaria. On 7 April 1919, radical socialists declared Bavaria a 'Soviet Republic'. As they waited for troops sent by Berlin to arrive and restore the deposed government, those opposed to the revolutionaries organised themselves into military-style companies. Heisenberg and some friends joined one of these. His duties were largely confined to writing reports and running errands. 'Our adventures were over after a few weeks,' Heisenberg recalled later, 'then the shooting died down and military service became increasingly monotonous.'[2] By the end of the first week in May the 'Soviet Republic' had been ruthlessly crushed, leaving over a thousand dead.

The harsh post-war reality led young middle-class teenagers like Heisenberg to embrace the romantic ideals of an earlier age as they flocked to join youth organisations such as the Pathfinders, the German equivalent of the Boy Scouts. Others, wanting more independence, set up their own groups and clubs. Heisenberg led one such group formed by younger pupils at his school. Gruppe Heisenberg, as they styled themselves, went hiking and camping in the Bavarian countryside and discussed the new world their generation would create.

In the summer of 1920, after graduating from the Gymnasium with such ease that he won a prestigious scholarship, Heisenberg wanted to study mathematics at Munich University. When a disastrous interview ended any chance of doing so, a despondent Heisenberg sought his father for advice. He made an appointment for his son to see an old friend, Arnold Sommerfeld. Although the 'small squat man with his martial dark moustache looked rather austere', Heisenberg did not feel intimidated.[3] He sensed that despite his appearance, here was a man with a 'genuine concern for young people'.[4] August Heisenberg had already told Sommerfeld that his son was particularly interested in relativity and atomic physics. 'You are much too demanding', he told Werner.[5] 'You can't possibly start with the most difficult part and hope that the rest will automatically fall into your lap.' Always eager to encourage and recruit raw talent to mould, he softened: 'It may be that you know something; it may be that you know nothing. We shall see.'[6]

Sommerfeld allowed the eighteen-year-old to attend the research seminar intended for more advanced students. Heisenberg was lucky. Together with Bohr's institute in Copenhagen and Born's group in Göttingen, Sommerfeld's institute would form the golden triangle of quantum research in the years to come. When Heisenberg attended his first seminar he 'spotted a dark-haired student with a somewhat secretive face in the third row'.[7] It was Wolfgang Pauli. Sommerfeld had already introduced him to the portly Viennese during a tour around the institute on his first visit. The professor had been quick to tell Heisenberg, once Pauli was out of earshot, that he considered the boy to be his most talented student. Recalling Sommerfeld's advice that he could learn a great deal from him, Heisenberg sat down next to Pauli.

'Doesn't he look the typical Hussar officer?' whispered Pauli as Sommerfeld entered.[8] It was the beginning of a lifelong professional relationship that never quite blossomed into a closer personal friendship.

They were simply too different. Heisenberg was quieter, friendlier, less outspoken and critical than Pauli. He romanticised nature and loved nothing more than hiking and camping with his friends. Pauli was drawn to cabarets, taverns and cafes. Heisenberg had done half a day's work while Pauli still slept soundly in his bed. Yet Pauli exerted a strong influence on Heisenberg and never passed up a chance to tell him, with tongue in cheek: 'You are a complete fool.'[9]

In the middle of writing his dazzling review of relativity, it was Pauli who steered Heisenberg away from Einstein's theory and towards the quantum atom as a more fertile area of research in which to make his name. 'In atomic physics we still have a wealth of uninterpreted experimental results,' he told Heisenberg; 'nature's evidence in one place seems to contradict that in another, and so far it has not been possible to draw an even halfway coherent picture of the relationship involved.'[10] It was likely, thought Pauli, that everyone would still be 'groping about in a thick mist' for years to come.[11] As Heisenberg listened, he was inexorably drawn into the realm of the quantum.

Sommerfeld soon assigned Heisenberg a 'little problem' in atomic physics. He asked him to analyse some new data on the splitting of spectral lines in a magnetic field and to construct a formula that replicated the splitting. Pauli warned Heisenberg that Sommerfeld hoped that deciphering such data would lead to new laws. It was an attitude that for Pauli bordered on 'a kind of number mysticism', but then he admitted, 'no one has been able to suggest anything better'.[12] The exclusion principle and electron spin still lay in the future.

Heisenberg's ignorance of the accepted rules and regulations of quantum physics allowed him to tread where others, wedded to a more cautious and rational approach, feared to. It enabled him to construct a theory that appeared to explain the anomalous Zeeman effect. Having dismissed an earlier version, Heisenberg was relieved when Sommerfeld sanctioned the publication of his latest effort. Although it was later shown to be incorrect, his first scientific paper brought Heisenberg to the attention of Europe's leading physicists. Bohr was one of those who sat up and took notice.

They first met in Göttingen in June 1922 when Sommerfeld took some of his students to hear Bohr's series of lectures on atomic physics. What struck Heisenberg was how precise Bohr was in his choice of words: 'Each one of his carefully formulated sentences revealed a long chain of underlying thoughts, of philosophical reflections, hinted at but never fully expressed.'[13] He was not alone in sensing that Bohr reached his conclusions more by intuition and inspiration than by detailed calculations. At the end of the third lecture, Heisenberg rose to point out some difficulties that remained in a published paper that Bohr had praised. As people began to mingle after the question-and-answer session, Bohr sought out Heisenberg and asked the twenty-year-old if he would like to accompany him on a walk later that day. Their hike to a nearby mountain lasted some three hours, and Heisenberg later wrote 'that my real scientific career only started that afternoon'.[14] For the first time, he saw 'that one of the founders of quantum theory was deeply worried by its difficulties'.[15] When Bohr invited him to Copenhagen for a term, Heisenberg suddenly saw his future as one 'full of hope and new possibilities'.[16]

Copenhagen would have to wait. Sommerfeld was due to go to America and in his absence had arranged for Heisenberg to study with Max Born in Göttingen. Although he looked 'like a simple farm boy, with short fair hair, clear bright eyes, and a charming expression', Born quickly discovered that there was much more to him than met the eye.[17] He was 'easily as gifted as Pauli', Born wrote to Einstein.[18] When he returned to Munich, Heisenberg finished his doctoral thesis on turbulence. Sommerfeld had chosen the topic to broaden his knowledge and understanding of physics. During the oral examination his inability to answer simple questions, such as the resolving power of a telescope, almost cost him his doctorate. Wilhelm Wien, the head of experimental physics, was dismayed when Heisenberg struggled to explain how a battery worked. He wanted to fail the upstart theorist, but reached a compromise with Sommerfeld. Heisenberg would get his doctorate, but would be awarded the second-lowest mark – grade III. Pauli had passed with grade I.

Feeling humiliated, that evening he packed his bags and caught the overnight train. He could not bear to stay in Munich a minute longer and fled to Göttingen. 'I was astonished when, one morning long before the appointed time, he suddenly appeared before me with an expression of embarrassment on his face', recalled Born later.[19] Heisenberg anxiously recounted the tale of his oral exam, worried that his services would no longer be required as an assistant. Eager to cement Göttingen's growing reputation for theoretical physics, Born was confident that Heisenberg would bounce back and told him so.

Born was convinced that physics had to be rebuilt from the ground up. The mish-mash of quantum rules and classical physics that was at the heart of the Bohr-Sommerfeld quantum atom had to give way to a logically consistent new theory that Born called 'quantum mechanics'. None of this was new for physicists trying to disentangle the problems of atomic theory. However, it signalled the awareness of a creeping sense of crisis in 1923 at the inability of physicists to cross the atomic Rubicon. Pauli was already loudly proclaiming to anyone who would listen that the failure to explain the anomalous Zeeman effect was evidence 'that we must create something fundamentally new'.[20] After meeting him, Heisenberg believed that Bohr was the one most likely to make the breakthrough.

Pauli had been in Copenhagen as Bohr's assistant since the autumn of 1922. He and Heisenberg kept each other informed about the latest developments at their respective institutes through a regular exchange of letters. Heisenberg, like Pauli, had also been working on the anomalous Zeeman effect. Just before Christmas 1923, he wrote to Bohr about his latest efforts and received an invitation to spend a few weeks in Copenhagen. On Saturday, 15 March 1924, Heisenberg stood in front of the three-storey neo-classical building with its red tiled roof at Blegdamsvej 17. Above the main entrance he saw the sign that greeted every visitor: 'Universitetets Institut for Teoretisk Fysik'. Better known as the Bohr Institute.

Heisenberg soon discovered that only half of the building, the basement and the ground floor, was used for physics. The rest was set aside for accommodation. Bohr and his growing family lived in an elegantly furnished flat that occupied the entire first floor. The family maid, the caretaker, and honoured guests were housed on the top floor. On the ground floor, besides the lecture hall with its six long rows of wooden benches, was a well-stocked library and offices for Bohr and his assistant. There was also a modest-sized workroom for visitors. Despite its name, the institute had two small laboratories on the first floor, with the main laboratory housed in the basement.

The institute was struggling for space with a permanent staff of six and almost a dozen visitors. Bohr was already making plans to expand. Over the next two years the adjacent land was bought and two new buildings were added that doubled the capacity of the institute. Bohr and his family moved out of their flat into a large purpose-built house next door. The extension meant a substantial renovation of the old building that included more office space, a dining room, and a new self-contained three-room flat on the top floor. It was here that Pauli and Heisenberg often stayed in later years.

There was one thing that no one at the institute wanted to miss: the arrival of the morning post. Letters from parents and friends were always welcome, but it was correspondence from far-flung colleagues and the journals that were seized upon for the latest breaking news from the frontiers of physics. However, not everything revolved around physics, even if much of the talking did. There were musical evenings, games of table tennis, hiking trips, and outings to watch the latest motion picture.

Heisenberg had arrived with such high hopes, but his first few days at the institute left him feeling frustrated. Expecting to spend time with Bohr almost as he stepped through the front door, he had hardly seen him. Used to being the best, Heisenberg was suddenly faced with Bohr's international posse of brilliant young physicists. He was intimidated. They all spoke several languages, while he sometimes struggled to express himself clearly in German. Enjoying nothing more than walks in the countryside with his friends, Heisenberg thought that everyone at the institute possessed a worldliness that he did not. However, nothing left him as despondent as the realisation that they understood much more of atomic physics than he did.

As he tried to shake off the blows to his self-esteem, Heisenberg wondered if he would ever get the chance to work with Bohr. He had been sitting in his room when there was a knock on the door and in strode Bohr. After apologising for being so busy, he proposed that the two of them go on a short walking tour. There was little chance, Bohr explained, of him being left alone long enough at the institute for the pair of them to talk at any length. What better way of getting to know one another than a few days of walking and talking? It was Bohr's favourite pastime.

Early the following morning they caught the tram to the northern outskirts of the city and began their walk. Bohr asked Heisenberg about his childhood and what he remembered about the outbreak of war ten years earlier. As they headed north, instead of physics they talked about the pros and cons of war, Heisenberg's involvement in the youth movement, and Germany. After spending the night at an inn, they walked to Bohr's country cottage in Tisvilde, before heading back to the institute on the third day. The 100-mile walk had the effect that Bohr desired and Heisenberg craved. They got to know each other more quickly.

They had talked about atomic physics, yet when they finally returned to Copenhagen, it was Bohr the man, rather than the physicist, that had captivated Heisenberg. 'I am, of course, absolutely enchanted with the days I am spending here', he wrote to Pauli.[21] He had never before met a man like Bohr with whom he could discuss just about anything. Despite his genuine concern for the welfare of everyone at his institute, Sommerfeld upheld the traditional German role of professor, one step removed from his subordinates. In Göttingen, Heisenberg would not have dared to broach with Born the range of subjects he and Bohr had discussed so freely. Unknown to him, it was Pauli, in whose footsteps he always seemed to be following, who was behind Bohr's warm reception.

Pauli always took a keen interest in what Heisenberg was doing, as the pair kept each other informed about their latest ideas. Pauli had returned to Hamburg University when he learnt that Heisenberg was going to spend a few weeks in Copenhagen, and he wrote to Bohr. For a man already notorious for his scathing wit, the fact that he described Heisenberg as a 'gifted genius' who would 'one day advance science greatly' made a deep impression on Bohr.[22] But before that day arrived, Pauli was sure that Heisenberg's physics had to be underpinned by a more coherent philosophical approach.

Pauli believed that to overcome the problems besetting atomic physics it was necessary to stop making arbitrary ad hoc assumptions whenever experiments yielded data in conflict with existing theory. Such an approach could only paper over the problems without ever leading to their solution. Given his deep understanding of relativity, Pauli was an ardent admirer of Einstein and the way in which he had constructed the theory using a few guiding principles and assumptions. Believing that it was the correct approach to adopt in atomic physics too, Pauli wanted to emulate Einstein by setting up the underlying philosophical and physical principles before moving on to develop the necessary formal mathematical nuts and bolts that held the theory together. By 1923 it was an approach that had left Pauli in despair. Having avoided introducing assumptions that could not be justified, he nevertheless failed to find a consistent and logical account of the anomalous Zeeman effect.

'Hopefully you will then take atomic theory forward in good measure and solve several of the problems with which I have tormented myself in vain and which are too difficult for me', Pauli wrote to Bohr.[23] 'I hope also that Heisenberg will then bring back home a philosophical attitude in his thinking.' By the time the young German arrived, Bohr had been well briefed. Throughout the two-week visit, the principles of physics rather than any particular problem was the focus of their discussions as Bohr and Heisenberg strolled through Faelledpark next to the institute or chatted over a bottle of wine in the evenings. Many years later, Heisenberg described his time in Copenhagen in March 1924 as a 'gift from heaven'.[24]

'I shall, of course, miss him (he is a charming, worthy, very bright man, who has become very dear to my heart), but his interest precedes mine, and your wish is decisive for me', Born wrote to Bohr after Heisenberg received an invitation for an extended stay in Copenhagen.[25] Due to spend the forthcoming winter semester teaching in America, Born would not need the services of his assistant until May the following year. At the end of July 1924, having successfully completed his *habilitation* thesis and gained the right to teach at German universities, Heisenberg left for a three-week hiking tour around Bavaria.

When he returned to Bohr's institute on 17 September 1924, Heisenberg was still only 22 years old, but had already written or co-written an impressive dozen papers on quantum physics. He still had much to learn and knew that Bohr was the man to teach him. 'From Sommerfeld I learned optimism, in Göttingen mathematics, from Bohr physics', he said later.[26] For the next seven months, Heisenberg was exposed to Bohr's approach to overcoming the problems that plagued quantum theory. While Sommerfeld and Born were also troubled by the same inconsistencies and difficulties, neither man was haunted like Bohr by them. He could hardly bring himself to talk of anything else.

From these intense discussions, Heisenberg 'realized how difficult it was to reconcile the results of one experiment with those of another'.[27] Among these experiments was Compton's scattering of X-rays by electrons that supported Einstein's light-quanta. The difficulties just seemed to multiply with de Broglie's extension of wave-particle duality to encompass all matter. Bohr, having taught Heisenberg all that he could, had great hopes for his young protégé: 'Now everything is in Heisenberg's hands – to find a way out of the difficulties.'[28]

By the end of April 1925, Heisenberg was back in Göttingen, thanking Bohr for his hospitality and 'sad about the fact that I must carry on wretchedly alone by myself in the future'.[29] Nevertheless, he had learned a valuable lesson from discussions with Bohr and in his ongoing dialogue with Pauli: something fundamental had to give. Heisenberg believed he knew what that might be as he tried to solve a long-standing problem: the intensities of the spectral lines of hydrogen. The Bohr-Sommerfeld quantum atom could account for the frequency of hydrogen's spectral lines, but not how bright or dim they were. Heisenberg's idea was to separate what was observable and what was not. The orbit of an electron around the nucleus of a hydrogen atom was not observable. So Heisenberg decided to abandon the idea of electrons orbiting the nucleus of an atom. It was a bold step, but one he was now ready to take, having long detested attempts at pictorial representations of the unobservable.

As a teenager in Munich, Heisenberg 'was enthralled by the idea that the smallest particles of matter might reduce to some mathematical form'.[30] At about the same time he came across an illustration in one of his textbooks that he found appalling. To explain how one atom of carbon and two atoms of oxygen formed a carbon dioxide molecule, the atoms were drawn with hooks and eyes by which they could hang together. Heisenberg found the idea of orbiting electrons inside the quantum atom similarly far-fetched. He now abandoned any attempt to visualise what was going on inside an atom. Anything that was unobservable he decided to ignore, focusing his attention only on those quantities that could be measured in the laboratory: the frequencies and intensities of the spectral lines associated with the light emitted or absorbed as an electron jumped from one energy level to another.

Even before Heisenberg adopted this new strategy, Pauli had already expressed his doubts about the usefulness of electron orbits more than a year earlier. 'The most important question seems to me to be this: *to what extent may definite orbits of electrons in stationary states be spoken of at all*', he had written in italics to Bohr in February 1924.[31] Even though he was well on the road that led to the exclusion principle, and concerned about the closure of electron shells, Pauli nevertheless answered his own question in another letter to Bohr in December: 'We must not bind atoms in the chains of our prejudices – to which, in my opinion, also belongs the assumption that electron orbits exist in the sense of ordinary mechanics – but we must, on the contrary, adapt our concepts to experience.'[32] They had to stop making compromises and cease trying to accommodate quantum concepts within the comfortable and familiar framework of classical physics. Physicists had to break free. The first to do so was Heisenberg when he pragmatically adopted the positivist credo that science should be based on observable facts, and attempted to construct a theory based solely on the observable quantities.

In June 1925, a little more than a month after returning from Copenhagen, Heisenberg was miserable in Göttingen. He was struggling to make headway in calculating the intensities of the spectral lines of hydrogen and admitted as much in a letter to his parents. He complained that 'everyone here is doing something different and no one anything worthwhile'.[33] A very severe attack of hay fever contributed to his low spirits. 'I couldn't see from my eyes, I just was in a terrible state', Heisenberg said later.[34] Unable to cope, he had to get away and a sympathetic Born granted him a two-week holiday. On Sunday, 7 June, Heisenberg caught the night train to the port of Cuxhaven on the coast. Arriving early in the morning, tired and hungry, Heisenberg went in search of breakfast at an inn and then boarded a ferry to the island of Helgoland, an isolated barren rock in the North Sea. Originally owned by the British until it was traded for Zanzibar in 1890, Helgoland was 30 miles from the German mainland and less than a square mile in size. It was here that Heisenberg hoped to find relief amid the bracing pollen-free sea air.

'On my arrival, I must have looked quite a sight with my swollen face; in any case, my landlady took one look at me, concluded that I had been in a fight and promised to nurse me through the after effects', Heisenberg recalled when he was 70.[35] The guesthouse was high on the southern edge of the distinctive island carved out of red sandstone rock. From the balcony of his second-floor room Heisenberg had a wonderful view of the village below, the beach, and the dark brooding sea beyond. In the days that followed he had time to think about 'Bohr's remark that part of infinity seems to lie within the grasp of those who look across the sea'.[36] It was in such reflective mood that he relaxed by reading Goethe, taking daily walks around the small resort, and swimming. Soon he was feeling much better. With little to distract him, Heisenberg's thoughts turned once more to problems of atomic physics. But here on Helgoland he felt none of the anxiety that had recently plagued him. Relaxed and carefree, he quickly jettisoned the mathematical ballast he had brought from Göttingen as he tried to solve the riddle of the intensities of the spectral lines.[37]

In his quest for a new mechanics for the quantised world of the atom, Heisenberg concentrated on the frequencies and relative intensities of the spectral lines produced when an electron instantaneously jumped from one energy level to another. He had no other choice; it was the only available data about what was happening inside an atom. Despite the imagery conjured up by all the talk of quantum jumps and leaps, an electron did not 'jump' through space as it moved between energy levels like a boy jumping off a wall onto the pavement below. It was simply in one place and an instant later it popped up in another without being anywhere in between. Heisenberg accepted that all observables, or anything connected with them, were associated with the mystery and magic of the quantum jump of an electron between two energy levels. Lost forever was the picturesque miniature solar system in which each electron orbited a nuclear sun.

On the pollen-free haven of Helgoland, Heisenberg devised a method of book-keeping to track all possible electron jumps, or transitions, that could occur between the different energy levels of hydrogen. The only way he could think of recording each observable quantity, associated with a unique pair of energy levels, was to use an array:

$$
\begin{array}{cccccc}
V_{11} & V_{12} & V_{13} & V_{14} & \cdots & V_{1n} \\
V_{21} & V_{22} & V_{23} & V_{24} & \cdots & V_{2n} \\
V_{31} & V_{32} & V_{33} & V_{34} & \cdots & V_{3n} \\
V_{41} & V_{42} & V_{43} & V_{44} & \cdots & V_{4n} \\
\vdots & \vdots & \vdots & \vdots & & \vdots \\
V_{m1} & V_{m2} & V_{m3} & V_{m4} & \cdots & V_{mn}
\end{array}
$$

This was the array for the entire set of possible frequencies of the spectral lines that could theoretically be emitted by an electron when it jumps between two different energy levels. If an electron quantum jumps from the energy level E_2 to the lower energy level E_1, a spectral line is emitted with a frequency designated by v_{21} in the array. The spectral line of frequency v_{12} would only be found in the absorption spectrum, since it is associated with an electron in energy level E_1 absorbing a quantum of energy sufficient to jump to energy level E_2. A spectral line of frequency v_{mn} would be emitted when an electron jumps between any two levels whose energies are E_m and E_n, where m is greater than n. Not all the frequencies v_{mn} are exactly observed. For example, measurement of v_{11} is impossible, since it would be the frequency of the spectral line emitted in a 'transition' from energy level E_1 to energy level E_1 – a physical impossibility. Hence v_{11} is zero, as are all potential frequencies when m=n. The collection of all non-zero frequencies, v_{mn}, would be the lines actually present in the emission spectrum of a particular element.

Another array could be formed from the calculation of transition rates between the various energy levels. If the probability for a particular transition, a_{mn}, from energy level E_m to E_n, is high, then the transition is more likely than one with a lower probability. The resulting spectral line with frequency v_{mn} would be more intense than for the less probable transition. Heisenberg realised that the transition probabilities a_{mn} and the frequencies v_{mn} could, after some deft theoretical manipulation, lead to a quantum counterpart for each observable quantity known in Newtonian mechanics such as position and momentum.

Of all things, Heisenberg began by thinking about electrons' orbits. He imagined an atom in which an electron was orbiting the nucleus at a great distance – more like Pluto orbiting the sun rather than Mercury. It was to prevent an electron spiralling into the nucleus at it radiated away energy that Bohr had introduced the concept of stationary orbits. However, in accordance with classical physics, the orbital frequency of an electron in such an exaggerated orbit, the number of complete orbits it makes per second, is equal to the frequency of the radiation it emits.

This was no flight of fancy, but a skilful use of the correspondence principle – Bohr's conceptual bridge between the quantum and classical realms. Heisenberg's hypothetical electron orbit was so large that it was on the border that divided the kingdoms of the quantum and the classical. Here in this borderland, the electron's orbital frequency was equal to the frequency of the radiation it emitted. Heisenberg knew that such an electron in an atom was akin to a hypothetical oscillator that could produce all the frequencies of the spectrum. Max Planck had adopted a similar approach a quarter of a century earlier. However, while Planck had used brute force and ad hoc assumptions to generate a formula that he already knew to be correct, Heisenberg was being guided by the correspondence principle onto the familiar landscape of classical physics. Once it was set into motion, he could calculate properties of the oscillator such as its momentum p, the displacement from its equilibrium position q, and its frequency of oscillation. The spectral line with a frequency v_{mn} would be emitted by one of a range of individual oscillators. Heisenberg knew that once he worked out the physics in this territory where the quantum and the classical met, he could extrapolate to explore the unknown interior of the atom.

Late one evening on Helgoland, all the pieces began falling into place. The theory built completely out of observables appeared to reproduce everything, but did it contravene the law of the conservation of energy? If it did, then it would collapse like a house of cards. Excited and nervous as he edged ever closer to proving that his theory was both physically and mathematically consistent, the 24-year-old physicist began making simple errors of arithmetic as he checked his calculations. It was almost three in the morning before Heisenberg could put down his pen, satisfied that the theory did not violate one of the most fundamental laws of physics. He was elated, but troubled. 'At first, I was deeply alarmed', Heisenberg recalled later.[38] 'I had the feeling that, through the surface of atomic phenomena, I was looking at a strangely beautiful interior, and felt almost giddy at the thought that I now had to probe this wealth of mathematical structures nature had so generously spread out before me.' Sleep was impossible – he was too excited. So as a new day dawned, Heisenberg walked to the southern tip of the island, where for days he had been longing to climb a rock jutting out into the sea. Fuelled by the adrenaline of discovery, he climbed it 'without too much trouble and waited for the Sun to rise'.[39]

In the cold light of day, Heisenberg's initial euphoria and optimism faded. His new physics appeared to work only with the help of a strange kind of multiplication where X times Y did not equal Y times X. With ordinary numbers it did not matter in which order they were multiplied: 4×5 gives exactly the same answer as 5×4, 20. Mathematicians called this property, where the ordering in multiplication is unimportant, commutation. Numbers obey the commutative law of multiplication, so (4×5)–(5×4) is always zero. It was a rule of mathematics that every child learned and Heisenberg was deeply troubled by the discovery that when he multiplied two arrays together, the answer was dependent on the order in which they were multiplied. (A×B)–(B×A) was not always zero.[40]

As the meaning of the peculiar multiplication he had been forced to use continued to elude him, on Friday, 19 June, Heisenberg travelled back to the mainland and headed straight to Hamburg and Wolfgang Pauli. A few hours later, having received words of encouragement from his severest critic, Heisenberg left for Göttingen and the task of refining and writing up what he had discovered. Only two days later, expecting to make quick progress, he wrote to Pauli that 'attempts to fabricate a quantum mechanics advance only slowly'.[41] As the days passed, his frustration grew as he failed to apply his new approach to the hydrogen atom.

Whatever doubts he harboured, there was one thing Heisenberg was certain about. In any calculation, only relationships between 'observable' quantities, or those that could be measured in principle if not in reality, were permissible. He had given the observability of all quantities in his equations the status of a postulate and devoted his 'entire meagre efforts' to 'killing off and suitably replacing the concept of the orbital paths that one cannot observe'.[42]

'My own works are at the moment not going especially well', Heisenberg wrote to his father at the end of June. A little more than a week later, he had finished the paper that ushered in a new era in quantum physics. Still uncertain about what he had done and its true significance, Heisenberg sent a copy to Pauli. Apologising, he asked him to read and return the paper within two or three days. The reason for the haste was that Heisenberg was due to give a lecture at Cambridge University on 28 July. With other commitments he was unlikely to return to Göttingen until late September and wanted 'either to complete it in the last days of my presence here or to burn it'.[43] Pauli greeted the paper 'with jubilation'.[44] It offered, he wrote to a colleague, 'a new hope, and a renewed enjoyment of life'.[45] 'Although it is not the solution to the riddle,' Pauli added, 'I believe that it is now once again possible to move forward.' The man who took those steps in the right direction was Max Born.

He had little inkling of what Heisenberg had been doing since returning from the little island in the North Sea. Born was therefore surprised when Heisenberg gave him the paper and requested that he decide whether it was worth publishing or not. Tired by his own exertions, Born put the paper to one side. When a couple of days later he sat down to read it and pass judgement on what Heisenberg had described as a 'crazy paper', Born was immediately captivated. He realised that Heisenberg was being uncharacteristically hesitant in what he was putting forward. Was it a consequence of having to employ a strange multiplication rule? Heisenberg was still groping even at the conclusion of the paper: 'Whether a method to determine quantum-mechanical data using relations between observable quantities, such as that proposed here, can be regarded as satisfactory in principle, or whether this method after all represents far too rough an approach to the physical problem of constructing a theoretical quantum mechanics, an obviously very involved problem at the moment, can be decided only by a more intensive mathematical investigation of the method which has been very superficially employed here.'[46]

What was the meaning of the mysterious multiplication law? It was a question that so obsessed Born, he could think of little else during the days and nights that followed. He was troubled by the fact that there was something vaguely familiar about it, but he could not pinpoint exactly what. 'Heisenberg's latest paper, soon to be published, appears rather mystifying, but is certainly true and profound', Born wrote to Einstein, even though he was still unable to explain the origin of the strange multiplication.[47] Praising the young physicists at his institute, especially Heisenberg, Born admitted 'that merely to keep up with their thoughts demands at times considerable effort on my part'.[48] After days of considering nothing else, the effort on this occasion was rewarded. One morning, Born suddenly recalled a long-forgotten lecture he had attended as a student and realised that Heisenberg had accidentally stumbled across matrix multiplication in which X times Y does not always equal Y times X.

On being told that the mystery of his strange multiplication rule had been solved, Heisenberg complained that 'I do not even know what a matrix is'.[49] A matrix is nothing more than an array of numbers placed in a series of rows and columns, just like the arrays that Heisenberg constructed in Helgoland. In the mid-nineteenth century the British mathematician Arthur Cayley had worked out how to add, subtract, and multiply matrices. If A and B are both matrices, then A×B can yield a different answer from B×A. Just like Heisenberg's array of numbers, matrices do not necessarily commute. Although they were established features of the mathematical landscape, matrices were unfamiliar territory for the theoretical physicists of Heisenberg's generation.

Once Born had correctly identified the roots of the strange multiplication, he knew that he needed help to turn Heisenberg's original scheme into a coherent theoretical framework that embraced all the multifarious aspects of atomic physics. He knew the perfect man for the job, one well versed in the intricacies of both quantum physics and mathematics. As luck would have it, he too would be in Hanover, where Born was due to attend a meeting of the German Physical Society. Once there, he immediately sought out Wolfgang Pauli. Born asked his former assistant to collaborate with him. 'Yes, I know you are fond of tedious and complicated formalisms', came the reply as Pauli refused. He wanted no part in Born's plans: 'You are only going to spoil Heisenberg's physical ideas by your futile mathematics.'[50] Feeling unable to make progress alone, he turned in desperation to one of his students for help.

In choosing 22-year-old Pascual Jordan, Born had unwittingly found the perfect collaborator for the task ahead. Entering the Technische Hochschule in Hanover in 1921 with the intention of studying physics, Jordan found the lectures rather poor and turned instead to mathematics. A year later he transferred to Göttingen to study physics. However, he rarely attended the lectures because they were too early in the morning, starting at either 7am or 8am. Then he met Born. Under his supervision, Jordan began to study physics seriously for the first time. 'He was not only my teacher, who in my student days introduced me to the wide world of physics – his lectures were a wonderful combination of intellectual clarity and horizon widening overview', Jordan later said of Born. 'But he was also, I want to assert, the person, who next to my parents, exerted the deepest, longest lasting influence on my life.'[51]

With Born as his guide, Jordan soon began concentrating on problems of atomic structure. Somewhat insecure and with a stutter, he appreciated Born's patience whenever they discussed the latest papers touching on atomic theory. Fortuitously, he had moved to Göttingen in time to attend the Bohr Festspiele and, like Heisenberg, was inspired by the lectures and the discussions that followed. After his doctoral dissertation in 1924, Jordan worked briefly with others before being asked by Born to collaborate with him on an attempt to explain the width of spectral lines. Jordan is 'exceptionally intelligent and astute and can think far more swiftly and confidently than I', Born wrote to Einstein in July 1925.[52]

By then Jordan had already heard of Heisenberg's latest ideas. Before he left Göttingen at the end of July, Heisenberg gave a talk to a small circle of students and friends about his attempt to construct a quantum mechanics based solely on the relations between observable properties. When Born asked him to collaborate, Jordan jumped at the chance to recast and extend Heisenberg's original ideas into a systematic theory of quantum mechanics. Unknown to Born, as he sent Heisenberg's paper to the journal *Zeitschrift für Physik*, Jordan was well versed in matrix theory through his background in mathematics. Applying these methods to quantum physics, in two months Born and Jordan laid the foundations for a new quantum mechanics that others would call matrix mechanics.[53]

Once Born identified Heisenberg's multiplication rule as a rediscovery of matrix multiplication, he quickly found a matrix formula that connected position q and momentum p using an expression that included Planck's constant: $pq-qp=(ih/2\pi)I$, where I is what mathematicians call a unit matrix. It allowed the right-hand side of the equation to be written as a matrix. It was from this fundamental equation using the methods of matrix mathematics that all of quantum mechanics was constructed in the months that followed. Born was proud to be 'the first person to write a physical law in terms of non-commuting symbols'.[54] But it 'was only a guess, and my attempts to prove it failed', he recalled later.[55] Within days of being shown the formula, Jordan came up with the rigorous mathematical derivation. No wonder Born was soon telling Bohr that, aside from Heisenberg and Pauli, he considered Jordan 'to be the most gifted of the younger colleagues'.[56]

In August, Born went on his summer holiday to Switzerland with his family while Jordan stayed in Göttingen to write up a paper by the end of September for publication. Before it appeared in print they sent a copy to Heisenberg, who was in Copenhagen at the time. 'Here, I got a paper from Born, which I cannot understand at all', Heisenberg said to Bohr as he handed him the paper.[57] 'It is full of matrices, and I hardly know what they are.'

Heisenberg was hardly alone in not being familiar with matrices, but he set about learning the new mathematics with gusto and mastered enough to begin collaborating with Born and Jordan while still in Copenhagen. Heisenberg returned to Göttingen in the middle of October in time to help write the final version of what became known as the *Drei-Männer-Arbeit*, the 'three-man paper' in which he, Born and Jordan presented the first logically consistent formulation of quantum mechanics – the long-sought-after new physics of the atom.

However, there were already reservations being expressed about Heisenberg's initial work. Einstein wrote to Paul Ehrenfest: 'In Göttingen they believe it (I don't).'[58] Bohr believed it was 'a step probably of fundamental importance' but 'it has not yet been possible to apply [the] theory to questions of atomic structure'.[59] While Heisenberg, Born and Jordan had been concentrating on developing the theory, Pauli had been busy using the new mechanics to do just that. By early November, while the 'three-man paper' was still being written, he had successfully applied matrix mechanics in a stunning tour de force. Pauli had done for the new physics what Bohr had done for the old quantum theory – reproduced the line spectrum of the hydrogen atom. For Heisenberg, to add insult to injury, Pauli had also calculated the Stark effect – the influence of an external electric field on the spectrum. 'I myself had been a bit unhappy that I could not succeed in deriving the hydrogen spectrum from the new theory', Heisenberg recalled.[60] Pauli had provided the first concrete vindication of the new quantum mechanics.

'The Fundamental Equations of Quantum Mechanics' read the title. Born had been in Boston for a nearly a month, as part of a five-month lecture tour of the United States, when one December morning he opened his post and received 'one of the greatest surprises' of his scientific life.[61] As he read the paper by one P.A.M. Dirac, a senior research student at Cambridge University, Born realised that 'everything was perfect in its way'.[62] Even more remarkably, Born soon discovered that Dirac had sent his paper to the *Proceedings of the Royal Society* containing the nuts and bolts of quantum mechanics a whole nine days before the 'three-man paper' was finished. Who was Dirac and how had he done it, wondered Born?

Paul Adrien Maurice Dirac was 23 years old in 1925. The son of a Swiss, French-speaking father, Charles, and an English mother, Florence, he was the second of three children. His father was such an overbearing and dominant figure that when he died in 1935, Dirac wrote: 'I feel much freer now.'[63] It was the trauma of having to remain silent in the presence of his father, a teacher of French, as he grew up that made Dirac a man of few words. 'My father made the rule that I should only talk to him in French.

He thought it would be good for me to learn French in that way. Since I found that I couldn't express myself in French, it was better for me to stay silent than to talk in English.'[64] Dirac's preference for silence, the legacy of a deeply unhappy childhood and adolescence, would become legendary.

Although interested in science, in 1918, Dirac acted on his father's advice and enrolled to study electrical engineering at the University of Bristol. Three years later, despite graduating with a first-class honours degree, he could not find a job as an engineer. With his employment prospects looking bleak as Britain's post-war depression continued, Dirac accepted the offer of free tuition for two years to study mathematics back at his old university. He would rather have gone to Cambridge, but the scholarship he had won did not cover all the expenses of studying at the university. However, in 1923, after gaining his mathematics degree and receiving a government grant, he finally arrived in Cambridge as a PhD student. His supervisor was Ralph Fowler, Rutherford's son-in-law.

Dirac had a thorough grasp of Einstein's theory of relativity, which had generated a firestorm of publicity around the world in 1919 while he was still an engineering student, but he knew very little about Bohr's decade-old quantum atom. Until his arrival in Cambridge, Dirac always considered atoms 'as very hypothetical things', hardly worth bothering about.[65] He soon changed his mind and set about making up for lost time.

The quiet, secluded life of a budding Cambridge theoretical physicist was tailor-made for the shy and introverted Dirac. Research students were largely left to work alone in either their college rooms or in the library. While others might have struggled with a lack of human contact day after day, Dirac was perfectly happy to be left alone in his room to think. Even on a Sunday as he relaxed by walking in the Cambridgeshire countryside, Dirac preferred to do it alone.

Like Bohr, whom he met for the first time in June 1925, Dirac chose his words, written or spoken, very carefully. If he gave a lecture and was asked to explain a point that had not been understood, Dirac would often repeat word for word what he had said before. Bohr had gone to Cambridge to lecture on the problems of quantum theory and Dirac had been impressed by the man, but not by his arguments. 'What I wanted was statements which could be expressed in terms of equations,' he said later, 'and Bohr's work very seldom provided such statements.'[66] Heisenberg, on the other hand, arrived from Göttingen to give a lecture having spent months doing just the sort of physics that Dirac would have found stimulating. But he did not hear about it from Heisenberg, who chose not to mention it as he spoke about atomic spectroscopy.

It was Ralph Fowler who alerted Dirac to Heisenberg's work by giving him a proof copy of the German's soon-to-be-published paper. Heisenberg had been Fowler's house-guest during his brief visit and had discussed his latest ideas with his host, who asked for a copy of the paper. When it arrived, Fowler had little time to study it thoroughly and so passed it on to Dirac, asking him for his opinion. When he first read it in early September, he found it difficult to follow and failed to appreciate what a breakthrough it represented. Then, as one week turned into two, Dirac suddenly realised that the fact that A×B did not equal B×A lay at the very heart of Heisenberg's new approach and 'provided the key to the whole mystery'.[67]

Dirac developed a mathematical theory that also led him to the formula $pq-qp=(ih/2\pi)I$ by distinguishing between what he called q-numbers and c-numbers, between those quantities that do not commute (AB does not equal BA) and those that do (AB=BA). Dirac showed that quantum mechanics differs from classical mechanics in that the variables, q and p, representing the position and momentum of a particle, do not commute with one another but obey the formula that he had found independently of Born, Jordan and Heisenberg. In May 1926, he received his PhD with the first-ever thesis on the subject of 'quantum mechanics'. By then physicists were beginning to breathe a little easier after being confronted by matrix mechanics, which was difficult to use and impossible to visualise, even though it generated the right answers.

'The Heisenberg-Born concepts leave us all breathless, and have made a deep impression on all theoretically orientated people', Einstein wrote in March 1926. 'Instead of dull resignation, there is now a singular tension in us sluggish people.'[68] They were roused out of their stupor by an Austrian physicist who found time while conducting an affair to produce an entirely different version of quantum mechanics that avoided what Einstein called Heisenberg's 'veritable calculation by magic'.[69]

Chapter 9

'A LATE EROTIC OUTBURST'

'I do not even know what a matrix is', Heisenberg had lamented when told of the origins of the strange multiplication rule that lay at the heart of his new physics. It was a reaction widely shared among physicists when they were presented with his matrix mechanics. Within a matter of months, however, Erwin Schrödinger offered them an alternative that they eagerly embraced. His friend, the great German mathematician Hermann Weyl, later described Schrödinger's astonishing achievement as the product of 'a late erotic outburst'.[1] A serial womaniser, the 38-year-old Austrian discovered wave mechanics while enjoying a secret tryst during Christmas 1925 at the Swiss ski resort of Arosa. Later, after fleeing Nazi Germany, he first scandalised Oxford and then Dublin when he set up home with his wife and yet another mistress under the same roof.

'His private life seemed strange to bourgeois people like ourselves', Born wrote some years after Schrödinger's death in 1961. 'But all this does not matter. He was a most lovable person, independent, amusing, temperamental, kind and generous, and he had a most perfect and efficient brain.'[2]

Erwin Rudolf Josef Alexander Schrödinger was born in Vienna on 12 August 1887. His mother wanted to name him Wolfgang, after Goethe, but allowed her husband to honour an older brother of his who had died in childhood. This brother was the reason why Schrödinger's father inherited the thriving family business manufacturing linoleum and oilcloth, ending his hopes of a being a scientist after studying chemistry at Vienna University. Schrödinger knew that the comfortable and carefree life he enjoyed before the First World War was possible only because his father had sacrificed his personal desires on the altar of duty.

Even before he could read or write, Schrödinger kept a record of the day's activities by dictating it to a willing adult. Precocious, he was educated at home by private tutors until the age of eleven when he began attending the Akademisches Gymnasium. Almost from the very first day until he left eight years later, Schrödinger excelled at the school. He was always first in his class without appearing to make much of an effort. A classmate recalled that 'especially in physics and mathematics, Schrödinger had a gift for understanding that allowed him, without any homework, immediately and directly to comprehend all the material during the class hours and to apply it'.[3] In truth, he was a dedicated student who worked hard in the privacy of his own study at home.

Schrödinger, like Einstein, had an intense dislike of rote learning and being forced to memorise useless facts. Nevertheless, he enjoyed the strict logic that underpinned the grammar of Greek and Latin. With a maternal grandmother who was English, he began learning the language early and spoke it almost as fluently as German. Later he learnt French and Spanish and was able to lecture in these languages whenever the occasion demanded. Well versed in literature and philosophy, he also loved the theatre, poetry and art. Schrödinger was just the sort of person to leave Werner Heisenberg feeling inadequate. Paul Dirac, when asked once if he played an instrument, replied that he did not know. He had never tried. Nor had Schrödinger, who shared his father's dislike of music.

After graduating from the Gymnasium in 1906, Schrödinger looked forward to studying physics at Vienna University under Ludwig Boltzmann. Tragically, the legendary theoretician committed suicide weeks before Schrödinger started his course. With his grey-blue eyes and shock of swept-back hair, Schrödinger made quite an impression despite being only 5ft 6in. Having shown himself to be an exceptional student at the Gymnasium, much was now expected from him. He did not disappoint, coming top of the class in one exam after another. Surprisingly, given his interest in theoretical physics, Schrödinger gained his doctorate in May 1910 with a dissertation entitled 'On the conduction of electricity on the surface of insulators in moist air'. It was an experimental investigation, showing that Schrödinger was, unlike Pauli and Heisenberg, perfectly at ease in the laboratory. Twenty-three-year-old Dr Schrödinger had a summer of freedom before reporting for military service on 1 October 1910.

All able-bodied young men in Austria-Hungary were required to do three years of military service. But as a university graduate he was able to choose a year's officer training, leading to a commission in the reserve ranks. When he returned to civilian life in 1911, Schrödinger secured a position as an assistant to the professor of experimental physics at his old university. He knew he was not cut out to be an experimenter, but never regretted the experience. 'I belong to those theoreticians who know by direct observation what it means to make a measurement', he later wrote.[4] 'Methinks it were better if there were more of them.'

In January 1914, Schrödinger, aged 26, became a *privatdozent*. Like everywhere else, opportunities in theoretical physics in Austria were few. The road to the professorship he desired seemed a long and difficult one. So he toyed with the idea of abandoning physics. Then in August that year the First World War began and he was called up to fight. He had luck on his side from the very beginning. As an artillery officer, he served in fortified positions high on the Italian front. The only real danger he faced during his various postings was boredom. Then he began receiving books and scientific journals that helped to relieve the tedium. 'Is this a life: to sleep, to eat, and to play cards?' he wrote in his diary before the first consignment arrived.[5] Philosophy and physics were the only things that kept Schrödinger from total despair: 'I no longer ask when will the war be over? But: will it be over?'[6]

Relief came when he was transferred back to Vienna in the spring of 1917 to teach physics at the university and meteorology at an anti-aircraft school. Schrödinger ended the war, as he wrote later, 'without getting wounded and without illness and with little distinction'.[7] As for most others, the early post-war years were difficult for Schrödinger and his parents, with the family business ruined. As the Habsburg Empire fell apart, the situation was made worse as the victorious allies maintained a blockade that cut off food supplies. As thousands starved and froze during the winter of 1918–19 in Vienna, with little money to buy food on the black market, the Schrödingers were often forced to eat at a local soup kitchen. Things began to improve slowly after March 1919 when the blockade was lifted and the emperor went into exile. Salvation for Schrödinger arrived early the following year with the offer of a job at the University of Jena. The salary was just enough for him to marry 23-year-old Annemarie Bertel.

Arriving in Jena in April, the couple stayed just six months before Schrödinger was appointed to an extraordinary professorship in October at the Technische Hochschule in Stuttgart. The money was better, and after the experiences of the past few years that mattered to him. By spring 1921 the universities of Kiel, Hamburg, Breslau and Vienna were all looking to appoint theoretical physicists. Schrödinger, who had by then earned a solid reputation, was being seriously considered by all of them. He accepted the offer of a professorship at Breslau.

At the age of 34, Schrödinger might have achieved the ambition of every academic; however, in Breslau he had the title but not the salary to go with it, and he left when the University of Zurich came calling. Not long after arriving in Switzerland in October 1921, Schrödinger was diagnosed with bronchitis and possibly tuberculosis. Negotiations surrounding his future, and the deaths of his parents during the previous two years, had taken their toll. 'I was actually so *kaput* that I could no longer get any sensible ideas', he later told Wolfgang Pauli.[8] On doctor's orders, Schrödinger went to a sanatorium in Arosa. It was in this high-altitude Alpine resort not far from Davos that he spent the next nine months recuperating. He was not idle during this time, but found the energy and enthusiasm to publish several papers.

As the years passed, Schrödinger began to wonder if he would ever make a major contribution that would establish him among the first rank of contemporary physicists. At the beginning of 1925 he was 37, long having celebrated the 30th birthday that was said to be the watershed in the creative life of a theorist. Doubts over his worth as a physicist were compounded by a marriage in trouble because of affairs on both sides. By the end of the year Schrödinger's marriage was shakier than ever, but he made the breakthrough that would ensure his place in the pantheon of physics.

Schrödinger was taking an ever more active interest in the latest developments in atomic and quantum physics. In October 1925, he read a paper that Einstein had written earlier in the year. A footnote that flagged up Louis de Broglie's thesis on wave-particle duality caught his eye. As with most footnotes, virtually everyone ignored it. Intrigued by Einstein's stamp of approval, Schrödinger set about acquiring a copy of the thesis, unaware that papers by the French prince had been in print for nearly two years. A couple of weeks later, on 3 November, he wrote to Einstein: 'A few days ago I read with the greatest interest the ingenious thesis of de Broglie, which I finally got hold of.'[9]

Others were also beginning to take note, but in the absence of any experimental support, few were as receptive to de Broglie's ideas as Einstein and Schrödinger. In Zurich, every fortnight, physicists from the university got together with those from the Eidgenossische Technische Hochschule (ETH), for a joint colloquium. Pieter Debye, the ETH professor of physics, ran the meetings and asked Schrödinger to give a talk on de Broglie's work. In the eyes of his colleagues, Schrödinger was an accomplished and versatile theoretician who had made solid but unremarkable contributions in his 40-odd papers that spanned areas as diverse as radioactivity, statistical physics, general relativity and colour theory. Among these were a number of well-received review articles that demonstrated his ability to absorb, analyse and organise the work of others.

On 23 November Felix Bloch, a 21-year-old student, was present when 'Schrödinger gave a beautifully clear account of how de Broglie associated a wave with a particle and how he could obtain the quantization rules of Niels Bohr and Sommerfeld by demanding that an integer number of waves should be fitted along a stationary orbit'.[10] With no experimental confirmation of wave-particle duality, which would come in 1927, Debye found it all far-fetched and 'rather childish'.[11] The physics of a wave – any wave, from sound to electromagnetic, even a wave travelling along a violin string – has an equation that describes it. In what Schrödinger had out-lined there was no 'wave equation'; de Broglie had never tried to derive one for his matter waves. Nor had Einstein after he read the French prince's thesis. Debye's point 'sounded quite trivial and did not seem to make a great impression', Bloch still remembered 50 years later.[12]

Schrödinger knew that Debye was right: 'You cannot have waves with-out a wave equation.'[13] Almost at once he decided to find the missing equation for de Broglie's matter waves. After returning from his Christmas holiday, Schrödinger was able to announce at the next colloquium held early in the New Year: 'My colleague Debye suggested that one should have a wave equation; well, I have found one!'[14] Between one meeting and the next, Schrödinger had taken de Broglie's nascent ideas and developed them into a fully-blown theory of quantum mechanics.

Schrödinger knew exactly where to start and what he had to do. De Broglie had tested his idea of wave-particle duality by reproducing the allowed electron orbits in the Bohr atom as those in which only a whole number of standing electron wavelengths could fit. Schrödinger knew that the elusive wave equation he sought would have to reproduce the three-dimensional model of the hydrogen atom with three-dimensional standing waves. The hydrogen atom would be the litmus test for the wave equation he needed to find.

Not long after starting the hunt, Schrödinger thought he had bagged just such an equation. However, when he applied it to the hydrogen atom, the equation churned out the wrong answers. The root of the failure lay in the fact that de Broglie had developed and presented wave-particle duality in a manner consistent with Einstein's theory of special relativity. Following de Broglie's lead, Schrödinger started out by looking for a wave equation that was 'relativistic' in form, and found one. In the meantime, Uhlenbeck and Goudsmit had discovered the concept of electron spin, but their paper did not appear in print until the end of November 1925. Schrödinger had found a relativistic wave equation, but unsurprisingly it did not include spin and therefore failed to agree with experiments.[15]

With the Christmas vacation fast approaching, Schrödinger began to concentrate his efforts on finding a wave equation without worrying about relativity. He knew that such an equation would fail for electrons travelling at speeds close to that of light where relativity could not be ignored. But for his purposes such a wave equation would do. Soon, however, there was more than just physics on his mind. He and his wife Anny were having another of their sustained bouts of marital turbulence, one that was lasting longer than most. Despite the affairs and talk of divorce, each seemed incapable and unwilling to permanently part from the other. Schrödinger wanted to escape for a couple of weeks. Whatever excuse he gave his wife, he left Zurich for the winter wonderland of his favourite Alpine resort, Arosa, and a rendezvous with an ex-lover.

Schrödinger was delighted to be back in the familiar and comfortable surroundings of the Villa Herwig. It was here that he and Anny had spent the previous two Christmas holidays, but there was hardly time enough over the next two weeks to feel guilty as Schrödinger spent his passion with his mysterious lady. However distracted he may have been, Schrödinger made time to continue the search for his wave equation. 'At the moment I am struggling with a new atomic theory', he wrote on 27 December.[16] 'If only I knew more mathematics! I am very optimistic about this thing and expect that if I can only … solve it, it will be very *beautiful*.' Six months of sustained creativity were to follow during this 'late erotic outburst' in his life.[17] Inspired by his unnamed Muse, Schrödinger had discovered *a* wave equation, but was it *the* wave equation he was seeking?

Schrödinger did not 'derive' his wave equation; there was just no way to do it from classical physics that was logically rigorous. Instead he constructed it out of de Broglie's wave-particle formula that linked the wavelength associated with a particle to its momentum, and from well-established equations of classical physics. As simple as it sounds, it required all of Schrödinger's skill and experience to be the first to write it down. It was the foundation on which he built the edifice of wave mechanics in the months ahead. But first he had to prove that it was *the* wave equation. When applied to the hydrogen atom, would it generate the correct values for the energy levels?

After returning to Zurich in January, Schrödinger found that his wave equation did reproduce the series of energy levels of the Bohr-Sommerfeld hydrogen atom. More complicated than de Broglie's one-dimensional standing electron waves fitted into circular orbits, Schrödinger's theory obtained their three-dimensional analogues – electron orbitals. Their associated energies were generated as part and parcel of the acceptable solutions of Schrödinger's wave equation. Banished once and for all were the ad hoc additions required by the Bohr-Sommerfeld quantum atom – all the previous tinkering and tweaking that sat uneasily now emerged naturally from within the framework of Schrödinger's wave mechanics. Even the mysterious quantum jumping between orbits by an electron appeared to be eliminated by the smooth and continuous transitions from one permitted three-dimensional electron standing wave to another. 'Quantization as an Eigenvalue Problem' was received by the *Annalen der Physik* on 27 January 1926.[18] Published on 13 March, it presented Schrödinger's version of quantum mechanics and its application to the hydrogen atom.

In a career that spanned some 50 years, Schrödinger's average annual output of research papers amounted to 40 printed pages. In 1926 he published 256 pages in which he demonstrated how wave mechanics could successfully solve a range of problems in atomic physics. He also came up with a time-dependent version of his wave equation that could tackle 'systems' that changed with time. Among them were processes involving

i. The fifth Solvay conference, 24 to 29 October 1927, devoted to the new quantum mechanics and to questions connected with it.
Back row, left to right: Auguste Piccard; E. Henriot; Paul Ehrenfest; E. Herzen; T. de Donder; Erwin Schrödinger;
J.E. Verschaffelt; Wolfgang Pauli; Werner Heisenberg; Ralph Fowler; Léon Brillouin.
Middle row, left to right: Pieter Debye; Martin Knudsen; William L. Bragg; Hendrik Kramers; Paul Dirac;
Arthur H. Compton; Louis de Broglie; Max Born; Niels Bohr.
Front row, left to right: Irving Langmuir; Max Planck; Marie Curie; Hendrik Lorentz; Albert Einstein;
Paul Langevin; Charles-Eugène Guye; C.T.R. Wilson; Owen Richardson.
(Photograph by Benjamin Couprie, Institut International de Physique Solvay, courtesy AIP Emilio Segrè Visual Archives)

ii. Max Planck, the conservative theorist who unwittingly started the quantum revolution in December 1900 when he unveiled his derivation for the distribution of electromagnetic radiation emitted by a blackbody.
(AIP Emilio Segrè Visual Archives, W. F. Meggers Collection)

iii. Ludwig Boltzmann, the Austrian physicist and foremost advocate of the atom until his suicide in 1906. (University of Vienna, courtesy AIP Emilio Segrè Visual Archives)

iv. 'The Olympia Academy'. Left to right: Conrad Habicht, Maurice Solovine and Albert Einstein. (© Underwood & Underwood/CORBIS)

v. Albert Einstein in 1912, seven years after the *annus mirabilis* in which he published five papers, including his quantum solution to the photoelectric effect and his special theory of relativity. (© Bettmann/CORBIS)

vi. The first Solvay conference, Brussels, 30 October to 3 November 1911 – a summit meeting on the quantum. Left to right seated: Walther Nernst; Marcel-Louis Brillouin; Ernest Solvay; Hendrik Lorentz; Emil Warburg; Jean-Baptiste Perrin; Wilhelm Wien; Marie Curie; Henri Poincaré. Left to right standing: Robert B. Goldschmidt; Max Planck; Heinrich Rubens; Arnold Sommerfeld; Frederick Lindemann; Maurice de Broglie; Martin Knudsen; Friedrich Hasenohrl; G. Hostelet; E. Herzen; Sir James Jeans; Ernest Rutherford; Heike Kamerlingh-Onnes; Albert Einstein; Paul Langevin.
(Photograph by Benjamin Couprie, Institut International de Physique Solvay, courtesy AIP Emilio Segrè Visual Archives)

vii. Niels Bohr, the 'golden Dane' who introduced the quantum into the atom.
This photo was taken in 1922, the year he won the Nobel Prize.
(Emilio Segrè Visual Archives, W. F. Meggers Collection)

viii. Ernest Rutherford, the charismatic New Zealander whose inspirational style motivated Bohr to run his own institute in Copenhagen along similar lines. Eleven of Rutherford's students would win the Nobel Prize.
(AIP Emilio Segrè Visual Archives)

ix. Always known as the Bohr Institute, the Universitetets Institut for Teoretisk Fysik was formally opened on 3 March 1921.
(Niels Bohr Archive, Copenhagen)

x. Einstein and Bohr walking together in Brussels during the 1930 Solvay conference. They are almost certainly discussing Einstein's light box thought experiment, which temporarily got the better of Bohr, leading him to fear the 'end of physics' if Einstein's ideas proved correct. (Photograph by Paul Ehrenfest, courtesy AIP Emilio Segrè Visual Archives, Ehrenfest Collection)

xi. Einstein and Bohr at Paul Ehrenfest's home in Leiden sometime after the 1930 Solvay conference. (Photograph by Paul Ehrenfest, courtesy AIP Emilio Segrè Visual Archives)

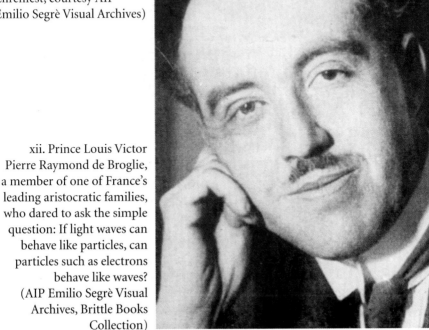

xii. Prince Louis Victor Pierre Raymond de Broglie, a member of one of France's leading aristocratic families, who dared to ask the simple question: If light waves can behave like particles, can particles such as electrons behave like waves? (AIP Emilio Segrè Visual Archives, Brittle Books Collection)

xiii. Wolfgang Pauli, the discoverer of the exclusion principle, was noted for his acerbic wit, but was also regarded as 'a genius comparable only with Einstein'. (©CERN, Geneva)

xiv. A moment to relax at the 'Bohr Festspiele', Göttingen University, June 1922. Left to right standing: Carl Wilhelm Oseen, Niels Bohr, James Franck and Oskar Klein. Max Born is seated. (AIP Emilio Segrè Visual Archives, Archive for the History of Quantum Physics)

xv. Left to right: Oskar Klein and the two 'spin doctors', George Uhlenbeck and Samuel Goudsmit, at Leiden University, summer 1926. (AIP Emilio Segrè Visual Archives)

xvi. Werner Heisenberg, aged 23. Two years later, he was responsible for one of the greatest and most profound achievements in the history of the quantum – the uncertainty principle.
(AIP Emilio Segrè Visual Archives/Gift of Jost Lemmerich)

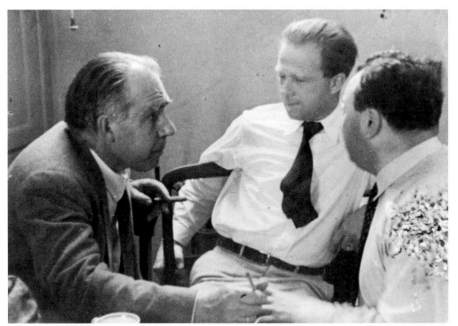

xvii. Bohr, Heisenberg and Pauli deep in discussion over lunch at the Bohr Institute in the mid-1930s. (Niels Bohr Institute, courtesy AIP Emilio Segrè Visual Archives)

xviii. The quiet Englishman, Paul Dirac, who helped to reconcile Heisenberg's matrix mechanics and Schrödinger's wave mechanics. (AIP Emilio Segrè Visual Archives)

xix. Erwin Schrödinger, whose discovery of wave mechanics was described as the product of 'a late erotic outburst'. (AIP Emilio Segrè Visual Archives)

xx. Left to right: Heisenberg's mother, Schrödinger's wife, Dirac's mother, Dirac, Heisenberg, Schrödinger at Stockholm train station in 1933. It was the year that Schrödinger and Dirac shared the Nobel Prize, and Heisenberg was awarded the deferred prize for 1932. (AIP Emilio Segrè Visual Archives)

xxi. Albert Einstein seated in his book-filled study at home in Princeton in 1954. (© Bettmann/CORBIS)

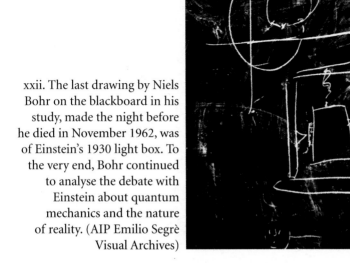

xxii. The last drawing by Niels Bohr on the blackboard in his study, made the night before he died in November 1962, was of Einstein's 1930 light box. To the very end, Bohr continued to analyse the debate with Einstein about quantum mechanics and the nature of reality. (AIP Emilio Segrè Visual Archives)

xxiii. David Bohm, who produced an alternative to the Copenhagen interpretation, seen here after refusing to testify whether or not he was a member of the Communist party before the House Un-American Activities Committee. (Library of Congress, New York World-Telegram and Sun Collection, courtesy AIP Emilio Segrè Visual Archives)

xxiv. John Stewart Bell, the Irish physicist who discovered what Einstein and Bohr could not: a mathematical theorem that could decide between their two opposing philosophical worldviews. (©CERN, Geneva)

the absorption and emission of radiation and the scattering of radiation by atoms.

On 20 February, as the first paper was being readied for the printers, Schrödinger used the name *Wellenmechanik*, wave mechanics, for the first time to describe his new theory. In stark contrast to the cold and austere matrix mechanics that proscribed even the hint of visualisability, Schrödinger offered physicists a familiar and reassuring alternative that offered to explain the quantum world in terms closer to those of nineteenth-century physics than Heisenberg's highly abstract formulation. In place of the mysterious matrices, Schrödinger came bearing differential equations, an essential part of every physicist's mathematical toolbox. Heisenberg's matrix mechanics gave them quantum jumps and discontinuity, and nothing to picture in their mind's eye as they sought to glimpse the inner workings of the atom. Schrödinger told physicists they no longer needed to 'suppress intuition and to operate only with abstract concepts such as transition probabilities, energy levels, and the like'.[19] It was hardly surprising that they greeted wave mechanics with enthusiasm and quickly rushed to embrace it.

As soon as he received complimentary copies of his paper, Schrödinger sent them out to colleagues whose opinions mattered most to him. Planck wrote back on 2 April that he had read the paper 'like an eager child hearing the solution to a riddle that had plagued him for a long time'.[20] Two weeks later, Schrödinger received a letter from Einstein, who told him 'the idea of your work springs from true genius'.[21] 'Your approval and Planck's mean more to me than that of half the world', Schrödinger wrote back.[22] Einstein was convinced that Schrödinger had made a decisive advance, 'just as I am convinced that the Heisenberg-Born method is misleading'.[23]

Others took longer to fully appreciate the product of Schrödinger's 'late erotic outburst'. Sommerfeld initially believed that wave mechanics was 'totally crazy', before changing his mind and declaring: 'although the truth of matrix mechanics is indubitable, its handling is extremely intricate and frighteningly abstract. Schrödinger has now come to our rescue.'[24] Many others also breathed easier as they learnt and began using the more familiar ideas embodied in wave mechanics rather than having to struggle with the abstract and alien formulation of Heisenberg and his Göttingen colleagues. 'The Schrödinger equation came as a great relief,' wrote the young spin doctor George Uhlenbeck, 'now we did not any longer have to learn the strange mathematics of matrices.'[25] Instead Ehrenfest, Uhlenbeck and the others in Leiden spent weeks 'standing for hours at a time in front of the blackboard' in order to learn all the splendid ramifications of wave mechanics.[26]

Pauli may have been close to the Göttingen physicists, but he recognised the significance of what Schrödinger had done and was deeply impressed. Pauli had strained every ounce of grey matter he possessed as he successfully applied matrix mechanics to the hydrogen atom. Everyone was later amazed by the speed and virtuosity with which he had done so. Pauli sent his paper to the *Zeitschrift für Physik* on 17 January, only ten days before Schrödinger posted his first paper. When he saw the relative ease with which wave mechanics allowed Schrödinger to tackle the hydrogen atom, Pauli was astonished. 'I believe that the work counts among the most significant recently written', he told Pascual Jordan. 'Read it carefully and with devotion.'[27] Not long afterwards, in June, Born described wave mechanics 'as the deepest form of the quantum laws'.[28]

Heisenberg was 'not very pleased', he told Jordan, by Born's apparent defection to wave mechanics.[29] Although he acknowledged that Schrödinger's paper was 'incredibly interesting' with its use of more familiar mathematics, Heisenberg firmly believed that when it came to physics, his matrix mechanics was a better description of the way things were at the atomic level.[30] 'Heisenberg from the very beginning did not share my opinion that your wave mechanics is physically more significant than our quantum mechanics', Born confided to Schrödinger in May 1927.[31] By then it was hardly a secret. Nor did Heisenberg want it to be. There was too much at stake.

As spring had given way to summer in 1925 there was still no quantum mechanics, a theory that would do for atomic physics what Newtonian mechanics did for classical physics. A year later there were two competing theories that were as different as particles and waves. They both gave identical answers when applied to the same problems. What, if any, was the connection between matrix and wave mechanics? It was a question that Schrödinger began to ponder almost as soon as he finished his first ground-breaking paper. After two weeks of searching he found no link. 'Consequently,' Schrödinger wrote to Wilhelm Wien, 'I have given up looking any further myself.'[32] He was hardly disappointed, as he confessed that 'matrix calculus was already unbearable to me long before I even distantly thought of my theory'.[33] But he was unable to stop digging until he unearthed the connection at the beginning of March.

The two theories that appeared to be so different in form and content, one employing wave equations and the other matrix algebra, one describing waves and the other particles, were mathematically equivalent.[34] No wonder they both gave exactly the same answers. The advantages of having two different but equivalent formalisms of quantum mechanics quickly became apparent. For most problems physicists encountered, Schrödinger's wave mechanics provided the easiest route to the solution. Yet for others, such as those involving spin, it was Heisenberg's matrix approach that proved its worth.

With any possible arguments about which of the two theories was correct smothered even before they could begin, attention turned from the mathematical formalism to the physical interpretation. The two theories might technically be equivalent, but the nature of physical reality that lay beyond the mathematics was altogether different: Schrödinger's waves and continuity versus Heisenberg's particles and discontinuity. Each man was convinced that his theory captured the true nature of physical reality. Both could not be right.

———

At the beginning there was no personal animosity between Schrödinger and Heisenberg as they began to question each other's interpretation of quantum mechanics. But soon emotions began to run high. In public and in their papers both managed, on the whole, to rein in their true feelings.

In their letters, however, there was no need for tact and restraint. When he initially tried but failed to prove the equivalence of wave and matrix mechanics, Schrödinger was somewhat relieved that there might be none, since 'the mere thought makes me shudder, if I later had to present the matrix calculus to a young student as describing the true nature of the atom'.[35] In his paper, 'On the Relation Between Heisenberg-Born-Jordan Quantum Mechanics and My Own', Schrödinger was at pains to distance wave mechanics from matrix mechanics. 'My theory was inspired by L. de Broglie and by brief but infinitely far-seeing remarks of A. Einstein', he explained. 'I was absolutely unaware of any genetic relationship with Heisenberg.'[36] Schrödinger concluded that, 'because of the lack of visualization' in matrix mechanics, 'I felt deterred by it, if not to say repelled'.[37]

Heisenberg was even less diplomatic about the continuity that Schrödinger was trying to restore to the atomic realm where, as far as he was concerned, discontinuity ruled. 'The more I think about the physical portion of the Schrödinger theory, the more repulsive I find it', he told Pauli in June.[38] 'What Schrödinger writes about the visualizability of his theory "is probably not quite right", in other words it's crap.' Two months earlier, Heisenberg had appeared more conciliatory when he described wave mechanics as 'incredibly interesting'.[39] But those who knew Bohr recognised that Heisenberg was employing exactly the sort of language favoured by the Dane, who always called an idea or an argument 'interesting' when in fact he disagreed with it. Increasingly frustrated as more of his colleagues abandoned matrix mechanics for the easier-to-use wave mechanics, Heisenberg finally snapped. He could hardly believe it when Born, of all people, started using Schrödinger's wave equation. In a fit of anger, Heisenberg called him a 'traitor'.

He may have been envious of the growing popularity of Schrödinger's alternative, but after its discovery it was Heisenberg who was responsible for the next great triumph of wave mechanics. He might have been annoyed at Born, but Heisenberg had also been seduced by the mathematical ease with which Schrödinger's approach could be applied to atomic problems. In July 1926 he used wave mechanics to account for the line spectra of helium.[40] Just in case anyone read too much into his adoption of the rival formulation, Heisenberg pointed out that it was nothing more than expediency. The fact that the two theories were mathematically equivalent meant he could use wave mechanics while ignoring the 'intuitive pictures' Schrödinger painted with it. However, even before Heisenberg posted his paper, Born had used Schrödinger's palette to paint an entirely different picture on the same canvas when he discovered that probability lay at the heart of wave mechanics and quantum reality.

Schrödinger was not trying to paint a new picture, but attempting to restore an old one. For him there were no quantum jumps between different energy levels in an atom, but only smooth, continuous transitions from one standing wave into another, with the emission of radiation being the product of some exotic resonance phenomenon. He believed that wave mechanics allowed the restoration of a classical, 'intuitive' picture of physical reality, one of continuity, causality and determinism. Born disagreed. 'Schrödinger's achievement reduces itself to something purely mathematical,' he told Einstein, 'his physics is wretched.'[41] Born used wave mechanics to paint a surreal picture of a reality with discontinuity, acausality and probability, instead of Schrödinger's attempt at a Newtonian-inspired old master. These two pictures of reality hang on different interpretations of the so-called wave function, symbolised by the Greek letter psi, ψ, in Schrödinger's wave equation.

Schrödinger had known from the very beginning that there was a problem with his version of quantum mechanics. According to Newton's laws of motion, if the position of an electron is known at a certain time together with its velocity, then it is theoretically possible to determine exactly where it will be at some later time. However, waves are much more difficult to pin down than a particle. Dropping a stone into a pond sends ripples of waves across its surface. Exactly where is the wave? Unlike a particle, a wave is not localised at a single place, but is a disturbance that carries energy through a medium. Like people taking part in a 'Mexican wave', a water wave is just individual water molecules bobbing up and down.

All waves, whatever their size and shape, can be described by an equation that mathematically maps their motion, just as Newton's equations do for a particle. The wave function, ψ, represents the wave itself and describes its shape at a given time. The wave function of a wave rippling across the surface of a pond specifies the size of the disturbance, the so-called amplitude, of the water at any point x at time t. When Schrödinger discovered the wave equation for de Broglie's matter waves, the wave function was the unknown part. Solving the equation for a particular physical situation, such as the hydrogen atom, would yield the wave function. However, there was a question that Schrödinger was finding difficult to answer: what was doing the waving?

In the case of water or sound waves, it was obvious: water or air molecules. Light had perplexed physicists in the nineteenth century. They had been forced to invoke the mysterious 'ether' as the necessary medium through which light travelled, until it was discovered that light was an electromagnetic wave with interlocked electric and magnetic fields doing the waving. Schrödinger believed that matter waves were as real as any of these more familiar types of waves. However, what was the medium through which an electron wave travelled? The question was akin to asking what does the wave function in Schrödinger's wave equation represent? In the summer of 1926 a witty little ditty summed up the situation that confronted Schrödinger and his colleagues:

Erwin with his psi can do
Calculations quite a few.
But one thing has not been seen:
Just what does psi really mean?[42]

Schrödinger finally proposed that the wave function of an electron, for example, was intimately connected to the cloud-like distribution of its electric charge as it travelled through space. In wave mechanics the wave function was not a quantity that could be directly measured because it was what mathematicians call a complex number. 4+3i is one example of such a number, and it consists of two parts: one 'real' and the other 'imaginary'. 4 is an ordinary number and is the 'real' part of the complex number 4+3i. The 'imaginary' part, 3i, has no physical meaning because i is the square root of −1. The square root of a number is just another number that multiplied by itself will give the original number. The square root of 4 is 2 since 2×2 equals 4. There is no number that multiplied by itself equals −1. While 1×1=1, −1×−1 is also equal to 1, since by the laws of algebra, a minus times a minus generates a plus.

The wave function was unobservable; it was something intangible that could not be measured. However, the square of a complex number gives a real number that is associated with something that can actually be measured in the laboratory.[43] The square of 4+3i is 25.[44] Schrödinger believed that the square of the wave function of an electron, $|\psi(x,t)|^2$, was a measure of the smeared-out density of electric charge at location x at time t.

As part of his interpretation of the wave function, Schrödinger introduced the concept of a 'wave packet' to represent the electron as he challenged the very idea that particles existed. He argued that an electron only 'appeared' to be particle-like but was not actually a particle, despite the overwhelming experimental evidence in favour of it being so. Schrödinger believed that a particle-like electron was an illusion. In reality there were only waves. Any manifestation of a particle electron was due to a group of matter waves being superimposed into a wave packet. An electron in motion would then be nothing more than a wave packet that moved like a pulse sent, with a flick of the wrist, travelling down the length of a taut rope tied at one end and held at the other. A wave packet that gave the appearance of a particle required a collection of waves of different wavelengths that interfered with one another in such a way that they cancelled each other out beyond the wave packet.

If giving up particles and reducing everything to waves rid physics of discontinuity and quantum jumps, then for Schrödinger it was a price worth paying. However, his interpretation soon ran into difficulties as it failed to make physical sense. Firstly, the wave packet representation of the electron began to unravel when it was discovered that the constituent

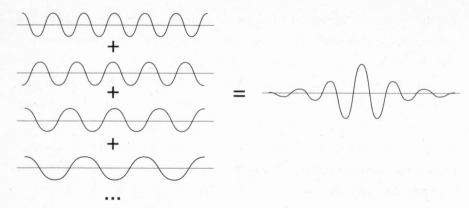

Figure 11: A wave packet formed from the superposition of a group of waves

waves would spread out across space to such a degree that they would have to travel faster than the speed of light if they were to be connected with the detection of a particle-like electron in an experiment.

Try as he might, there was no way for Schrödinger to prevent this dispersal of the wave packet. Since it was made up of waves that varied in wavelength and frequency, as the wave packet travelled through space it would soon begin to spread out as individual waves moved at different velocities. An almost instantaneous coming together, a localisation at one point in space, would have to take place every time an electron was detected as a particle. Secondly, when attempts were made to apply the wave equation to helium and other atoms, Schrödinger's vision of the reality that lay beneath his mathematics disappeared into an abstract, multi-dimensional space that was impossible to visualise.

The wave function of an electron encodes everything there is to know about its single three-dimensional wave. Yet the wave function for the two electrons of the helium atom could not be interpreted as two three-dimensional waves existing in ordinary three-dimensional space. Instead the mathematics pointed to a single wave inhabiting a strange six-dimensional space. In each move across the periodic table from one element to the next, the number of electrons increased by one and an additional three dimensions were required. If lithium, third in the table, required a nine-dimensional space, then uranium had to be accommodated in a space with

276 dimensions. The waves that occupied these abstract multi-dimensional spaces could not be the real, physical waves that Schrödinger hoped would restore continuity and eliminate the quantum jump.

Nor could Schrödinger's interpretation account for the photoelectric and Compton effects. There were unanswered questions: how could a wave packet possess electric charge? Could wave mechanics incorporate quantum spin? If Schrödinger's wave function did not represent real waves in everyday three-dimensional space, then what were they? It was Max Born who provided the answer.

Born was nearing the end of his five-month stay in America when Schrödinger's first paper on wave mechanics appeared in March 1926. Reading it on his return to Göttingen in April, he was taken completely 'by surprise' as others had been.[45] The terrain of quantum physics had dramatically changed during his absence. Almost out of nowhere, Born immediately recognised, Schrödinger had constructed a theory of 'fascinating power and elegance'.[46] He was quick to acknowledge the 'superiority of wave mechanics as a mathematical tool', as demonstrated by the relative ease with which it solved 'the fundamental atomic problem' – the hydrogen atom.[47] After all, it had taken someone of Pauli's prodigious talent to apply matrix mechanics to the hydrogen atom. Born might have been taken by surprise but he was already familiar with the idea of matter waves long before Schrödinger's paper was published.

'A letter from Einstein directed my attention to de Broglie's thesis shortly after its publication, but I was too much involved in our speculations to study it carefully', Born admitted more than half a century later.[48] By July 1925 he had made time to study de Broglie's work and wrote to Einstein that 'the wave theory of matter could be of very great importance'.[49] Enthused, he had already begun 'speculating a little about de Broglie's waves', Born told Einstein.[50] But just then he shoved de Broglie's ideas aside to make sense of the strange multiplication rule in a paper given to him by Heisenberg. Now, almost a year later, Born solved some of the problems encountered by wave mechanics, but at a price far higher than Schrödinger demanded with his sacrifice of particles.

The rejection of particles and quantum jumps that Schrödinger advocated was too much for Born. He witnessed regularly in Göttingen what he called 'the fertility of the particle concept' in experiments on atomic collisions.[51] Born accepted the richness of Schrödinger's formalism but rejected the Austrian's interpretation. 'It is necessary,' Born wrote late in 1926, 'to drop completely the physical pictures of Schrödinger which aim at a revitalization of the classical continuum theory, to retain only the formalism and to fill that with a new physical content.'[52] Already convinced 'that particles could not simply be abolished', Born found a way to weave them together with waves using probability as he came up with a new interpretation of the wave function.[53]

Born had been working on applying matrix mechanics to atomic collisions while in America. Back in Germany with Schrödinger's wave mechanics suddenly at his disposal, he returned to the subject and produced two seminal papers bearing the same title, 'Quantum mechanics of collision phenomena'. The first, only four pages long, was published on 10 July in Zeitschrift für Physik. Ten days later the second paper, more polished and refined than the first, was finished and in the post.[54] While Schrödinger renounced the existence of particles, Born in his attempt to save them put forward an interpretation of the wave function that challenged a fundamental tenet of physics – determinism.

The Newtonian universe is purely deterministic with no room for chance. In it, a particle has a definite momentum and position at any given time. The forces that act on the particle determine the way its momentum and position vary in time. The only way that physicists such as James Clerk Maxwell and Ludwig Boltzmann could account for the properties of a gas that consists of many such particles was to use probability and settle for a statistical description. The forced retreat into a statistical analysis was due to the difficulties in tracking the motion of such an enormous number of particles. Probability was a consequence of human ignorance in a deterministic universe where everything unfolded according to the laws of nature. If the present state of any system and the forces acting upon it are known, then what happens to it in the future is already determined. In classical physics, determinism is bound by an umbilical cord to causality – the notion that every effect has a cause.

Like two billiard balls colliding, when an electron slams into an atom it can be scattered in almost any direction. However, that is where the similarity ends, argued Born as he made a startling claim. When it comes to atomic collisions, physics could not answer the question 'What is the state after collision?', but only 'How probable is a given effect of the collision?'[55] 'Here the whole problem of determinism arises', admitted Born.[56] It was impossible to determine exactly where the electron was after the collision. The best that physics could do, he said, was to calculate the probability that the electron would be scattered through a certain angle. This was Born's 'new physical content', and it all hinged on his interpretation of the wave function.

The wave function itself has no physical reality; it exists in the mysterious, ghost-like realm of the possible. It deals with abstract possibilities, like all the angles by which an electron could be scattered following a collision with an atom. There is a real world of difference between the possible and the probable. Born argued that the square of the wave function, a real rather than a complex number, inhabits the world of the probable. Squaring the wave function, for example, does not give the actual position of an electron, only the probability, the odds that it will found here rather than there.[57] For example, if the value of the wave function of an electron at X is double its value at Y, then the probability of it being found at X is four times greater than the probability of finding it at Y. The electron could be found at X, Y or somewhere else.

Niels Bohr would soon argue that until an observation or measurement is made, a microphysical object like an electron does not exist anywhere. Between one measurement and the next it has no existence outside the abstract possibilities of the wave function. It is only when an observation or measurement is made that the 'wave function collapses' as one of the 'possible' states of the electron becomes the 'actual' state and the probability of all the other possibilities becomes zero.

For Born, Schrödinger's equation described a probability wave. There were no real electron waves, only abstract waves of probability. 'From the point of view of our quantum mechanics there exists no quantity which in an individual case causally determines the effect of a collision', wrote Born.[58] And he confessed, 'I myself tend to give up determinism in the atomic world.'[59] Yet while the 'motion of particles follows probability rules', he pointed out, 'probability itself propagates according to the law of causality'.[60]

It took Born the time between his two papers to fully grasp that he had introduced a new kind of probability into physics. 'Quantum probability', for want of a better term, was not the classical probability of ignorance that could in theory be eliminated. It was an inherent feature of atomic reality. For example, the fact that it was impossible to predict when an individual atom would decay in a radioactive sample, amid the certainty that one would do so, was not due to a lack of knowledge but was the result of the probabilistic nature of the quantum rules that dictate radioactive decay.

Schrödinger dismissed Born's probability interpretation. He did not accept that a collision of an electron or an alpha particle with an atom is 'absolutely accidental', i.e. 'completely undetermined'.[61] Otherwise, if Born was right, then there was no way to avoid quantum jumps and causality was once again threatened. In November 1926, he wrote to Born: 'I have, however, the impression that you and others, who essentially share your opinion, are too deeply under the spell of those concepts (like stationary states, quantum jumps, etc.), which have obtained civic rights in our thinking in the last dozen years; hence, you cannot do full justice to an attempt to break away from this scheme of thought.'[62] Schrödinger never relinquished his interpretation of wave mechanics and the attempt at a visualisability of atomic phenomena. 'I can't imagine that an electron hops about like a flea', he once memorably said.[63]

Zurich lay well outside the golden quantum triangle of Copenhagen, Göttingen and Munich. As the new physics of wave mechanics spread like wildfire through Europe's physics community in the spring and summer of 1926, many were eager to hear Schrödinger discuss his theory in person. When the invitation arrived from Arnold Sommerfeld and Wilhelm Wien to give two lectures in Munich, Schrödinger readily accepted. The first, on 21 July, to Sommerfeld's 'Wednesday Colloquium', was routine and well-received. The second, on 23 July, to the Bavarian section of the German Physical Society, was not. Heisenberg, who at the time was based in Copenhagen as Bohr's assistant, had returned to Munich in time to hear both of Schrödinger's lectures before going on a hiking tour.

As he sat in the packed lecture theatre for a second time, Heisenberg listened quietly until the end of Schrödinger's talk, entitled 'New results of wave mechanics'. During the question-and-answer session that followed, he became increasingly agitated until he could no longer remain silent. As he rose to speak, all eyes were on him. Schrödinger's theory, he pointed out, could not explain Planck's radiation law, the Frank-Hertz experiment, the Compton effect, or the photoelectric effect. None could be explained without discontinuity and quantum jumps – the very concepts that Schrödinger wanted to eliminate.

Before Schrödinger could reply, with some in the audience already expressing their disapproval at the remarks of the 24-year-old, an annoyed Wien stood up and intervened. The old physicist, Heisenberg told Pauli later, 'almost threw me out of the room'.[64] The pair had a history going back to Heisenberg's days as a student in Munich and his poor showing during the oral examination for his doctorate on anything connected to experimental physics. 'Young man, Professor Schrödinger will certainly take care of all these questions in due time', Wien told Heisenberg as he motioned for him to sit down.[65] 'You must understand that we are now finished with all that nonsense about quantum jumps.' Schrödinger, unfazed, replied that he was confident that all remaining problems would be overcome.

Heisenberg could not stop himself from lamenting later that Sommerfeld, who had witnessed the whole incident, had 'succumbed to the persuasive force of Schrödinger's mathematics'.[66] Shaken and dejected at being forced to retire from the arena vanquished before battle had been properly joined, Heisenberg needed to regroup. 'A few days ago I heard two lectures here by Schrödinger,' he wrote to Jordan, 'and I am rock-solid convinced of the incorrectness of the physical interpretation of QM presented by Schrödinger.'[67] He already knew that conviction alone was not enough, given that 'Schrödinger's mathematics signifies a great progress'.[68] After his disastrous intervention, Heisenberg had sent a dispatch to Bohr from the front line of quantum physics.

After reading Heisenberg's version of events in Munich, Bohr invited Schrödinger to Copenhagen to give a lecture and participate in 'some discussions for the narrower circle of those who work here at the Institute, in which we can deal more deeply with the open questions of atomic theory'.[69] When Schrödinger stepped off the train on 1 October 1926, Bohr was waiting for him at the station. Remarkably, it was the first time they had ever met.

After the exchange of pleasantries, battle began almost at once, and according to Heisenberg, 'continued daily from early morning until late at night'.[70] There was to be little respite for Schrödinger from Bohr's continual probing in the days ahead. He installed Schrödinger in the guest room at his home to maximise their time together. Although usually the most kind and considerate of hosts, in his desire to convince Schrödinger that he was in error, Bohr appeared even to Heisenberg to act as a 'remorseless fanatic, one who was not prepared to make the least concession or grant that he could ever be mistaken'.[71] Each man passionately defended his deeply-rooted convictions concerning the physical interpretation of the new physics. Neither was prepared to concede a single point without putting up a fight. Each pounced on any weakness or lack of precision in the argument of the other.

During one discussion Schrödinger called 'the whole idea of quantum jumps a sheer fantasy'. 'But it does not prove that there are no quantum jumps', Bohr countered. All it proved, he continued, was that 'we cannot imagine them'. Emotions soon ran high. 'You can't seriously be trying to cast doubt on the whole basis of quantum theory!' asked Bohr. Schrödinger conceded there was much that still needed to be fully explained, but that Bohr had also 'failed to discover a satisfactory physical interpretation of quantum mechanics'. As Bohr continued to press, Schrödinger finally snapped. 'If all this damned quantum jumping were really here to stay, I should be sorry I ever got involved with quantum theory.' 'But the rest of us are extremely grateful that you did,' Bohr replied, 'your wave mechanics has contributed so much to mathematical clarity and simplicity that it represents a gigantic advance over all previous forms of quantum mechanics.'[72]

After a few days of these relentless discussions, Schrödinger fell ill and took to his bed. Even as his wife did all she could to nurse their houseguest, Bohr sat on the edge of the bed and continued the argument. 'But surely Schrödinger, you must see ...' He did see, but only through the glasses that he had long worn, and he was not about to change them for ones prescribed by Bohr. There had been little, if any, chance of the two men ever reaching a concord. Each remained unconvinced by the other. 'No real understanding could be expected since, at the time, neither side was able to offer a complete and coherent interpretation of quantum mechanics', Heisenberg later wrote.[73] Schrödinger did not accept that quantum theory represented a complete break with classical reality. As far as Bohr was concerned, there was no going back to the familiar ideas of orbits and continuous paths in the atomic realm. The quantum jump was here to stay whether Schrödinger liked it or not.

As soon as he arrived back in Zurich, Schrödinger recounted Bohr's 'really remarkable' approach to atomic problems in a letter to Wilhelm Wien. 'He is completely convinced that any understanding in the usual sense of the word is impossible', he told Wien. 'Therefore the conversation is almost immediately driven into philosophical questions, and soon you no longer know whether you really take the position he is attacking, or whether you really must attack the position that he is defending.'[74] Yet despite their theoretical differences, Bohr and 'especially' Heisenberg had behaved 'in a touchingly kind, nice, caring and attentive manner', and all 'was totally, cloudlessly amiable and cordial'.[75] Distance and a few weeks had made it seem less of an ordeal.

A week before Christmas 1926, Schrödinger and his wife travelled to America, where he had accepted an invitation from the University of Wisconsin to give a series of lectures for which he would receive the princely sum of $2,500. Afterwards he criss-crossed the country, giving nearly 50 lectures. By the time he arrived back in Zurich in April 1927, Schrödinger had turned down several job offers. He had his eye on a far greater prize, Planck's chair in Berlin.

Having been appointed in 1892, Planck was due to retire on 1 October 1927 to an emeritus professorship. Heisenberg, 24, was too young for such an elevated position. Arnold Sommerfeld had been first choice, but at 59, he decided to stay in Munich. It was now either Schrödinger or Born. Schrödinger was appointed as Planck's successor and it was the discovery of wave mechanics that had clinched it. In August 1927, Schrödinger moved to Berlin and found someone there who was just as unhappy with Born's probabilistic interpretation of the wave function as he was – Einstein.

Einstein had been the first to introduce probability into quantum physics in 1916 when he provided the explanation for the spontaneous emission of light-quanta as an electron jumped from one atomic energy level to another. Ten years later, Born had put forward an interpretation of the wave function and wave mechanics that could account for the probabilistic character of quantum jumps. It came with a price tag that Einstein did not want to pay – the renunciation of causality.

In December 1926, Einstein had expressed his growing disquiet at the rejection of causality and determinism in a letter to Born: 'Quantum mechanics is certainly imposing. But an inner voice tells me that it is not yet the real thing. The theory says a lot, but does not really bring us any closer to the secret of the "old one". I, at any rate, am convinced that He is not playing at dice.'[76] As the battle lines were being drawn, Einstein was unwittingly the inspiration for a stunning breakthrough, one of the greatest and profoundest achievements in the history of the quantum – the uncertainty principle.

Chapter 10

UNCERTAINTY IN COPENHAGEN

s Werner Heisenberg stood in front of the blackboard, with his
notes spread out on the table before him, he was nervous. The
brilliant 25-year-old physicist had every reason to be. It was
Wednesday, 28 April 1926, and he was about to deliver a lecture on matrix
mechanics to the famed physics colloquium at Berlin University. Whatever
the merits of Munich or Göttingen, it was Berlin that Heisenberg rightly
called 'the stronghold of physics in Germany'.[1] His eyes scanned the faces
in the audience and settled on four men sitting in the front row, each with
a Nobel Prize to his name: Max von Laue, Walter Nernst, Max Planck, and
Albert Einstein.

Any nerves at this 'first chance to meet so many famous men' quickly
subsided as Heisenberg, by his own reckoning, presented 'a clear account
of the concepts and mathematical foundations of what was then a most
unconventional theory'.[2] As the audience drifted away after the lecture,
Einstein invited Heisenberg back to his apartment. During the half-hour
stroll to Haberlandstrasse, Einstein asked Heisenberg about his family, education and early research. It was only when they were comfortably seated
in his apartment that the real conversation began, recalled Heisenberg,
as Einstein probed 'the philosophical background of my recent work'.[3]
'You assume the existence of electrons inside the atom, and you are probably right to do so', said Einstein. 'But you refuse to consider their orbits,

even though we can observe electron tracks in a cloud chamber. I should very much like to hear more about your reasons for making such strange assumptions.'[4] This was just what he had hoped for, a chance to win over the 47-year-old quantum master.

'We cannot observe electron orbits inside the atom,' replied Heisenberg, 'but the radiation which an atom emits during discharges enables us to deduce the frequencies and corresponding amplitudes of its electrons.'[5] Warming to his theme, he explained that 'since a good theory must be based on directly observable magnitudes, I thought it more fitting to restrict myself to these, treating them, as it were, as representatives of the electron orbits'.[6] 'But you don't seriously believe,' Einstein protested, 'that none but observable magnitudes must go into a physical theory?'[7] It was a question that struck at the very foundations on which Heisenberg had constructed his new mechanics. 'Isn't that precisely what you have done with relativity?' he countered.

A 'good trick should not be tried twice', smiled Einstein.[8] 'Possibly I did use this kind of reasoning,' he conceded, 'but it is nonsense all the same.' Although it might be heuristically useful to bear in mind what one has actually observed, in principle, he argued, 'it is quite wrong to try founding a theory on observable magnitudes alone'. 'In reality the very opposite happens. It is the theory which decides what we can observe.'[9] What did Einstein mean?

Almost a century before, in 1830, the French philosopher Auguste Comte had argued that, while every theory has to be based on observation, the mind also needs a theory in order to make observations. Einstein tried to explain that observation was a complex process, involving assumptions about phenomena that are used in theories. 'The phenomenon under observation produces certain events in our measuring apparatus', said Einstein.[10] 'As a result, further processes take place in the apparatus, which eventually and by complicated paths produce sense impressions and help fix the effects in our consciousness.' These effects, Einstein maintained, depend on our theories. 'And in your theory,' he told Heisenberg, 'you quite obviously assume that the whole mechanism of light transmission from the vibrating atom to the spectroscope or to the eye works just as one has always supposed it does, that is, essentially according to Maxwell's law. If that were no longer the case, you could not possibly observe any of the magnitudes you call observable.'[11] Einstein continued to press: 'Your claim that you are introducing none but observable magnitudes is therefore an assumption about a property of the theory that you are trying to formulate.'[12] 'I was completely taken aback by Einstein's attitude, though I found his arguments convincing', Heisenberg later admitted.[13]

While Einstein was still a patent clerk he had studied the work of the Austrian physicist Ernst Mach, for whom the goal of science was not to discern the nature of reality, but to describe experimental data, the 'facts', as economically as possible. Every scientific concept was to be understood in terms of its operational definition – a specification of how it could be measured. It was while under the influence of this philosophy that Einstein had challenged the established concepts of absolute space and time. But he had long since abandoned Mach's approach because, as he told Heisenberg, it 'rather neglects the fact that the world really exists, that our sense impressions are based on something objective'.[14]

As he left the apartment disappointed at his failure to persuade Einstein, Heisenberg needed to make a decision. In three days' time, on 1 May, he was due in Copenhagen to begin his dual appointment as Bohr's assistant and as a lecturer at the university. However, he had just been offered an ordinary professorship at Leipzig University. Heisenberg knew it was a tremendous honour for one so young, but should he accept? Heisenberg told Einstein of the difficult choice he had to make. Go and work with Bohr, was his advice. The next day, Heisenberg wrote to his parents that he was turning down the Leipzig offer. 'If I continue to produce good papers,' he reassured himself and them, 'I will always receive another call; otherwise I don't deserve it.'[15]

'Heisenberg is now here and we are all very much occupied with discussions about the new development of the quantum theory and the great prospects it holds out', Bohr wrote to Rutherford in the middle of May 1926.[16] Heisenberg lived at the institute in a 'cosy little attic flat with slanting walls' and a view of Faelled Park.[17] Bohr and his family had moved into the plush and spacious director's villa next door. Heisenberg was such a regular visitor that he soon felt 'half at home with the Bohrs'.[18] The enlargement and renovation of the institute had taken far longer than expected and Bohr was exhausted. Sapped of energy, he suffered a severe case of flu. As Bohr spent the next two months recovering, Heisenberg successfully used wave mechanics to account for the line spectrum of helium.

Once Bohr was back to his old self, living next door to him was something of a mixed blessing. 'After 8 or 9 o'clock in the evening Bohr, all of a sudden, would come up to my room and say, "Heisenberg, what do you think about this problem?" And then we would start talking and talking and quite frequently we went on till twelve or one o'clock at night.'[19] Or he would invite Heisenberg over to the villa for a chat that lasted long into the evening, fuelled by glasses of wine.

As well as working with Bohr, Heisenberg gave two lectures a week on theoretical physics at the university in Danish. He was not much older than his students, and one of them could barely believe 'he was so clever since he looked like a bright carpenter's apprentice just returned from technical school'.[20] Heisenberg quickly adapted to the rhythm of life at the institute and with his new colleagues enjoyed sailing, horse riding, and walking tours at the weekends. But there was less and less time for such activities after Schrödinger's visit at the beginning of October 1926.

Schrödinger and Bohr had failed to reach any sort of accord over the physical interpretation of either matrix or wave mechanics. Heisenberg saw how 'terribly anxious' Bohr was 'to get to the bottom of things'.[21] In the months that followed, the interpretation of quantum mechanics was all that Bohr and his young apprentice talked about as they tried to reconcile theory and experiment. 'Bohr often came up to my room late at night to talk to me of the difficulties in quantum theory which tortured both of us', Heisenberg said later.[22] Nothing caused them more pain than wave-particle duality. As Einstein told Ehrenfest: 'On the one hand waves, on the other quanta! The reality of both is firm as a rock. But the devil makes a verse out of this (which really rhymes).'[23]

In classical physics something can be either a particle or a wave; it cannot be both. Heisenberg had used particles and Schrödinger waves as they discovered their respective versions of quantum mechanics. Even the demonstration that both matrix and wave mechanics were mathematically equivalent had not yielded any deeper understanding of wave-particle duality. The crux of the whole problem, Heisenberg said, was that no one could answer the questions: 'Is an electron now a wave or is it a particle, and how does it behave if I do this and that and so on?'[24] The harder Bohr and Heisenberg thought about wave-particle duality, the worse things seemed to become. 'Like a chemist who tries to concentrate his poison more and more from some kind of solution,' remembered Heisenberg, 'we tried to concentrate the poison of the paradox.'[25] As they did so there was an increasing tension between the two men, as each adopted a different approach in an attempt to resolve the difficulties.

In the search for a physical interpretation of quantum mechanics, what the theory revealed about the nature of reality at the atomic level, Heisenberg was totally committed to particles, quantum jumps, and dis-continuity. For him the particle aspect was dominant in wave-particle duality. He was not prepared to make room to accommodate anything remotely linked to Schrödinger's interpretation. To Heisenberg's horror, Bohr wanted to 'play with both schemes'.[26] Unlike the young German, he was not wedded to matrix mechanics and had never been enthralled by any mathematical formalism. While Heisenberg's first port of call was always the mathematics, Bohr weighed anchor and sought to understand the physics behind the mathematics. In probing quantum concepts such as wave-particle duality, he was more interested in grasping the physical content of an idea rather than the mathematics it came wrapped in. Bohr believed that a way had to be found to allow for the simultaneous existence of both particles and waves in any complete description of atomic proc-esses. Reconciling these two contradictory concepts was for him the key that would open the door leading to a coherent physical interpretation of quantum mechanics.

Ever since Schrödinger's discovery of wave mechanics it was understood that there was one quantum theory too many. What was needed was a single formulation, especially given that the two were mathematically the same. It was Paul Dirac and Pascual Jordan, independently of each other, who came up with just such a formalism that autumn. Dirac, who had arrived in Copenhagen in September 1926 for a six-month stay, showed that matrix and wave mechanics were just special cases of an even more abstract formulation of quantum mechanics called transformation theory. All that was missing was a physical interpretation of the theory, and the search for it was beginning to take its toll.

'Since our talks often continued till long after midnight and did not produce a satisfactory conclusion despite protracted efforts over several months,' recalled Heisenberg, 'both of us became utterly exhausted and rather tense.'[27] Bohr decided that enough was enough and went on a four-week skiing holiday in Guldbrandsdalen, Norway in February 1927. Heisenberg was glad to see him go, so that he 'could think about these hopelessly complicated problems undisturbed'.[28] None was more pressing than the trajectory of an electron in a cloud chamber.

When Bohr met Rutherford at the research students' Christmas party in Cambridge in 1911, he was struck by the New Zealander's generous praise for the recent invention of the cloud chamber by C.T.R. Wilson. The Scotsman had managed to create clouds in a small glass chamber that contained air saturated with water vapour. Cooling the air by allowing it to expand caused the vapour to condense into minuscule water droplets on particles of dust, producing a cloud. Before long, Wilson was able to create a 'cloud' even after removing all traces of dust from the chamber. The only explanation he could offer was that the cloud was formed by condensation on ions present in the air within the chamber. However, there was another possibility. Radiation passing through the chamber could rip electrons from atoms in the air, forming ions, thereby leaving a trail of tiny water droplets in its wake. It was soon discovered that radiation did exactly that.

Wilson appeared to have given physicists a tool for observing the trajectories of alpha and beta particles emitted from radioactive substances.

Particles followed well-defined paths, while waves, because they spread out, did not. However, quantum mechanics did not allow for the existence of the particle trajectories that were clearly visible for all to see in a cloud chamber. The problem seemed insurmountable. But it ought to be possible, Heisenberg was convinced, to establish a connection between what was observed in the cloud chamber and quantum theory, 'hard though it appeared to be'.[29]

Working late one evening in his small attic flat at the institute, Heisenberg's mind began to wander as he pondered the riddle of electron tracks in a cloud chamber where matrix mechanics said there should be none. All of a sudden he heard the echo of Einstein's rebuke that 'it is the theory that decides what we can observe'.[30] Convinced that he was on to something, Heisenberg needed to clear his head. Although it was well past midnight, he went for a walk in the neighbouring park.

Barely feeling the chill, he began to focus on the precise nature of the electron track left behind in a cloud chamber. 'We had always said so glibly that the path of the electron in the cloud chamber could be observed', he wrote later.[31] 'But perhaps what we really observed was something much less. Perhaps we merely saw a series of discrete and ill-defined spots through which the electron had passed. In fact, all we do see in the cloud chamber are individual water droplets which must certainly be much larger than the electron.'[32] There was no continuous, unbroken path, Heisenberg believed. He and Bohr had been asking the wrong questions. The one to answer was: 'Can quantum mechanics represent the fact that an electron finds itself approximately in a given place and that it moves approximately with a given velocity?'

Hurrying back to his desk, Heisenberg began manipulating the equations he knew so well. Quantum mechanics apparently placed restrictions on what could be known and observed. But how did the theory decide what can and cannot be observed? The answer was the uncertainty principle.

Heisenberg had discovered that quantum mechanics forbids, at any given moment, the precise determination of both the position and the momentum of a particle. It is possible to measure exactly either where an electron is or how fast it is moving, but not both simultaneously. It was nature's price for knowing one of the two exactly. In a quantum dance of give-and-take, the more accurately one is measured the less accurately the other can be known or predicted. If he was right, then Heisenberg knew that it meant no experiment probing the atomic realm would ever succeed in overcoming the limits imposed by the uncertainty principle. It was, of course, impossible to 'prove' such a claim, but Heisenberg was certain it must be so, given that all processes involved in any such experiment 'had necessarily to satisfy the laws of quantum mechanics'.[33]

In the days that followed he tested the uncertainty principle, or as he preferred to call it, the indeterminacy principle. In the laboratory of the mind, he conducted one imaginary 'thought experiment' after another in which it might be possible to measure position and momentum simultaneously with an accuracy that the uncertainty principle said was impossible. As calculation after calculation revealed that the uncertainty principle had not been violated, one particular thought experiment convinced Heisenberg that he had successfully demonstrated that 'It is the theory which decides what we can and cannot observe'.

Heisenberg had once discussed with a friend the difficulties surrounding the concept of electron orbits. His friend had argued that it should, in principle, be possible to construct a microscope that allowed electron paths inside the atom to be observed. However, such an experiment was now ruled out because, according to Heisenberg, 'not even the best microscope could cross the limits set by the uncertainty principle'.[34] All he had to do was prove it theoretically by trying to determine the exact position of a moving electron.

To 'see' an electron required a special kind of microscope. Ordinary microscopes use visible light to illuminate an object and then focus the reflected light into an image. The wavelengths of visible light are much larger than an electron and therefore could not be used to determine its exact position as they washed over it like waves over a pebble. What was required was a microscope that used gamma rays, 'light' of extremely short wavelength and high frequency, to pinpoint its position. Arthur Compton, in 1923, had investigated X-rays striking electrons and found conclusive evidence for the existence of Einstein's light-quanta. Heisenberg imagined that, like two billiard balls colliding, when a gamma ray photon hits the electron, it is scattered into the microscope as the electron recoils.

There is, however, a discontinuous shove rather than a smooth transition in the electron's momentum due to the impact of the gamma ray photon. Since the momentum that an object possesses is its mass multiplied by its velocity, any change in its velocity causes a corresponding change in its momentum.[35] When the photon hits the electron it jolts its velocity. The only way to minimise the discontinuous change in the electron's momentum is by reducing the energy of the photon, thereby lessening the impact of the collision. To do so entails using light of a longer wavelength and lower frequency. However, such a switch in wavelength means that it is no longer possible to pin down the exact position of the electron. The more precisely the electron's position is measured, the more uncertain or imprecise any measurement of its momentum and vice versa.[36]

Heisenberg showed that if Δp and Δq (where Δ is the Greek letter delta) are the 'imprecision' or 'uncertainty' with which the momentum and the position are known, then Δp multiplied by Δq is always greater than or equal to $h/2\pi$: $\Delta p \Delta q \geq h/2\pi$, where h is Planck's constant.[37] This was the mathematical form of the uncertainty principle or the 'imprecision in knowledge of simultaneous measurements' of position and momentum. Heisenberg also discovered another 'uncertainty relation' involving a different pair of so-called conjugate variables, energy and time. If ΔE and Δt are the uncertainties with which the energy E of a system can be determined and the time t at which E is observed, then $\Delta E \Delta t \geq h/2\pi$.

At first there were some who thought that the uncertainty principle was the result of the technological shortcomings of the equipment used in an experiment. If the equipment could be improved, they believed, then the uncertainty would disappear. This misunderstanding arose because of

Heisenberg's use of thought experiments to draw out the significance of the uncertainty principle. However, thought experiments are imaginary experiments employing perfect equipment under ideal conditions. The uncertainty discovered by Heisenberg is an intrinsic feature of reality. There could be no improvement, he argued, on the limits set by the size of Planck's constant and enforced by the uncertainty relations on the precision of what is observable in the atomic world. Rather than 'uncertain' or 'indeterminate', 'unknowable' may have been a more apt description of his remarkable discovery.

Heisenberg believed it was the act of measuring the position of the electron that made the precise determination of its momentum at the same time impossible. The reason appeared, as far as he was concerned, to be straightforward. The electron is disturbed unpredictably when struck by the photon used to 'see it' in order to locate its position. It was this unavoidable disturbance during the act of measurement that Heisenberg identified as the origin of uncertainty.[38]

It was an explanation that he believed was supported by the fundamental equation of quantum mechanics: $pq-qp=-ih/2\pi$, where p and q are the momentum and position of a particle. It was the inherent uncertainty of nature that lay behind non-commutativity – the fact that p×q does not equal q×p. If an experiment to locate an electron were followed by one measuring its velocity (and therefore its momentum) they would give two precise values. Multiplying the two values together yields an answer A. However, repeating the experiments in reverse order, measuring the velocity first and then the position, would lead to a completely different result, B. In each case the first measurement caused a disturbance that affected the outcome of the second. If there had been no disturbance, which was different in each experiment, then p×q would be the same as q×p. As pq–qp would then equal zero, there would be no uncertainty and no quantum world.

Heisenberg was delighted as he saw the pieces fit neatly together. His version of quantum mechanics was built out of matrices representing observables such as position and momentum that do not commute. Ever since he discovered the strange rule that made the order in which two arrays of numbers were multiplied an essential component of the mathematical scheme of his new mechanics, the physical reason why this was so had been shrouded in mystery. Now he had lifted the veil. It was, according to Heisenberg, 'only the uncertainty specified by $\Delta p \Delta q \geq h/2\pi$', that 'creates room for the validity of the relations' in $pq-qp=-ih/2\pi$.[39] It was uncertainty, he claimed, that 'makes possible this equation without requiring that the physical meaning of the quantities p and q be changed'.[40]

The uncertainty principle had exposed a deep fundamental difference between quantum and classical mechanics. In classical physics both the position and momentum of an object can in principle be simultaneously determined to any degree of accuracy. If the position and velocity were known precisely at any given moment, then the path of an object, past, present and future, could also be exactly mapped out. These long-established concepts of everyday physics 'can also be defined exactly for the atomic processes', said Heisenberg.[41] However, the limitations of these concepts are laid bare when attempts are made to measure simultaneously a pair of conjugate variables: position and momentum or energy and time.

For Heisenberg the uncertainty principle was the bridge between the observation of what appeared to be electron tracks in a cloud chamber and quantum mechanics. As he built that bridge between theory and experiment, he assumed that 'only such experimental situations can arise in nature as can be expressed in the mathematical formalism' of quantum mechanics.[42] He was convinced that if quantum mechanics said it could not happen, then it did not. 'The physical interpretation of quantum mechanics is still full of internal discrepancies,' Heisenberg wrote in his uncertainty paper, 'which show themselves in arguments about continuity versus discontinuity and particle versus wave.'[43]

It was a sorry state of affairs that arose because concepts that had been the foundation of classical physics ever since Newton 'fit nature only inaccurately' at the atomic level.[44] He believed that with a more precise analysis of concepts such as position, momentum, velocity, and the path of an electron or atom it might be possible to eliminate 'the contradictions evident up to now in the physical interpretations of quantum mechanics'.[45]

What is meant by 'position' in the quantum realm? Nothing more or less, Heisenberg answered, than the result of a specific experiment designed to measure, say, the 'position of the electron' in space at a given moment, 'otherwise this word has no meaning'.[46] For him there simply is no electron with a well-defined position or a well-defined momentum in the absence of an experiment to measure its position or momentum. A measurement of an electron's position creates an electron-with-a-position, while a measurement of its momentum creates an electron-with-a-momentum. The very idea of an electron with a definite 'position' or 'momentum' is meaningless prior to an experiment that measures it. Heisenberg had adopted an approach to defining concepts through their measurement that harked back to Ernst Mach and what philosophers called operationalism. But it was more than just a redefinition of old concepts.

With the track left behind by an electron passing through a cloud chamber firmly on his mind, Heisenberg examined the concept of the 'path of the electron'. A path is an unbroken, continuous series of positions taken up by the moving electron in space and time. Under his new criteria, to observe the path involves measuring the electron's position at each successive point. However, hitting the electron with a gamma ray photon in the act of measuring its position disturbs it, therefore its future trajectory cannot be predicted with certainty. In the case of an atomic electron 'orbiting' a nucleus, a gamma ray photon is energetic enough to knock it out of the atom, and only one point in its 'orbit' is measured and therefore known. Since the uncertainty principle forbids an exact measurement of both the position and velocity that define the path of an electron or its orbit in an atom, there simply is no path or orbit. The only thing that is known for certain, says Heisenberg, is one point along the path, and 'therefore here the word "path" has no definable meaning'.[47] It is measurement that defines what is being measured.

There is no way of knowing, argued Heisenberg, what happens between two consecutive measurements: 'It is of course tempting to say that the electron must have been somewhere between the two observations and that therefore the electron must have described some kind of path or orbit even if it may be impossible to know which path.'[48] Tempting or not, he maintained that the classical notion of an electron's trajectory being a continuous, unbroken path through space is unjustified. An electron track observed in a cloud chamber only 'looks' like a path, but is really nothing more than a series of water droplets left in its wake.

Heisenberg was desperately trying to understand the sort of questions that it was possible to answer experimentally after his discovery of the uncertainty principle. It was an unspoken basic tenet of classical physics that a moving object possessed both a precise location in space at a given time and a precise momentum, irrespective of whether it was measured or not. From the fact that the position and momentum of an electron cannot be measured with absolute accuracy at the same time, Heisenberg asserted that the electron does not possesses precise values of 'position' and 'momentum' simultaneously. To talk as if it did, or that it has a 'trajectory', is meaningless. To speculate about the nature of reality that lies beyond the realm of observation and measurement is pointless.

———

In later years, Heisenberg repeatedly chose to highlight the moment he remembered his talk with Einstein in Berlin as the crucial juncture on his journey to the uncertainty principle. Yet as he travelled the road to discovery that ended in the depths of a winter's night in Copenhagen, others had walked parts of the route with him. His most influential and valued companion was not Bohr, but Wolfgang Pauli.

As Schrödinger, Bohr and Heisenberg were locked in debate in Copenhagen in October 1926, Pauli was in Hamburg quietly analysing the collision of two electrons. He discovered, aided by Born's probabilistic interpretation, what he described in a letter to Heisenberg as a 'dark point'. Pauli had found that when electrons collide their respective momenta 'must be taken as controlled' and their positions 'uncontrolled'.[49] A probable change in momentum was accompanied by a simultaneous but indeterminable change in position. He had found that one could not 'ask simultaneously' about momentum (q) and position (p).[50] 'One can see the world with the p-eye and one can view it with the q-eye,' Pauli stressed, 'but if one opens both eyes together, then one goes astray.'[51] Pauli took it no further, but his 'dark point' lurked in the back of Heisenberg's mind as he and Bohr grappled with the problem of interpretation and wave-particle duality in the months before the discovery of the uncertainty principle.

On 23 February 1927, Heisenberg wrote a fourteen-page letter to Pauli summarising his work on the uncertainty principle. He relied on the critical judgement of the Viennese 'Wrath of God' more than most. 'Day is dawning in quantum theory', replied Pauli.[52] Any lingering doubts vanished and, by 9 March, Heisenberg had turned the contents of his letter into a paper for publication. It was only then that he wrote to Bohr in Norway: 'I believe that I have succeeded in treating the case where both [the momentum] p and [the position] q are given to a certain accuracy ... I have written a draft of a paper about these problems which yesterday I sent Pauli.'[53]

Heisenberg chose not to send Bohr either a copy of the paper or the details of what he had done. It was a sign of how strained their relationship had become. 'I wanted to get Pauli's reactions before Bohr was back because I felt again that when Bohr comes back he will be angry about my interpretation', he explained later.[54] 'So I first wanted to have some support, and see whether somebody else liked it.' Five days after Heisenberg posted his letter, Bohr was back in Copenhagen.

Refreshed after his month-long vacation, Bohr dealt with pressing institute business before carefully reading the uncertainty paper. When they met to discuss it, he told a stunned Heisenberg that it was 'not quite right'.[55] Bohr not only disagreed with Heisenberg's interpretation, but he had also spotted an error in the analysis of the gamma-ray microscope thought experiment. The workings of the microscope had nearly proved to be Heisenberg's undoing as a student in Munich. Only the intervention of Sommerfeld had secured his doctorate. Afterwards, a contrite Heisenberg had read up on microscopes, but he was about to discover that he still had some more to learn.

Bohr told Heisenberg it was wrong to place the origin of the uncertainty in the momentum of the electron in the discontinuous recoil it suffers due to the collision with the gamma-ray photon. What prohibits the precise measurement of the momentum of the electron is not the discontinuous and uncontrollable nature of the momentum change, Bohr argued, but the impossibility of measuring that change exactly. The Compton effect, he explained, allows the change in momentum to be calculated with pin-point accuracy as long as the angle by which the photon is scattered after the collision through the aperture of the microscope is known. However, it is impossible to fix the point where the photon enters the microscope. Bohr identified this as the source of the uncertainty in the momentum of the electron. The electron's position when it collides with the photon is uncertain, since the finite aperture of any microscope limits its resolving power and therefore its ability to locate any microphysical object exactly. Heisenberg had failed to take all this into account, and there was worse to come.

Bohr maintained that a wave interpretation of the scattered light-quantum was indispensable for the correct analysis of the thought experiment. It was the wave-particle duality of radiation and matter that was at the heart of quantum uncertainty for Bohr as he linked Schrödinger's wave packets with Heisenberg's new principle. If the electron is viewed as a wave packet, then for it to have a precise, well-defined position requires it to be localised and not spread out. Such a wave packet is formed from the superposition of a group of waves. The more tightly localised or confined the wave packet is, the greater the variety of waves needed, the greater the range of frequencies and wavelengths involved. A single wave has a precise momentum, but it was an established fact that a group of superimposed waves of differing wavelengths cannot have a well-defined momentum. Equally, the more precisely defined the momentum of a wave packet, the fewer component waves it has and the more spread out it is, thereby increasing the uncertainty in its position. The simultaneously precise measurement of position and momentum is impossible, as Bohr showed that the uncertainty relations could be derived from the wave model of the electron.

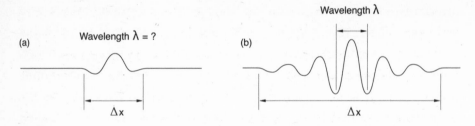

Figure 12: (a) Position of the wave can be precisely determined but not the wavelength (and hence momentum); (b) wavelength can be measured accurately but not the position, since the wave is spread out

What troubled Bohr was that Heisenberg had adopted an approach based exclusively on particles and discontinuity. The wave interpretation, Bohr believed, could not be ignored. He regarded Heisenberg's failure to accommodate wave-particle duality as a deep conceptual flaw. 'I did not know exactly what to say to Bohr's argument,' Heisenberg said later, 'so the discussion ended with the general impression that now Bohr has again shown that my interpretation is not correct.'[56] He was furious and Bohr upset at the reaction of his young protégé.

Living next to door to each other and with their offices on the ground floor of the institute separated only by a staircase, Bohr and Heisenberg did well to avoid one another for a few days before meeting again to discuss the uncertainty paper. Bohr hoped that, having had time to cool down, Heisenberg would see reason and rewrite it. He refused. 'Bohr tried to explain that it was not right and I shouldn't publish the paper', Heisenberg said later.[57] 'I remember that it ended by my breaking out in tears because I just couldn't stand this pressure from Bohr.'[58] There was too much at stake for him to simply make the changes being demanded.

Heisenberg's reputation as the wunderkind of physics rested on his discovery of matrix mechanics aged just 24. The growing popularity of Schrödinger's wave mechanics threatened to overshadow, even undermine, that astonishing achievement. Before long he was complaining about the number of papers being written that simply reworked into the language of wave mechanics results first obtained using matrix methods. Although he too had employed the alternative to matrix mechanics as a handy set of mathematical tools with which to calculate the spectrum of helium, Heisenberg harboured hopes of slamming the door on Schrödinger's wave mechanics and the Austrian's claims at having restored continuity. With the discovery of the uncertainty principle, and his interpretation of it based on particles and discontinuity, Heisenberg thought he had closed the door and locked it. He wept tears of frustration as he tried to prevent Bohr from opening it again.

Heisenberg believed that his future was intimately bound to whether it was particles or waves, discontinuity or continuity that ruled in the atomic domain. He wanted to publish as quickly as possible and challenge Schrödinger's claim that matrix mechanics was *unanschaulich*, unvisualisable, and therefore untenable. Schrödinger disliked discontinuity and a particle-based physics as much as Heisenberg loathed a physics of continuity and waves. Armed with the uncertainty principle and what he deemed to be *the* correct interpretation of quantum mechanics, Heisenberg went on the attack as he consigned his rival to a footnote in his paper: 'Schrödinger describes quantum mechanics as a formal theory of frightening, indeed repulsive, abstractness and lack of visualizability. Certainly one cannot overestimate the value of the mathematical (and to that extent physical) mastery of the quantum-mechanical laws that Schrödinger's theory has made possible. However, as regards questions of physical interpretation and principle, the popular view of wave mechanics, as I see it, has actually deflected us from exactly those roads which were pointed out by the papers of Einstein and de Broglie on the one hand and by the papers of Bohr and by quantum mechanics [i.e. matrix mechanics] on the other hand.'[59]

On 22 March 1927, Heisenberg posted his paper, 'On the perceptual content of quantum theoretical kinematics and mechanics', to the *Zeitschrift für Physik*, the quantum theorist's journal of choice.[60] 'I quarrel with Bohr', he wrote to Pauli two weeks later.[61] 'By exaggerating one side or the other', protested Heisenberg, 'one can discuss a lot without saying anything new.' Believing that he had dealt with Schrödinger and his wave mechanics once and for all, Heisenberg now faced a far more tenacious opponent.

While Heisenberg was busy exploring the consequences of the uncertainty principle in Copenhagen, on the ski slopes in Norway, Bohr came up with complementarity. It was for him no mere theory or a principle, but the necessary conceptual framework hitherto missing for describing the strange nature of the quantum world. Complementarity, Bohr believed, could accommodate the paradoxical nature of wave-particle duality. The wave and particle properties of electrons and photons, matter and radiation, were mutually exclusive yet complementary aspects of the same phenomenon. Waves and particles were two sides of the same coin.

Complementarity neatly sidestepped the difficulties that arose from having to use two disparate classical descriptions, waves and particles, to describe a non-classical world. Both particles and waves were, according to Bohr, indispensable for a complete description of quantum reality. Either description by itself is only partially true. Photons paint one picture of light, waves another. Both hang side by side. But to avoid contradictions, there were limitations. The observer can look at only one of them at any given time. No experiment would ever reveal a particle and a wave at the same time. Bohr argued that 'evidence obtained under different conditions cannot be comprehended within a single picture, but must be regarded as *complementary* in the sense that only the totality of the phenomena exhausts the possible information about the objects'.[62]

Bohr found support for his emerging ideas when he saw something in the uncertainty relations, $\Delta p \Delta q \geq h/2\pi$ and $\Delta E \Delta t \geq h/2\pi$, that Heisenberg, blinded by his intense dislike of waves and continuity, did not. The Planck-Einstein equation $E = h\nu$ and de Broglie's formula $p = h/\lambda$ embodied wave-particle duality. Energy and momentum are properties commonly associated with particles, whereas frequency and wavelength are both characteristics of waves. Each equation contained one particle-like and one wave-like variable. The meaning of this combination of particle and wave characteristics in the same equation was something that niggled Bohr. After all, a particle and a wave are two wholly distinct physical entities.

As he corrected Heisenberg's analysis of the microscope thought experiment, Bohr spotted that the same was true for the uncertainty relations. It was a finding that led him to interpret the uncertainty principle as revealing the extent to which two complementary but mutually exclusive classical concepts, either particles and waves or momentum and position, could be applied simultaneously without contradiction in the quantum world.[63]

The uncertainty relations also implied that a choice has to be made between what Bohr called a 'causal' description based on the conservation laws of energy and momentum (E and p in the uncertainty relations), and a 'space-time' description in which events are followed in space and time (q and t). The two descriptions were mutually exclusive but complementary so as to account for the results of all possible experiments. To Heisenberg's dismay, Bohr had reduced the uncertainty principle to a special rule exposing the limits inherent in nature on any simultaneous measurements of complementary pairs of observables such as position and momentum or on the simultaneous use of two complementary descriptions.

There was another difference of opinion. Whereas the uncertainty principle led Heisenberg to question the extent to which classical concepts such as 'particle', 'wave', 'position', 'momentum' and 'trajectory' were applicable in the atomic realm, Bohr argued that the 'interpretation of the experimental material rests essentially upon the classical concepts'.[64] While Heisenberg insisted upon an operational definition of these concepts, a sort of meaning through measurement, Bohr argued that their meanings were already fixed by how they were used in classical physics. 'Every description of natural processes,' he had written in 1923, 'must be based on ideas which have been introduced and defined by the classical theory.'[65] Regardless of any limitations imposed by the uncertainty principle, they could not be replaced for the simple reason that all experimental data, its discussion and interpretation, by which theories are put to the test in the laboratory, is of necessity expressed in the language and concepts of classical physics.

Heisenberg suggested that since classical physics was found wanting at the atomic level, why should these concepts be retained? 'Why should we not simply say that we cannot use these concepts with a very high precision, therefore the uncertainty relations, and therefore we have to abandon these concepts to a certain extent', he argued in the spring of 1927.[66] When it comes to the quantum, 'we must realize that our words don't fit'. If words fail, then the only sensible option for Heisenberg was to retreat into the formalism of quantum mechanics. After all, he maintained, 'a new mathematical scheme is just as good as anything because the new mathematical scheme then tells what may be there and what may not be there'.[67]

Bohr was unconvinced. The gathering of every piece of information about the quantum world, he pointed out, involves performing an experiment the results of which are recorded as fleeting flashes of light on a screen, or as clicks of a Geiger counter, or registered by the movement of needles on voltmeters and the like. Such instruments belong to the everyday world of the physics laboratory, but they are the only means by which an event at the quantum level can be magnified, measured, and recorded. It is the interaction between a piece of laboratory equipment and a microphysical object, an alpha particle or an electron, which triggers the click of a Geiger counter or causes the needle of a voltmeter to move.

Any such interaction involves the exchange of at least one quantum of energy. The consequence of this, Bohr said, is the 'impossibility of any sharp distinction between the behaviour of atomic objects and the interactions with the measuring instruments which serve to define the conditions under which the phenomena appear'.[68] In other words, it was no longer possible to make the separation that existed in classical physics between the observer and the observed, between the equipment used to make a measurement and what was being measured.

Bohr was adamant that it was the specific experiment being performed that revealed either the particle or wave aspects of an electron or a beam of light, of matter or radiation. Since particle and wave were complementary but mutually exclusive facets of one underlying phenomenon, in no actual or imaginary experiment could both be revealed. When equipment was set up to investigate the interference of light, as in Young's famous two-slits experiment, it was the wave nature of light that was manifest. If it was an experiment to study the photoelectric effect by shining a beam of light onto a metal surface, then it was light as a particle that would be observed.

To ask whether light is either a wave or a particle is meaningless. In quantum mechanics, said Bohr, there is no way of knowing what light 'really is'. The only question worth asking is: Does the light 'behave' like a particle or a wave? The answer is that sometimes it behaves like a particle and at others like a wave, depending upon the choice of experiment.

Bohr assigned a pivotal role to the act of choosing which experiment to perform. Heisenberg identified the act of measurement to determine, for example, the exact position of an electron as the origin of a disturbance that ruled out a simultaneously precise measurement of its momentum. Bohr agreed that there was a physical disturbance. 'Indeed, our usual [classical] description of physical phenomena is based entirely on the idea that the phenomena concerned may be observed without disturbing them appreciably', he said during a lecture delivered in September 1927.[69] It was a statement implying that such a disturbance is caused by the act of observing phenomena in the quantum world. A month later he was more explicit when, in a draft of a paper, he wrote 'that no observation of atomic phenomena is possible without their essential disturbance'.[70] However, he believed that the origin of this irreducible and uncontrollable disturbance lay not in the act of measurement but in the experimenter having to choose one side of the wave-particle duality in order to perform that measurement. Uncertainty, Bohr argued, was nature's price for making that choice.

In the middle of April 1927, as he worked on formulating a consistent interpretation of quantum mechanics within the conceptual framework provided by complementarity, Bohr sent a copy of the uncertainty paper to Einstein at Heisenberg's request. In the accompanying letter he wrote that it was a 'very important contribution to the discussion of the general problems of quantum theory'.[71] In spite of their ongoing and often heated arguments, Bohr informed Einstein that 'Heisenberg shows in an exceedingly brilliant manner how his uncertainty relations may be utilized not only in the actual development of quantum theory, but also for the judgement of its visualizable content'.[72] He went on to outline some of his own emerging ideas that would throw light on 'the difficulties of the quantum theory [that] are connected with the concepts, or rather with the words that are used in the customary description of nature, and which always have their origin in the classical theories'.[73] Einstein, for some unknown reason, chose not to reply.

If he was hoping to elicit a response from Einstein, then Heisenberg must have been disappointed when he returned to Copenhagen after spending Easter in Munich. It was a much-needed break from the constant pressure to yield to Bohr's interpretation. 'So I have come to be in a fight for the matrices and against the waves', Heisenberg wrote to Pauli on 31 May, the very day his 27-page paper appeared in print. 'In the ardour of this struggle I have often criticized Bohr's objections to my work too sharply and, without realizing or intending it, have in this way personally wounded him. When I now reflect on these discussions, I can very well understand that Bohr was angry about them.'[74] The reason for such contrition was that two weeks earlier, he had finally admitted to Pauli that Bohr was right.

The scattering of gamma rays into the aperture of the hypothetical microscope was the basis of the uncertainty relation for momentum and position. 'Thus the relation $\Delta p \Delta q \approx h$ indeed comes out naturally, but not entirely as I had thought.'[75] Heisenberg went on to concede that 'certain points' were easier to handle using Schrödinger's wave description, but he remained utterly convinced that in quantum physics 'only discontinuities are interesting' and they could never be emphasised enough. It was still not too late to withdraw the paper, but it was a step too far. 'All results of the paper are correct after all,' he told Pauli, 'and I am also in agreement with Bohr concerning these.'[76]

As a compromise, Heisenberg added a postscript. 'After the conclusion of the foregoing paper,' it began, 'more recent investigations of Bohr have led to a point of view which permits an essential deepening and sharpening of the analysis of quantum-mechanical correlations attempted in this work.'[77] Heisenberg acknowledged that Bohr had brought to his attention crucial points that he had overlooked – uncertainty was a consequence of wave-particle duality. He closed by thanking Bohr, and with the publication of the paper, months of wrangling and 'gross personal misunderstandings', though not entirely forgotten, were firmly pushed aside.[78] Whatever their differences, as Heisenberg said later, 'all that mattered now was to present the facts in such a way that despite their novelty they could be grasped and accepted by all physicists'.[79]

'I am very ashamed to have given the impression of being quite ungrateful', Heisenberg wrote to Bohr in the middle of June, not long after Pauli had visited Copenhagen.[80] Two months later, still full of remorse, he explained to Bohr how he reflected 'almost every day on how that came about and am ashamed that it could not have gone otherwise'.[81] Future job prospects had been a major determining factor in the rush to publish. When he turned down the Leipzig professorship in favour of Copenhagen, Heisenberg was certain that if he continued producing 'good papers', then universities would come calling.[82] After the publication of the uncertainty paper, the job offers came. Anxious that Bohr might think otherwise, he was quick to explain that he had not encouraged potential suitors because of their recent dispute over uncertainty. Not yet 26, Heisenberg became Germany's youngest ordinary professor when he accepted a new offer from Leipzig University. He left Copenhagen at the end of June. By then life at the institute was back to normal, as Bohr continued the painfully slow business of dictating the paper on complementarity and its implications for the interpretation of quantum mechanics.

He had been hard at work on it since April, and Oskar Klein, a 32-year-old Swede based at the institute, was the person Bohr turned to for help. As the argument over uncertainty and complementarity raged, Hendrik Kramers, Bohr's former assistant, warned Klein: 'Do not enter this conflict, we are both too kind and gentle to participate in that kind of struggle.'[83] When Heisenberg first learnt that Bohr was writing a paper aided by Klein on the basis that 'there exists waves and particles', he wrote rather disparagingly to Pauli that 'when one starts like that, then one can of course make everything consistent'.[84]

As one draft followed another and the title changed from 'The philosophical foundations of the quantum theory' to 'The quantum postulate and the recent development of atomic theory', Bohr tried hard to finish the paper so he could present it at a forthcoming conference. But it turned out to be yet another draft. For the time being, it would have to do.

———

The International Physics Congress from 11 to 20 September 1927 in Como, Italy was held to commemorate the 100th anniversary of the death of the Italian Alessandro Volta, the inventor of the battery. With the conference in full swing, Bohr was still finalising his notes until the day of the lecture on 16 September. Among the audience at the Istituto Carducci eager to hear what he had to say were Born, de Broglie, Compton, Heisenberg, Lorentz, Pauli, Planck, and Sommerfeld.

It was impossible for some in the audience to catch every softly spoken word that followed as Bohr outlined for the first time his new framework of complementarity, followed by an exposition of Heisenberg's uncertainty principle and the role of measurement in quantum theory. Bohr stitched each of these elements together, including Born's probabilistic interpretation of Schrödinger's wave function, so that they constituted the foundations of a new physical understanding of quantum mechanics. Physicists would later call this fusion of ideas the 'Copenhagen interpretation'.

Bohr's lecture was the culmination of what Heisenberg later described as 'an intensive study of all questions concerning the interpretation of quantum theory in Copenhagen'.[85] At first even the young quantum magician was uneasy with the Dane's answers. 'I remember discussions with Bohr which went through many hours till very late at night and ended almost in despair,' Heisenberg wrote later, 'and when at the end of the discussion I went alone for a walk in the neighbouring park I repeated to myself again and again the question: Can nature possibly be as absurd as it seemed to us in these atomic experiments?'[86] Bohr's answer was an unequivocal yes. The central role given to measurement and observation vitiated all attempts to unearth regular patterns in nature or any causal connections.

It was Heisenberg, in his uncertainty paper, who first advocated in print the rejection of one of the central tenets of science: 'But what is wrong in the sharp formulation of the law of causality, "When we know the present precisely, we can predict the future," is not the conclusion but the assumption. Even in principle we cannot know the present in all detail.'[87] Not knowing simultaneously the exact initial position and velocity of an electron, for example, allows only probabilities of a 'plenitude of possibilities' of future positions and velocities to be calculated.[88] Therefore it is impossible to predict the exact result of any single observation or measurement of an atomic process. Only the probability of a given outcome among a range of possibilities can be precisely predicted.

The classical universe built on the foundations laid down by Newton was a deterministic, clockwork cosmos. Even after Einstein's relativistic remodelling, if the exact position and velocity of an object, particle or planet, are known at any given moment, then in principle its position and velocity can be completely determined for all time. In the quantum universe there was no room for the determinism of the classical, where all phenomena can be described as a causal unfolding of events in space and time. 'Because all experiments are subject to the laws of quantum mechanics, and therefore to equation $\Delta p \Delta q \approx h$,' Heisenberg boldly asserted in the last paragraph of his uncertainty paper, 'it follows that quantum mechanics establishes the final failure of causality.'[89] Any hope of restoring it was as 'fruitless and senseless' as any lingering belief in a 'real' world hidden behind what Heisenberg called 'the perceived statistical world'.[90] It was a view shared by Bohr, Pauli and Born.

At Como two physicists were noticeable by their absence. Schrödinger had only weeks earlier moved to Berlin as Planck's successor and was busy settling in. Einstein refused to set foot in fascist Italy. Bohr would have to wait just a month before they met in Brussels.

PART III

TITANS CLASH OVER REALITY

'There is no quantum world. There is only an abstract quantum
mechanical description.'
—Niels Bohr

'I still believe in the possibility of a model of reality – that is to say,
of a theory that represents things themselves and not merely
the probability of their occurrence.'
—Albert Einstein

Chapter 11

SOLVAY 1927

'Now, I am able to write to Einstein', Hendrik Lorentz wrote on 2 April 1926.[1] Earlier that day this elder statesman of physics had been granted a private audience with the King of the Belgians. Lorentz had sought and received royal approval for Einstein's election to the scientific committee of the International Institute of Physics set up by industrialist Ernest Solvay. Once described by Einstein as 'a marvel of intelligence and exquisite tact', Lorentz had also obtained the king's permission to invite German physicists to the fifth Solvay conference scheduled for October 1927.[2]

'His Majesty expressed the opinion that, seven years after the war, the feelings which they aroused should be gradually damped down, that a better understanding between peoples was absolutely necessary for the future, and that science could help to bring this about', reported Lorentz.[3] Aware that Germany's brutal violation of Belgian neutrality in 1914 was still fresh in the memory, the king felt 'it necessary to stress that in view of all that the Germans had done for physics, it would be very difficult to pass them over'.[4] But passed over and isolated from the international scientific community they had been ever since the end of the war.

'The only German invited is Einstein who is considered for this purpose to be international', Rutherford told a colleague before the third Solvay conference in April 1921.[5] Einstein decided not to attend because Germans were excluded, and instead went on a lecture tour of America to raise funds for the founding of the Hebrew University in Jerusalem. Two years later he said he would decline any invitation to the fourth Solvay conference because of the continuing prohibition on German participation. 'In my opinion it is not right to bring politics into scientific matters,' he wrote to Lorentz, 'nor should individuals be held responsible for the government of the country to which they happen to belong.'[6]

Unable to attend the 1921 conference because of ill health, Bohr too declined an invitation to Solvay 1924. He feared that to go might be interpreted by some as tacit approval of the policy to exclude the Germans. When Lorentz became president of the League of Nations' Committee on Intellectual Cooperation in 1925, he saw little prospect of the ban on German scientists from international conferences being lifted in the near future.[7] Then, unexpectedly in October that same year, the door barring them was unlocked if not yet opened.

In an elegant palazzo in the small Swiss resort of Locarno, on the northern tip of Lake Maggiore, treaties were ratified that many hoped would ensure the future peace of Europe. Locarno was the sunniest place in Switzerland and an apt setting for such optimism.[8] It had taken months of intense diplomatic negotiations to arrange the meeting so that emissaries of Germany, France and Belgium could settle their post-war borders with one another. The Locarno treaties paved the way for Germany's acceptance, in September 1926, into the League of Nations, and membership brought with it an end to the exclusion of her scientists from the international stage. When the King of Belgium gave his consent, prior to the final moves on the diplomatic chessboard, Lorentz wrote to Einstein asking him attend the fifth Solvay conference and to accept his election to the committee responsible for planning it. Einstein agreed, and in the coming months the participants were selected, the agenda finalised, and the coveted invitations sent out.

All those invited fell into one of three groups. The first were members of the scientific committee: Hendrik Lorentz (president), Martin Knudsen (secretary), Marie Curie, Charles-Eugène Guye, Paul Langevin, Owen

Richardson and Albert Einstein.[9] The second group consisted of a scientific secretary, a Solvay family representative, and three professors from the Free University of Brussels, invited as a matter of courtesy. The American physicist Irving Langmuir, due to visit Europe at the time, would be present as a guest of the committee.

The invitation made clear that the 'conference will be devoted to the new quantum mechanics and to questions connected with it'.[10] This was reflected in the composition of the third group: Niels Bohr, Max Born, William L. Bragg, Léon Brillouin, Arthur H. Compton, Louis de Broglie, Pieter Debye, Paul Dirac, Paul Ehrenfest, Ralph Fowler, Werner Heisenberg, Hendrik Kramers, Wolfgang Pauli, Max Planck, Erwin Schrödinger and C.T.R. Wilson.

The old masters of quantum theory and the young turks of quantum mechanics would all travel to Brussels. Sommerfeld and Jordan were the most prominent of those not invited to what looked like the physicists' equivalent of a theological council convened to settle some disputed point of doctrine. During the conference, five reports would be presented: William L. Bragg on the intensity of X-ray reflection; Arthur Compton on disagreements between experiment and the electromagnetic theory of radiation; Louis de Broglie on the new dynamics of quanta; Max Born and Werner Heisenberg on quantum mechanics; and Erwin Schrödinger on wave mechanics. The last two sessions of the conference would be devoted to a wide-ranging general discussion concerning quantum mechanics.

Two names were missing from the agenda. Einstein had been asked, but decided he was 'not competent' enough to present a report. 'The reason,' he told Lorentz, 'is that I have not been able to participate as intensively in the modern development of quantum theory as would be necessary for that purpose. This is in part because I have on the whole too little receptive talent for fully following the stormy developments, in part also because I do not approve of the purely statistical way of thinking on which the new theory is founded.'[11] It was not an easy decision, since Einstein had wanted to 'contribute something of value in Brussels', but he confessed: 'I have now given up that hope.'[12]

In fact Einstein had closely monitored 'the stormy developments' of the new physics, and indirectly stimulated and encouraged the work of de Broglie and Schrödinger. However, from the very beginning he doubted that quantum mechanics was a consistent and complete description of reality. Bohr's name was also missing. He too had played no direct part in the theoretical development of quantum mechanics, but had exerted his influence through discussions with the likes of Heisenberg, Pauli and Dirac who did.

All those invited to the fifth Solvay conference on 'Electrons and Photons' knew it was designed to address the most pressing problem of the day, more philosophy than physics: the meaning of quantum mechanics. What did the new physics reveal about the nature of reality? Bohr believed he had found the answer. For many he arrived in Brussels as king of the quantum, but Einstein was the pope of physics. Bohr was anxious 'to learn his reaction to the latest stage of the development which, to our view, went far in clarifying the problems which he had himself from the outset elicited so ingeniously'.[13] What Einstein thought mattered deeply to Bohr.

So it was in a mood of great expectancy that most of the world's leading quantum physicists assembled at 10am on a grey, overcast Monday on 24 October 1927, at the Institute of Physiology in Léopold Park for the start of the first session. The conference had taken eighteen months to arrange and required the consent of a king and the ending of Germany's pariah status.

After a few brief words of welcome from Lorentz as president of the scientific committee and chair of the conference, the task of opening the proceedings fell to William L. Bragg, professor of physics at Manchester University. Now 37, Bragg was only 25 when he was awarded the Nobel Prize for physics in 1915, together with his father, William H. Bragg, for pioneering the use of X-rays to investigate the structure of crystals. He was the obvious choice to report on the latest data concerning the reflection of X-rays by crystals and how these results led to a better understanding of atomic structure. After Bragg's presentation, Lorentz invited questions and contributions from the floor. The agenda had been organised to allow ample time after each report for a thorough discussion. With Lorentz using his command of English, German and French to help those less fluent, Bragg, Heisenberg, Dirac, Born, de Broglie, and the old Dutch master himself were among those who took part in the discussion before the first session came to an end and everyone adjourned for lunch.

In the afternoon session, the American Arthur Compton reported on the failure of the electromagnetic theory of radiation to explain either the photoelectric effect or the increase in the wavelength of X-rays when they are scattered by electrons. Although awarded a share of the 1927 Nobel Prize only a few weeks earlier, genuine modesty prevented him from referring to this last phenomenon as the Compton effect, as it was universally known. Where James Clerk Maxwell's great nineteenth-century theory failed, Einstein's light-quantum, newly rebranded as the 'photon', succeeded in uniting theory and experiment. The reports presented by Bragg and Compton were intended to facilitate the discussion of theoretical concepts. At the end of the first day all the leading players had spoken bar one, Einstein.

After a leisurely reception on Tuesday morning at the Free University of Brussels, everyone reconvened in the afternoon to hear Louis de Broglie's paper on 'The new dynamics of quanta'. Speaking in French, de Broglie began by outlining his own contribution, the extension of wave-particle duality to matter, and how Schrödinger ingeniously developed it into wave mechanics. Then, treading carefully by conceding that Born's idea contained a great deal of truth, he offered an alternative to the probabilistic interpretation of Schrödinger's wave function.

In the 'pilot wave theory', as de Broglie later called it, an electron really exists both as a particle and a wave, in contrast to the Copenhagen interpretation where an electron behaves like *either* a particle *or* a wave depending on the type of experiment performed. Both particles and waves exist simultaneously, de Broglie argued, with the particle, akin to a surfer, riding a wave. The waves leading or 'piloting' the particles from one place to another were physically real rather than Born's abstract waves of probability. With Bohr and his associates determined to assert the primacy of the Copenhagen interpretation and Schrödinger still doggedly wanting to promote his views on wave mechanics, de Broglie's pilot wave proposal came under attack. Looking for support from the one man who might sway the neutrals, de Broglie was disappointed when Einstein remained silent.

On Wednesday, 26 October, the proponents of the two rival versions of quantum mechanics addressed the conference. During the morning session, Heisenberg and Born gave a joint report. It was divided into four broad sections: the mathematical formalism; the physical interpretation; the uncertainty principle; and the applications of quantum mechanics.

The presentation, like the writing of the report, was a double act. Born, the senior man, delivered the introduction and sections I and II before handing over to Heisenberg. 'Quantum mechanics,' they began, 'is based on the intuition that the essential difference between atomic physics and classical physics is the occurrence of discontinuities.'[14] Then came the metaphorical tipping of their hats to colleagues sitting only feet away as they pointed out that quantum mechanics was essentially 'a direct continuation of the quantum theory founded by Planck, Einstein, and Bohr'.[15]

After an exposition of matrix mechanics, the Dirac-Jordan transformation theory, and the probability interpretation, they turned to the uncertainty principle and the 'actual meaning of Planck's constant h'.[16] It was nothing less, they maintained, than the 'universal measure of the indeterminacy that enters the laws of nature through the dualism of waves and corpuscles'. In effect, if there were no wave-particle duality of matter and radiation there would be no Planck's constant and no quantum mechanics. In conclusion, they made the provocative statement that 'we consider quantum mechanics to be a closed theory, whose fundamental physical and mathematical assumptions are no longer susceptible of any modification'.[17]

Closure implied that no future developments would ever alter any of the fundamental features of the theory. Any such claim to the completeness and finality of quantum mechanics was something that Einstein could not accept. For him quantum mechanics was indeed an impressive achievement but not yet the real thing. Refusing to take the bait, Einstein took no part in the discussion that followed the report. Nor did any one else raise objections, as only Born, Dirac, Lorentz and Bohr spoke.

Paul Ehrenfest, sensing Einstein's disbelief at the boldness of the Born-Heisenberg assertion that quantum mechanics was a closed theory, scribbled a note and passed it to him: 'Don't laugh! There is a special section in purgatory for professors of quantum theory, where they will be obliged to listen to lectures on classical physics ten hours every day.'[18] 'I laugh only at their naiveté', Einstein replied. 'Who knows who would have the [last] laugh in a few years?'

After lunch it was Schrödinger who took centre stage as he delivered his report in English on wave mechanics. 'Under this name at present two theories are being carried on, which are indeed closely related but not identical', he said.[19] There was really only one theory, but it was effectively split in two. One part concerned waves in ordinary, everyday three-dimensional space, while the other required a highly abstract multi-dimensional space. The problem, Schrödinger explained, was that for anything other than a moving electron this was a wave that existed in a space with more than three dimensions. Whereas the single electron of the hydrogen atom could be accommodated in a three-dimensional space, helium with two electrons needed six dimensions. Nevertheless, Schrödinger argued that this multi-dimensional space, known as configuration space, was only a mathematical tool and ultimately whatever was being described, be it many electrons colliding or orbiting the nucleus of an atom, the process took place in space and time. 'In truth, however, a complete unification of the two conceptions has not yet been achieved', he admitted, before going on to outline both.[20]

Although physicists found it easier to use wave mechanics, no leading theorist agreed with Schrödinger's interpretation of the wave function of a particle as representing the cloud-like distribution of its charge and mass. Undeterred by the widespread acceptance of Born's alternative probability interpretation, Schrödinger highlighted his own and questioned the accepted notion of the 'quantum jump'.

From the moment he received the invitation to speak in Brussels, Schrödinger was acutely aware of the possibility of a clash with the 'matricians'. The discussion began with Bohr asking if a remark about 'difficulties' later in Schrödinger's report implied that a result he had stated earlier was incorrect. Schrödinger dealt with Bohr's inquiry comfortably, only to find Born challenging the correctness of another calculation. Somewhat annoyed, he said it was 'perfectly correct and rigorous and that this objection by Mr Born is unfounded'.[21]

After a couple of others had spoken, it was Heisenberg's turn: 'Mr Schrödinger says at the end of his report that the discussion he has given reinforces the hope that when our knowledge will be deeper it will be possible to explain and to understand in three dimensions the results provided by the multi-dimensional theory. I see nothing in Mr Schrödinger's calculations that would justify this hope.'[22] Schrödinger argued that his 'hope of achieving a three-dimensional conception is not quite utopian'.[23] A few minutes later the discussion ended and brought to a close the first part of the proceedings, the presentation of the commissioned reports.

When it was already too late to change the dates, it was discovered that the Académie des Sciences in Paris had chosen Thursday, 27 October to mark the centenary of the death of the French physicist Augustin Fresnel. It was decided that the Solvay meeting would be suspended for a day and a half to allow those wishing to attend the ceremonial event to do so and return for the climax of the conference, a wide-ranging general discussion spread over the last two sessions. Lorentz, Einstein, Bohr, Born, Pauli, Heisenberg and de Broglie were among the twenty who travelled to Paris to honour a kindred spirit.

Amid the distraction of German, French and English voices all seeking permission from Lorentz to speak next, Paul Ehrenfest suddenly got up and walked over to the blackboard and wrote: 'The Lord did there confound the languages of all the earth.' As he returned to his chair there was laughter as his colleagues realised that Ehrenfest was not just referring to the biblical Tower of Babel. The first session of the general discussion began on Friday afternoon, 28 October, with Lorentz making some introductory remarks as he tried to focus minds on the issues of causality, determinism, and probability. Were quantum events caused or not? Or as he put it: 'Could one not maintain determinism by making it an article of faith? Must one necessarily elevate indeterminism to a principle?'[24] Offering no further thoughts of his own, Lorentz invited Bohr to address the meeting. As he spoke about the 'epistemological problems confronting us in quantum physics', it was clear to all present that Bohr was attempting to convince Einstein about the correctness of the Copenhagen solutions.[25]

When the conference proceedings were published in French in December 1928, many mistook Bohr's contribution, then and later, as one of the official reports. When asked for an edited version of his comments for inclusion, Bohr requested that a much-expanded version of his Como lecture, which had been published the previous April, be reprinted in lieu of his remarks. Bohr being Bohr, his request was granted.[26]

Einstein listened as Bohr outlined his belief that wave-particle duality was an intrinsic feature of nature that was explicable only within the framework of complementarity, that complementarity underpinned the uncertainty principle which exposed the limits of applicability of classical concepts. However, the ability to communicate unambiguously the results of experiments probing the quantum world, Bohr explained, required the experimental set-up as well as the observations themselves to be expressed in a language 'suitably refined by the vocabulary of classical physics'.[27]

In February 1927, as Bohr was edging towards complementarity, Einstein had given a lecture in Berlin on the nature of light. He argued that instead of either a quantum or a wave theory of light, what was needed was 'a synthesis of both conceptions'.[28] It was a view he had first expressed almost twenty years earlier. Where he had long hoped to see some sort of 'synthesis', Einstein now heard Bohr imposing segregation through complementarity. It was either waves or particles depending on the choice of experiment.

Scientists had always conducted their experiments on the unspoken assumption that they were passive observers of nature, able to look without disturbing what they were looking at. There was a razor-sharp distinction between object and subject, between the observer and observed. According to the Copenhagen interpretation, this was not true in the atomic realm, as Bohr identified what he called the 'essence' of the new physics – the 'quantum postulate'.[29] It was a term he introduced to capture the existence of discontinuity in nature due to indivisibility of the quantum. The quantum postulate, said Bohr, led to no clear separation of the observer and the observed. When investigating atomic phenomena, the interaction between what is measured and the measuring equipment meant, according to Bohr, that 'an independent reality in the ordinary physical sense can neither be ascribed to the phenomenon nor to the agencies of observation'.[30]

The reality Bohr envisaged did not exist in the absence of observation. According to the Copenhagen interpretation, a microphysical object has no intrinsic properties. An electron simply does not exist at any place until an observation or measurement is performed to locate it. It does not have a velocity or any other physical attribute until it is measured. In between measurements it is meaningless to ask what is the position or velocity of an electron. Since quantum mechanics says nothing about a physical reality that exists independently of the measuring equipment, only in the act of measurement does the electron become 'real'. An unobserved electron does not exist.

'It is wrong to think that the task of physics is to find out how nature is', Bohr would argue later.[31] 'Physics concerns what we can say about nature.' Nothing more. He believed that science had but two goals, 'to extend the range of our experience and to reduce it to order'.[32] 'What we call science,' Einstein once said, 'has the sole purpose of determining what *is*.'[33] Physics for him was an attempt to grasp reality, as it is, independent of observation. It is in this sense, he said, that 'one speaks of "physical reality"'.[34] Bohr, armed with the Copenhagen interpretation, was not interested in what 'is', but in what we can say to each other about the world. As Heisenberg later stated, unlike objects in the everyday world, 'atoms or the elementary particles themselves are not as real; they form a world of potentialities or possibilities rather than one of things or facts'.[35]

For Bohr and Heisenberg, the transition from the 'possible' to the 'actual' took place during the act of observation. There was no underlying quantum reality that exists independently of the observer. For Einstein, a belief in the existence of an observer-independent reality was fundamental to the pursuit of science. At stake in the debate that was about to begin between Einstein and Bohr was the soul of physics and the nature of reality.

After Bohr's contribution, three others had already spoken when Einstein indicated to Lorentz that he wanted to break his self-imposed silence. 'Despite being conscious of the fact that I have not entered deeply enough into the essence of quantum mechanics,' he said, 'nevertheless I want to present here some general remarks.'[36] Quantum mechanics, Bohr had argued, 'exhausted the possibilities of accounting for observable phenomena'.[37] Einstein disagreed. A line had been drawn in the microphysical sands of the quantum realm. Einstein knew that the onus was on him to show that the Copenhagen interpretation was inconsistent and thereby wreck the claims of Bohr and his supporters that quantum mechanics was a closed and complete theory. He resorted to his favourite tactic – the hypothetical thought experiment conducted in the laboratory of the mind.

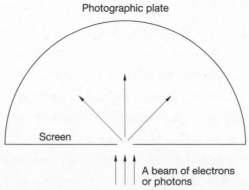

Figure 13: Einstein's single-slit thought experiment

Einstein went over to the blackboard and drew a line representing an opaque screen with a small slit in it. Just behind the screen he drew a semicircular curve representing a photographic plate. Using the sketch, Einstein outlined his experiment. When a beam of electrons or photons strikes the screen, some will pass through the slit and hit the photographic plate. Because of the narrowness of the slit, the electrons passing through it will diffract like waves in every possible direction. In keeping with the demands of quantum theory, Einstein explained, the electrons travelling outwards from the slit towards the photographic plate do so as spherical waves. Nonetheless, the electrons actually strike the plate as individual particles. There were, said Einstein, two distinct points of view concerning this thought experiment.

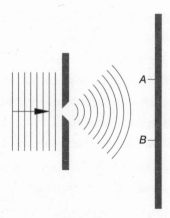

Figure 14: A later rendition by Bohr of Einstein's single-slit thought experiment

According to the Copenhagen interpretation, before any observation is made, and striking the photographic plate counts as such, there is a non-zero probability of detecting an individual electron at every point on the plate. Even though the wave-like electron is spread over a large region of space, the very moment a particular electron is detected at point A, the probability of finding it at point B or anywhere else on the plate instantly becomes zero. Since the Copenhagen interpretation maintains that quantum mechanics gives a complete description of individual electron events in the experiment, the behaviour of each electron is described by a wave function.

Here's the rub, said Einstein. If prior to the observation the probability of finding the electron was 'smeared' over the entire photographic plate, then the probability at B and everywhere else had to be instantaneously affected at the very moment the electron hit the plate at point A. Such an instantaneous 'collapse of the wave function' implied the propagation of some sort of faster-than-light cause and effect outlawed by his special theory of relativity. If an event at A is the cause of another at B, then there must be a time lapse between them to allow a signal to travel at light speed from A to B. Einstein believed the violation of this requirement, later called locality, indicated that the Copenhagen interpretation was inconsistent and quantum mechanics was not a complete theory of individual processes. Einstein proposed an alternative explanation.

Each electron that passes through the slit follows one of many possible trajectories until it hits the photographic plate. However, the spherical waves do not correspond to individual electrons, argued Einstein, but to 'a cloud of electrons'.[38] Quantum mechanics does not give any information about individual processes, but only about what he called an 'ensemble' of processes.[39] Though each individual electron of the ensemble follows its own distinct trajectory from slit to plate, the wave function does not represent an individual electron but the electron cloud. Therefore, the square of the wave function, $|\psi(A)|^2$, represents not the probability of finding a particular electron at A, but that of finding any member of the ensemble at that point.[40] It was, Einstein said, a 'purely statistical' interpretation, by which he meant that the statistical distribution of the large number of electrons striking the plate produced the characteristic diffraction pattern.[41]

Bohr, Heisenberg, Pauli and Born were not entirely sure what Einstein was driving at. He had not clearly stated his aim: to show that quantum mechanics was inconsistent and therefore an incomplete theory. Sure, the wave function collapses instantaneously, they thought, but it was an abstract wave of probability, not a real wave travelling in ordinary three-dimensional space. Nor was it possible to choose between the two viewpoints Einstein outlined on the basis of observing what happens to an individual electron. In both cases an electron passes through the slit and hits the plate at some point.

'I feel myself in a very difficult position because I don't understand what precisely is the point which Einstein wants to [make]', said Bohr.[42] 'No doubt it is my fault.' Remarkably, he then said: 'I do not know what quantum mechanics is. I think we are dealing with some mathematical methods which are adequate for [a] description of our experiments.'[43] Instead of responding to Einstein's analysis, Bohr simply went on to restate his own views. But in this game of quantum chess, the Danish grandmaster later recounted in a paper, written in 1949 to celebrate his opponent's 70th birthday, the reply he gave that evening and on the last day of the conference in 1927.[44]

According to Bohr, Einstein's analysis of his thought experiment tacitly assumed that the screen and photographic plate both had a well-defined position in space and time. However, maintained Bohr, this implied that both had an infinite mass, for only then would there be no uncertainty in either position or time as the electron emerged from the slit. As a result, the exact momentum and energy of the electron is unknown. This was the only possible scenario, argued Bohr, given that the uncertainty principle implies that the more precisely the electron's position is known, the more inexact any concurrent measurement of its momentum must be. The infinitely heavy screen in Einstein's imaginary experiment left no room for uncertainty in the space and time location of the electron at the slit. However, such precision came at a price: its momentum and energy were completely indeterminate.

It was more realistic, Bohr suggested, to assume that the screen did not have an infinite mass. Although still much heavier, the screen would now move when the electron passed through the slit. While any such movement would be so small as to be impossible to detect in the laboratory, its measurement presented no problem in the abstract world of the idealised thought experiment furnished, as it was, with measuring devices capable of perfect accuracy. Because the screen moves, the position of the electron in space and time is uncertain during the process of diffraction, resulting in a corresponding uncertainty in both its momentum and energy. However, compared to the case of an infinitely massive screen, it would lead to an improved prediction of where the diffracted electron will hit the photographic plate. Within the limits imposed by the uncertainty principle, argued Bohr, quantum mechanics was as complete a description of individual events as was possible.

Unimpressed by Bohr's reply, Einstein asked him to consider the possibility of controlling and measuring the transfer of momentum and energy between the screen and the particle, be it an electron or a photon, as it passed through the slit. Then, he argued, the state of the particle immediately afterwards could be determined with an accuracy greater than that allowed by the uncertainty principle. As the particle passes through the slit, said Einstein, it would be deflected and its trajectory towards the photographic plate would be determined by the law of conservation of momentum, which requires the sum total of the momenta of two bodies (particle and screen) that interact to remain constant. If the particle is deflected upwards, then the screen must be pushed downwards and vice versa.

Having used the moveable screen introduced by Bohr for his own ends, Einstein modified the imaginary experiment further by inserting a two-slit screen between the moveable screen and the photographic plate.

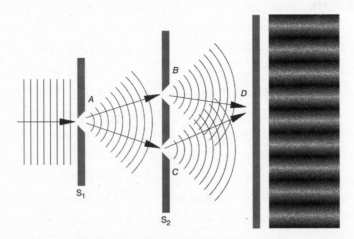

Figure 15: Einstein's two-slits thought experiment. At far right, the resulting interference pattern on the screen is shown

Einstein reduced the intensity of a beam until only one particle at a time passed through the slit in the first screen, S_1, and one of the two slits of the second screen, S_2, before hitting the photographic plate. As each particle left an indelible mark where it hit the plate, something remarkable would happen. What initially appeared to be a random sprinkling of specks was slowly transformed, as more and more particles left their imprint, by the laws of statistics into the characteristic interference pattern of light and dark bands. While each particle was responsible for only a single mark, it nevertheless contributed decisively through some statistical imperative to the overall interference pattern.

By controlling and measuring the transfer of momentum between the particle and the first screen it was possible, said Einstein, to determine if the particle was deflected towards the upper or lower slit in the second screen. From where it hit the photographic plate and the movement of the first screen, it was possible to trace through which of the two slits the particle had passed. It appeared that Einstein had devised an experiment in which it was possible to simultaneously determine the position and momentum of a particle with a greater precision than the uncertainty principle allowed. In the process he also seemed to have contradicted another fundamental tenet of the Copenhagen interpretation. Bohr's framework of complementarity posited that *either* particle-like *or* wave-like properties of an electron or a photon could be manifest in any given experiment.

There had to be a flaw in Einstein's argument, and Bohr set out to find it by sketching the sort of equipment needed to conduct the experiment. The apparatus he focused on was the first screen. Bohr realised that the control and measurement of the transfer of momentum between the particle and screen hinged on the screen's ability to move vertically. It is the observation of the screen moving either up or down as the particle passes through the slit that allows the determination of whether it passes through either the upper or lower slit in the second screen, after it strikes the photographic plate.

Einstein, despite his years at the Swiss Patent Office, had not considered the details of the experiment. Bohr knew that the quantum devil lay in the details. He replaced the first screen with one hanging by a pair of springs fixed to a supporting frame so that its vertical motion due to the transfer of momentum from a particle passing through the slit could be measured. The measuring device was simple: a pointer attached to the supporting frame and a scale engraved on the screen itself. It was crude, but sensitive enough to allow the observation of any individual interaction between screen and particle in an imaginary experiment.

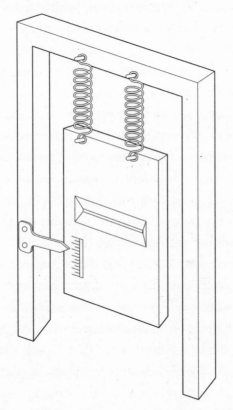

Figure 16: Bohr's design of a moveable first screen

Bohr argued that if the screen was already moving with an unknown velocity greater than any due to an interaction with a particle as it passed through the slit, then it would be impossible to ascertain the degree of momentum transfer and with it the trajectory of the particle. On the other hand, if it was possible to control and measure the transfer of momentum from particle to screen, the uncertainty principle implied a simultaneous uncertainty in the position of the screen and slit. However precise the measurement of the screen's vertical momentum, it was strictly matched, in accordance with the uncertainty principle, by a corresponding imprecision in the measurement of its vertical position.

Bohr went on to argue that the uncertainty in the position of the first screen destroys the interference pattern. For example, D on the photographic plate is a point of destructive interference, a dark spot in the interference pattern. A vertical displacement of the first screen would result in a change in the length of the two paths ABD and ACD. If the new lengths differed by half a wavelength, then instead of destructive interference there would be constructive interference and a bright spot at D.

To accommodate uncertainty in the vertical displacement of the first screen, S_1, requires an 'averaging' over all of its possible positions. This leads to interference somewhere between the extremes of total constructive and total destructive interference, resulting in a washed-out pattern on the photographic plate. Controlling the transfer of momentum from the particle to the first screen allows the trajectory of the particle through a slit in the second screen to be tracked; however, it destroys the interference pattern, argued Bohr. He concluded that Einstein's 'suggested control of momentum transfer would involve a latitude in the knowledge of the position of the diaphragm $[S_1]$ which would exclude the appearance of the interference phenomena in question'.[45] Bohr had not only defended the uncertainty principle but also the belief that the wave and particle aspects of a microphysical object cannot both appear in a single experiment, imaginary or not.

Bohr's rebuttal rested on the assumption that controlling and measuring the momentum transferred to S_1 accurately enough to determine the particle's direction afterwards results in an uncertainty in the position of S_1. The reason for this, Bohr explained, lay in reading the scale on S_1. To do so, it has to be illuminated, and that requires the scattering of photons from the screen and results in an uncontrollable transfer of momentum. This impedes the precise measurement of the momentum transferred from the particle to the screen as it passes through the slit. The only way to eliminate the impact of the photon is by not illuminating the scale at all, making it impossible to read. Bohr had resorted to employing the same concept of 'disturbance' that he had earlier criticised Heisenberg for using as an explanation of the origin of uncertainty in the microscope thought experiment.

There was another curious phenomenon associated with the two-slit experiment. If one of the two slits has a shutter that is closed, then the interference pattern disappears. Interference occurs only when both slits are open at the same time. But how was that possible? A particle can go through only one slit. How did the particle 'know' that the other slit was open or closed?

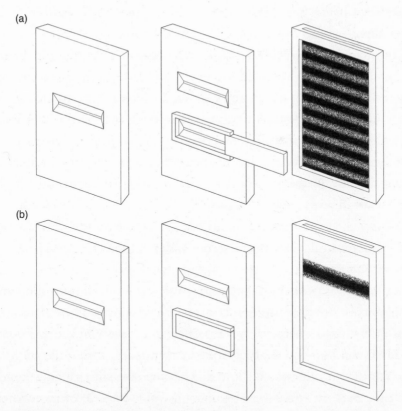

(a)

(b)

Figure 17: Two-slit experiment (a) with both slits open; (b) with one slit closed

Bohr had a ready answer. There was no such thing as a particle with a well-defined path. It was this lack of a definite trajectory that was behind the appearance of an interference pattern, even though it was particles, one at a time, which had passed through the two-slit set-up, and not waves. This quantum fuzziness enables a particle to 'sample' a variety of possible paths and so it 'knows' if one of the slits is open or closed. Whether it is open or not affects the particle's future path.

If detectors are placed in front of the two slits to sneak a look at which slit a particle is going to pass through, then it seems possible to close the other slit without affecting the particle's trajectory. When such a 'delayed-choice' experiment was later actually conducted, instead of an interference pattern there was an enlarged image of the slit. In trying to measure the position of the particle to establish through which slit it would pass, it is disturbed from its original course and the interference pattern fails to materialise.

The physicist has to choose, says Bohr, between '*either* tracing the path of a particle *or* observing interference effects'.[46] If one of the two slits of S_2 is closed, then the physicist knows through which slit the particle passed before hitting the photographic plate, but there will be no interference pattern. Bohr argues that this choice allows an 'escape from the paradoxical necessity of concluding that the behaviour of an electron or a photon should depend on the presence of a slit in the diaphragm [S_2] through which it could be proved not to pass'.[47]

The two-slit experiment was for Bohr 'a typical example' of the appearance of complementary phenomena under mutually exclusive experimental conditions.[48] Given the quantum mechanical nature of reality, he argued, light was neither a particle nor a wave. It was both, and sometimes it behaved like a particle and sometimes like a wave. On any given occasion, nature's answer to whether it was a particle or a wave simply depended on the question asked – on the type of experiment performed. An experiment to determine through which slit in S_2 a photon passed was a question that solicited a 'particle' answer and therefore no interference pattern. It was the loss of an independent, objective reality and not probability, God playing dice, that Einstein found unacceptable. Quantum mechanics, therefore, could not be the fundamental theory of nature that Bohr claimed it to be.

'Einstein's concern and criticism provided a most valuable incentive for us all to re-examine the various aspects of the situation as regards the description of atomic phenomena', recalled Bohr.[49] A major point of contention, he stressed, was 'the distinction between the objects under investigation and the measuring instruments which serve to define, in classical terms, the conditions under which the phenomena appear'.[50] In the Copenhagen interpretation the measuring instruments were inextricably linked with the object under investigation: no separation is possible.

While a microphysical object such as an electron was subject to the laws of quantum mechanics, the apparatus obeyed the laws of classical physics. Yet Bohr had to retreat in the face of Einstein's challenge as he applied the uncertainty principle to a macroscopic object, the first screen S_1. By doing so, Bohr had imperiously consigned an element of the large-scale world of the everyday to the realm of the quantum as he failed to establish where is 'the cut' between the classical and the quantum worlds, the border between the macro and micro. It would not be the last time that Bohr played a questionable move in his game of quantum chess with Einstein. The spoils for the victor were just too high.

Einstein spoke only once more during the general discussion, when he asked a question. De Broglie recalled later that 'Einstein said hardly anything beyond presenting a very simple objection to the probability interpretation' and then 'he fell back into silence'.[51] However, with all the participants staying at the Hotel Metropole, it was in its elegant art deco dining room that the keenest arguments took place, not in the conference room at the Institute of Physiology. 'Bohr and Einstein,' said Heisenberg, 'were in the thick of it all.'[52]

Surprisingly for an aristocrat, de Broglie spoke only French. He must have seen Einstein and Bohr deep in conversation in the dining room, with the likes of Heisenberg and Pauli listening closely. As they spoke in German, de Broglie did not realise that they were engaged in what Heisenberg called a 'duel'.[53] The acknowledged master of the thought experiment, Einstein would arrive at breakfast armed with a new proposal that challenged the uncertainty principle and with it the much-lauded consistency of the Copenhagen interpretation.

The analysis would begin over coffee and croissants. It continued as Einstein and Bohr headed to the Institute of Physiology, usually with Heisenberg, Pauli and Ehrenfest trailing alongside. As they walked and talked, assumptions were probed and clarified before the start of the morning session. 'During the meeting and particularly in the pauses we younger people, mostly Pauli and I, tried to analyse Einstein's experiment,' Heisenberg said later, 'and at lunch time the discussions continued between Bohr and the others from Copenhagen.'[54] Late in the afternoon, following further consultations among themselves, the collaborative effort would yield a rebuttal. During dinner back at the Metropole, Bohr would explain to Einstein why his latest thought experiment had failed to break the limits imposed by the uncertainty principle. Each time Einstein could find no fault with the Copenhagen response, but they knew, said Heisenberg, 'in his heart he was not convinced'.[55]

After several days, Heisenberg later recalled, 'Bohr, Pauli and I – knew that we could now be sure of our ground, and Einstein understood that the new interpretation of quantum mechanics cannot be refuted so simply'.[56] But Einstein refused to yield. Even if it failed to capture the essence of his rejection of the Copenhagen interpretation, he would say, 'God does not play dice'. 'But still, it cannot be for us to tell God, how he is to run the world', replied Bohr on one occasion.[57] 'Einstein, I am ashamed of you,' said Paul Ehrenfest only half-joking, 'you are arguing against the new quantum theory just as your opponents argue about relativity theory.'[58]

The only impartial witness to the private encounters between Einstein and Bohr at Solvay 1927 was Ehrenfest. 'Einstein's attitude gave rise to ardent discussions within a small circle, in which Ehrenfest, who through the years had been a close friend of us both,' recalled Bohr, 'took part in a most active and helpful way.'[59] A few days after the conference ended,

Ehrenfest wrote a letter to his students at Leiden University vividly describing the goings-on in Brussels: 'Bohr towering completely over everybody. At first not understood at all (Born was also there), then step by step defeating everybody. Naturally once again the awful Bohr incantation terminology. (Poor Lorentz as interpreter between the British and the French who were absolutely unable to understand each other. Summarizing Bohr. And Bohr responding with polite despair.) Every night at 1 a.m. Bohr came into my room just to say ONE SINGLE WORD to me, until 3 a.m. It was delightful for me to be present during the conversations between Bohr and Einstein. Like a game of chess. Einstein all the time with new examples. ... to break the UNCERTAINTY RELATION. Bohr from out of the philosophical smoke clouds constantly searching for the tools to crush one example after the other. Einstein like a jack-in-the-box, jumping out fresh every morning. Oh, that was priceless. But I am almost without reservation pro Bohr and contra Einstein.'[60] However, Ehrenfest admitted 'that he would not be able to find relief in his own mind before concord with Einstein was reached'.[61]

At Solvay 1927 the discussions with Einstein were conducted, Bohr said later, in 'a most humorous spirit'.[62] Yet he noted wistfully, 'a certain difference in attitude and outlook remained, since with his mastery for coordinating apparently contrasting experiences without abandoning continuity and causality, Einstein was perhaps more reluctant to renounce such ideals than someone for whom renunciation in this respect appeared to be the only way to proceed with the immediate task of coordinating the multifarious evidence regarding atomic phenomena, which accumulated from day to day in the exploration of this new field of knowledge.'[63] It was Einstein's very successes, implied Bohr, that kept him anchored in the past.

The fifth Solvay conference ended with Bohr, in the minds of those gathered in Brussels, having successfully argued for the logical consistency of the Copenhagen interpretation, but failing to convince Einstein that it was the only possible interpretation of what was a 'complete', closed theory.

On his journey home, Einstein travelled to Paris with a small group that included de Broglie. 'Carry on', he told the French prince as they parted company. 'You are on the right road.'[64] But de Broglie, disheartened at the lack of support in Brussels, would soon recant and accept the Copenhagen interpretation. When Einstein reached Berlin he was exhausted and subdued. Within a fortnight he wrote to Arnold Sommerfeld that quantum mechanics 'may be a correct theory of the statistical laws, but it is an inadequate conception of individual elementary processes'.[65]

While Paul Langevin later said that 'the confusion of ideas reached its zenith' at Solvay 1927, for Heisenberg this meeting of minds was the decisive turning point in establishing the correctness of the Copenhagen interpretation.[66] 'I am satisfied in every respect with the scientific results', he wrote as the conference ended.[67] 'Bohr's and my views have been generally accepted; at least serious objections are no longer being made, not even by Einstein and Schrödinger.' As far as Heisenberg was concerned, they had won. 'We could get anything clear by using the old words and limiting them by the uncertainty relations and still get a completely consistent picture', he recalled almost 40 years later. When asked whom he meant by 'we', Heisenberg replied: 'I could say that at that time it was practically Bohr, Pauli, and myself.'[68]

Bohr never used the term the 'Copenhagen interpretation', nor did anyone else until Heisenberg in 1955. Yet from a handful of adherents it quickly spread so that for most physicists the 'Copenhagen interpretation of quantum mechanics' became synonymous with quantum mechanics. Three factors lay behind this rapid dissemination and acceptance of the 'Copenhagen spirit'. The first was the pivotal role of Bohr and his institute. Inspired by his stay in Rutherford's laboratory in Manchester as a young postdoctoral student, Bohr had managed to create an institute of his own with the same zing in the air – the sense that anything was possible.

'Bohr's Institute quickly became the world centre of quantum physics, and to paraphrase the old Romans, "all roads lead to Blegdamsvej 17"', recalled the Russian George Gamov who arrived there in the summer of 1928.[69] The Kaiser Wilhelm Institute of Theoretical Physics of which

Einstein was the director existed only on paper, and he preferred it that way. While he usually worked alone, or later with an assistant who carried out the calculations, Bohr fathered many scientific children. The first to rise to prominence and positions of authority were Heisenberg, Pauli and Dirac. Though only young men, as Ralph Kronig later recalled, other young physicists did not dare to go against them. Kronig, for one, had not published the idea of electron spin after Pauli ridiculed it.

Secondly, around the time of Solvay 1927 a number of professorships became vacant. Those who had helped create the new physics filled nearly all of these. The institutes they headed quickly began to attract many of best and brightest students from Germany and across Europe. Schrödinger had secured the most prestigious position, as Planck's successor in Berlin. Immediately after the Solvay conference, Heisenberg arrived in Leipzig to take up his post as professor and director of the institute for theoretical physics. Within six months, in April 1928, Pauli moved from Hamburg to a professorship at the EHT in Zurich. Pascual Jordan, whose mathematical skills had been vital to the development of matrix mechanics, succeeded Pauli in Hamburg. Before long, through regular visits and the exchange of assistants and students between each other and Bohr's institute, Heisenberg and Pauli established Leipzig and Zurich as centres of quantum physics. With Kramers already installed at the University of Utrecht and Born at Göttingen, the Copenhagen interpretation soon became quantum dogma.

Lastly, despite their differences, Bohr and his younger associates always presented a united front against all challenges to the Copenhagen interpretation. The one exception was Paul Dirac. Appointed Lucasian Professor of Mathematics at Cambridge University in September 1932, a chair once occupied by Isaac Newton, Dirac was never interested in the question of interpretation. It seemed to him to be a pointless preoccupation that led to no new equations. Tellingly, he called himself a mathematical physicist, whereas neither his contemporaries Heisenberg and Pauli nor Einstein and Bohr ever described themselves as such. They were theoretical physicists to a man, as was Lorentz, the acknowledged elder statesman of the clan who died in February 1928. 'To me personally,' Einstein wrote later, 'he meant more than all the others encountered in my lifetime.'[70]

Soon Einstein's own health became a matter of concern. In April 1928 during a short visit to Switzerland he collapsed as he carried his suitcase up a steep hill. At first it was thought that he had suffered a heart attack, but then an enlargement of the heart was diagnosed. Later Einstein told his friend Michele Besso that he had felt 'close to croaking', before adding, 'which of course one shouldn't put off unduly'.[71] Once back in Berlin under Elsa's watchful eye, visits by friends and colleagues were strictly rationed. She was once more Einstein's gatekeeper and nurse, as she had been after he had fallen ill following his Herculean effort in formulating the general theory of relativity. This time Elsa needed help and hired a friend's unmarried sister. Helen Dukas was 32 and became Einstein's trusted secretary and friend.[72]

As he recuperated, a paper by Bohr was published in three languages: English, German and French. The English version, entitled 'The Quantum Postulate and the Recent Development of Atomic Theory', appeared on 14 April 1928. A footnote stated: 'The content of this paper is essentially the same as that of a lecture on the present state of quantum theory delivered on September 16, 1927, at the Volta celebrations in Como.'[73] In truth, Bohr had produced a more refined and advanced exposition of his ideas surrounding complementarity and quantum mechanics than he had presented in either Como or Brussels.

Bohr sent a copy to Schrödinger, who replied: 'if you want to describe a system, e.g. a mass point by specifying its [momentum] p and [position] q, then you find that this description is only possible with a limited degree of accuracy.'[74] What was therefore needed, Schrödinger argued, was the introduction of *new* concepts with respect to which this limitation *no longer* applies. 'However,' he concluded, 'it will no doubt be *very* difficult to invent this conceptual scheme, since – as you emphasize so impressively – the new-fashioning required touches upon the deepest levels of our experience: space, time and causality.'

Bohr wrote back thanking Schrödinger for his 'not altogether unsympathetic attitude', but he did not see the need for 'new concepts' in quantum theory since the *old* empirical concepts appeared inseparably linked to the 'foundations of the human means of visualization'.[75] Bohr restated his position that it was not a question of a more or less arbitrary limitation in the applicability of the classical concepts, but an inescapable feature of complementarity that emerges in an analysis of the concept of observation. He ended by encouraging Schrödinger to discuss the contents of his letter with Planck and Einstein. When Schrödinger informed him of the exchange with Bohr, Einstein replied that the 'Heisenberg-Bohr tranquilizing philosophy – or religion? – is so delicately contrived that for the time being, it provides a gentle pillow for the true believer from which he cannot very easily be aroused. So let him lie there.'[76]

Four months after collapsing, Einstein was still weak but no longer confined to his bed. To continue his convalescence he rented a house in the sleepy town of Scharbeutz on the Baltic coast. There he read Spinoza and enjoyed being away from the 'idiotic existence one leads in the city'.[77] It was almost a year before he was well enough to return to his office. He would work there all morning before going home for lunch and a rest until three o'clock. 'Otherwise he was always working,' recalled Helen Dukas, 'sometimes all through the night.'[78]

During the Easter vacation of 1929 Pauli went to see Einstein in Berlin. He found Einstein's 'attitude regarding modern quantum physics reactionary' because he continued to believe in a reality where natural phenomena unfolded according to the laws of nature, independently of an observer.[79] Shortly after Pauli's visit, Einstein made his views perfectly clear as he received the Planck medal from Planck himself. 'I admire to the highest degree the achievements of the younger generation of physicists which goes by the name quantum mechanics and believe in the deep level of truth of that theory,' he told the audience, 'but I believe that the restriction to statistical laws will be a passing one.'[80] Einstein had already embarked on his solitary journey in search of a unified field theory that he believed would save causality and an observer-independent reality. In the meantime

he would continue to challenge what was becoming the quantum orthodoxy, the Copenhagen interpretation. When they met again in Brussels
at the sixth Solvay conference in 1930, Einstein presented Bohr with an
imaginary box of light.

Chapter 12

EINSTEIN FORGETS RELATIVITY

Bohr was stunned. Einstein smiled.

Over the past three years, Bohr had re-examined the imaginary experiments Einstein had proposed at the Solvay conference in October 1927. Each was designed to show that quantum mechanics was inconsistent, but he had found the flaw in Einstein's analysis in every case. Not content to rest on his laurels, Bohr devised some thought experiments of his own involving an assortment of slits, shutters, clocks and the like as he probed his interpretation for any weaknesses. He found none. But Bohr never conjured up anything as simple and ingenious as the thought experiment that Einstein had just finished describing to him in Brussels at the sixth Solvay conference.

The theme of the six-day meeting that began on 20 October 1930 was the magnetic properties of matter. The format remained the same: a series of commissioned reports on various topics related to magnetism, each followed by a discussion. Bohr had joined Einstein as a member of the nine-strong scientific committee and both were therefore automatically invited to the conference. After the death of Lorentz, the Frenchman Paul Langevin had agreed to take on the demanding dual responsibilities of presiding over the committee and the conference. Dirac, Heisenberg, Kramers, Pauli and Sommerfeld were among the 34 participants.

As a meeting of minds it was a close second to Solvay 1927, with twelve current and future Nobel laureates present. It was the backdrop to the 'second round' of the ongoing struggle between Einstein and Bohr over the meaning of quantum mechanics and the nature of reality. Einstein had travelled to Brussels armed with a new thought experiment designed to deliver a fatal blow to the uncertainty principle and the Copenhagen interpretation. An unsuspecting Bohr was ambushed after one of the formal sessions.

———

Imagine a box full of light, Einstein asked Bohr. In one of its walls is a hole with a shutter that can be opened and closed by a mechanism connected to a clock inside the box. This clock is synchronised with another in the laboratory. Weigh the box. Set the clock to open the shutter at a certain time for the briefest of moments, but long enough for a single photon to escape. We now know, explained Einstein, precisely the time at which the photon left the box. Bohr listened unconcerned; everything Einstein had proposed appeared straightforward and beyond contention. The uncertainty principle applied only to pairs of complementary variables – position and momentum or energy and time. It did not impose any limit on the degree of accuracy with which any one of the pair could be measured. Just then, with a hint of smile, Einstein uttered the deadly words: weigh the box again. In a flash, Bohr realised that he and the Copenhagen interpretation were in deep trouble.

To work out how much light had escaped locked up in a single photon, Einstein used a remarkable discovery he had made while still a clerk at the Patent Office in Bern: energy is mass and mass is energy. This astonishing spin-off from his work on relativity was captured by Einstein in his simplest and most famous equation: $E=mc^2$, where E is energy, m is mass, and c is the speed of light.

By weighing the box of light before and after the photon escapes, it is easy to work out the difference in mass. Although such a staggeringly small change was impossible to measure using equipment available in 1930, in the realm of the thought experiment it was child's play. Using $E=mc^2$ to convert the quantity of missing mass into an equivalent amount of energy, it was possible to calculate precisely the energy of the escaped photon. The time of the photon's escape was known via the laboratory clock being synchronised with the one inside the light box controlling the shutter. It appeared that Einstein had conceived an experiment capable of measuring simultaneously the energy of the photon and the time of its escape with a degree of accuracy proscribed by Heisenberg's uncertainty principle.

'It was quite a shock for Bohr', recalled the Belgian physicist Léon Rosenfeld, who had recently begun what turned into a long-term collaboration with the Dane.[1] 'He did not see the solution at once.' While Bohr was desperately worried by Einstein's latest challenge, Pauli and Heisenberg were dismissive. 'Ah, well, it will be all right, it will be all right', they told him.[2] 'During the whole evening he was extremely unhappy, going from one to the other and trying to persuade them that it couldn't be true, that it would be the end of physics if Einstein were right,' recalled Rosenfeld, 'but he couldn't produce any refutation.'[3]

Rosenfeld was not invited to Solvay 1930, but had travelled to Brussels to meet Bohr. He never forgot the sight of the two quantum adversaries heading back to the Hotel Metropole that evening: 'Einstein, a tall majestic figure, walking quietly, with a somewhat ironical smile on his face, and Bohr trotting near him, very excited, ineffectually pleading that if Einstein's device would work, it would mean the end of physics.'[4] For Einstein it was neither an end nor a beginning. It was nothing more than a demonstration that quantum mechanics was inconsistent and therefore not the closed and complete theory that Bohr claimed. His latest thought experiment was simply an attempt to rescue the kind of physics that aimed to understand an observer-independent reality.

A photograph shows Einstein and Bohr walking together, but slightly out of step. Einstein is just ahead as if trying to flee. Bohr, mouth open, is hurrying to keep pace. He leans towards Einstein, desperate to make himself heard. Despite having his coat draped over his left arm, Bohr gestures with his left forefinger to emphasise whatever point he is trying to make. Einstein's hands are by his side, one clutching a briefcase and the other a possible victory cigar. As he listens, Einstein's moustache fails to hide the half-knowing smile of a man who thinks he has just gained the upper hand. That evening, said Rosenfeld, Bohr looked 'like a dog who has received a thrashing'.[5]

Bohr spent a sleepless night examining every facet of Einstein's thought experiment. He took the imaginary box of light apart to find the flaw that he hoped existed. Einstein did not picture, even in his mind's eye, either the details of the inner workings of the light box or how to weigh it. Bohr, desperate to get to grips with the device and the measurements that would have to be made, drew what he called a 'pseudorealistic' diagram of the experimental set-up to help him.

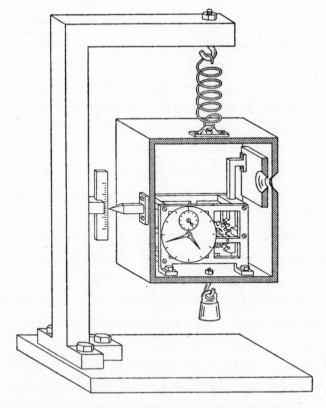

Figure 18: Bohr's later rendition of Einstein's 1930s light box
(Niels Bohr Archive, Copenhagen)

Given the need to weigh the light box before the shutter is opened at a pre-set time and after the photon has escaped, Bohr decided to focus on the weighing process. With mounting anxiety and little time, he chose the simplest possible method. He suspended the light box from a spring fixed to a supporting frame. To turn it into a weighing scale, Bohr attached a pointer to the light box so its position could be read on a scale attached to the vertical arm of what resembled a hangman's gallows. To ensure that the pointer was positioned at zero on the scale, Bohr attached a small weight to the bottom of the box. There was nothing whimsical in the construction, as Bohr included even the nuts and bolts used to fix the frame to a base, and drew the clockwork mechanism controlling the opening and closing of the hole through which the photon was to escape.

The initial weighing of the light box is simply the configuration with the attached weight chosen to ensure that the pointer is at zero. After the photon escapes, the light box is lighter and is pulled upwards by the spring. To reposition the pointer at zero, the attached weight has to be replaced by a slightly heavier one. There is no time limit on how long the experimenter can take to change the weights. The difference in the weights is the mass lost due to the escaped photon, and from $E=mc^2$ the energy of the photon can be calculated precisely.

From the arguments he deployed at Solvay 1927, Bohr held that any measurement of the position of the light box would lead to an inherent uncertainty in its momentum, because to read the scale would require it to be illuminated. The very act of measuring its weight would cause an uncontrollable transfer of momentum to the light box because of the exchange of photons between the pointer and the observer causing it to move. The only way to improve the accuracy of the position measurement was to carry out the balancing of the light box, the positioning of the pointer at zero, over a comparatively long time. However, Bohr argued that this would lead to a corresponding uncertainty in the momentum of the box. The more accurately the position of the box was measured, the greater the uncertainty attached to any measurement of its momentum.

Unlike at Solvay 1927, Einstein was attacking the energy–time uncertainty relation, not the position–momentum incarnation. It was now, in the early hours of the morning, that a tired Bohr suddenly saw the flaw in Einstein's *gedankenexperiment*. He reconstructed the analysis bit by bit until he was satisfied that Einstein had indeed made an almost unbelievable mistake. Relieved, Bohr went to sleep for a few hours, knowing that when he awoke it would be to savour his triumph over breakfast.

In his desperation to destroy the Copenhagen view of quantum reality, Einstein had forgotten to take into account his own theory of general relativity. He had ignored the effects of gravity on the measurement of time by the clock inside the light box. General relativity was Einstein's greatest achievement. 'The theory appeared to me then, and it still does, the greatest feat of human thinking about Nature, the most amazing combination of philosophic penetration, physical intuition, and mathematical skill', said Max Born.[6] He called it 'a great work of art, to be enjoyed and admired from a distance'. When the bending of light predicted by general relativity was confirmed in 1919, it made headlines around the world. J.J. Thomson told one British newspaper that Einstein's theory was 'a whole new continent of new scientific ideas'.[7]

One of these new ideas was gravitational time dilation. Two identical and synchronised clocks in a room with one fixed to the ceiling and the other on the floor would be out of step by 300 parts in a billion billion, because time flows more slowly at the floor than at the ceiling.[8] The reason was gravity. According to general relativity, Einstein's theory of gravity, the rate at which a clock ticks depends upon its position in a gravitational field. Also, a clock moving in a gravitational field ticks slower than one that is stationary. Bohr realised that this implied that weighing the light box affected the time-keeping of the clock inside.

The position of the light box in the earth's gravitational field is altered by the act of measuring the pointer against the scale. This change in position would alter the rate of the clock and it would no longer be synchronised with the clock in the laboratory, making it impossible to measure as accurately as Einstein presumed the precise time the shutter opened and the photon escaped from the box. The greater the accuracy in measuring the energy of the photon, via $E=mc^2$, the greater the uncertainty in the position of the light box within the gravitational field. This uncertainty of position prevents, due to gravity's ability to affect the flow of time, the determination of the exact time the shutter opens and the photon escapes. Through this chain of uncertainties Bohr showed that Einstein's light box experiment could not simultaneously measure exactly both the energy of the photon and the time of its escape.[9] Heisenberg's uncertainty principle remained intact, and with it the Copenhagen interpretation of quantum mechanics.

When Bohr came down to breakfast he was no longer looking 'like a dog who has received a thrashing' the night before. Now it was Einstein who was stunned into silence as he listened to Bohr explain why his latest challenge, like those of three years earlier, had failed. Later there would be those who questioned Bohr's refutation because he had treated macroscopic elements such as the pointer, the scale, and the light box as if they were quantum objects and therefore subject to limitations imposed by the uncertainty principle. To handle macroscopic objects in this way ran counter to his insistence that laboratory equipment be treated classically. But Bohr had never been particularly clear about where to draw the line between the micro and macro, since in the end every classical object is nothing but a collection of atoms.

Whatever reservations some had later, Einstein accepted Bohr's counter-arguments, as did the physics community at the time. As a result he ceased his attempts to circumvent the uncertainty principle to demonstrate that quantum mechanics was logically inconsistent. Instead Einstein would henceforth focus on exposing the theory as incomplete.

In November 1930 Einstein lectured in Leiden on the light box. Afterwards a member of the audience argued that there was no conflict within quantum mechanics. 'I know, this business is free of contradictions,' replied Einstein, 'yet in my view it contains a certain unreasonableness.'[10] In spite of this, in September 1931, he once again nominated Heisenberg and Schrödinger for a Nobel Prize. But after going two rounds with Bohr and his seconds at the Solvay conferences, one sentence in Einstein's letter of nomination was telling: 'In my opinion, this theory contains without doubt a piece of the ultimate truth.'[11] His 'inner voice' continued to whisper that quantum mechanics was incomplete, that it was not the 'whole' truth as Bohr would have everyone believe.

At the end of the 1930 Solvay conference, Einstein travelled to London for a few days. He was the guest of honour at a fundraising dinner on 28 October for the benefit of impoverished eastern European Jews. Held at the Savoy Hotel, and hosted by Baron Rothschild, the fundraiser drew almost a thousand people. With the great and the good elegantly dressed, Einstein willingly donned white tie and tails to play his part in what he called the 'monkey comedy' if it helped open wallets.[12] George Bernard Shaw was the master of ceremonies.

Although he occasionally departed from his prepared script, the 74-year-old Shaw gave a virtuoso performance that began with him complaining that he had to talk about 'Ptolemy and Aristotle, Kepler and Copernicus, Galileo and Newton, gravitation and relativity and modern astrophysics and Heavens knows what …'[13] Then, with his usual wit, Shaw summarised everything in three sentences: 'Ptolemy made a universe, which lasted 1,400 years. Newton, also, made a universe, which lasted for 300 years. Einstein has made a universe, and I can't tell you how long that will last.'[14] The guests laughed, none louder than Einstein. After comparing the achievements of Newton and Einstein, Shaw ended with a toast: 'I drink to the greatest of our contemporaries, Einstein!'[15]

It was a difficult act to follow, but Einstein was every bit as much the showman when the occasion demanded. He expressed his gratitude to Shaw for 'the unforgettable words which you have addressed to my mythical namesake who makes life so difficult for me'.[16] He offered words of praise to Jews and Gentiles alike 'of noble spirit and with a strong sense of justice, who had devoted their lives to uplifting human society and liberating the individual from degrading oppression'. 'To you all I say,' knowing that he was addressing a sympathetic audience, 'that the existence and destiny of our people depends less on external factors than on us remaining faithful to the moral traditions which have enabled us to survive for thousands of years despite the fierce storms that have broken over our heads.' 'In the service of life,' Einstein added, 'sacrifice becomes grace.'[17] Words said in hope would, for millions, soon be put to the test as the dark clouds of the coming Nazi storm gathered.

Six weeks earlier, on 14 September, the Nazis had gained 6.4 million votes in the Reichstag elections. The size of the Nazi vote stunned many. In May 1924 the party had won 32 seats, and in the December elections that same year, just fourteen. In May 1928 they did even worse, winning a mere twelve seats and 812,000 votes. The result seemed to confirm the Nazis as just another far-right fringe group. Now, little more than two years later, they had increased their share of the vote eight-fold and were the second-largest party in the Reichstag with 107 deputies.[18]

Einstein was not alone in believing that 'the Hitler vote is only a symptom, not necessarily of anti-Jewish hatred but of momentary resentment caused by economic misery and unemployment within the ranks of misguided German youths'.[19] However, only about one quarter of those who voted Nazi were young first-time voters. It was among the older generation of white-collar workers, shopkeepers, small businessmen, Protestant farmers in the north, craftsmen, and unskilled workers outside the industrial centres that Nazi support was strongest. What contributed decisively to the changed German political landscape between the elections of 1928 and 1930 was the Wall Street Crash in October 1929.

Germany was hardest hit by the financial shockwaves emanating from New York. The lifeblood of its fragile economic revival of the past five years had been short-term loans from the United States. With mounting losses, and in disarray, American financial institutions demanded immediate repayment of existing loans. The result was a rapid rise in unemployment from 1.3 million in September 1929 to over 3 million in October 1930. Einstein for the moment saw the Nazis as nothing more than a 'childish disease of the Republic' that would soon pass.[20] The disease, however, would kill off an already ailing Weimar Republic that had in all but name abandoned parliamentary democracy in favour of rule by decree.

'We are moving toward bad times', wrote a pessimistic Sigmund Freud on 7 December 1930.[21] 'I ought to ignore it with the apathy of old age, but I can't help feeling sorry for my seven grandchildren.' Five days earlier, Einstein had left Germany to spend two months at Caltech, the California Institute of Technology in Pasadena. Boltzmann, Schrödinger and Lorentz had all lectured at what had fast become one America's leading centres of scientific excellence. When his ship docked in New York, Einstein was persuaded to give a fifteen-minute press conference to the horde of waiting reporters. 'What do you think of Adolf Hitler?' shouted one. 'He is living on the empty stomach of Germany', replied Einstein. 'As soon as the economic conditions improve, he will no longer be important.'[22]

A year later, in December 1931, when he set off for a second stint at Caltech, Germany was in an even deeper economic depression and greater political turmoil. 'I decided today that I shall essentially give up my Berlin position and shall be a bird of passage for the rest of my life', Einstein wrote in his diary as he crossed the Atlantic.[23] While in California, Einstein happened to meet Abraham Flexner, who was in the process of establishing a unique research centre, the Institute for Advanced Study, in Princeton, New Jersey. Armed with a $5 million donation, Flexner wanted to create a 'society of scholars' devoted entirely to research, freed from the demands of teaching students. Serendipitously meeting Einstein, Flexner wasted little time in taking the first steps that eventually led to the recruitment of the world's most celebrated scientist.

Einstein agreed to spend five months a year at the institute and the remainder in Berlin. 'I am not abandoning Germany', he told the *New York Times*.[24] 'My permanent home will still be in Berlin.' The five-year arrangement would begin in the autumn of 1933 because Einstein had already committed himself to another spell at Caltech. He was fortunate that he had, for it was during this third visit to Pasadena that Hitler was appointed Chancellor on 30 January 1933. For Germany's half-million Jews, the exodus began slowly, with only 25,000 leaving by June. Einstein, safely in California, did not speak out, but acted as if he would return when the time came. He wrote to the Prussian Academy asking about his salary, but had already made his decision. 'In view of Hitler,' he wrote to a friend on 27 February, 'I don't dare step on German soil.'[25] That very day the Reichstag was set alight. It signalled the beginning of the first wave of state-sponsored Nazi terror.

In the midst of the violence unleashed by the Nazis, 17 million voted for them in the Reichstag election on 5 March. Five days later, on the eve of his planned departure from Pasadena, Einstein gave an interview and made public what he thought about events in Germany. 'As long as I have any choice in the matter,' he said, 'I shall live only in a country where civil liberty, tolerance and equality of all citizens before the law prevail. Civil liberty implies freedom to express one's political convictions, in speech and in writing; tolerance implies respect for the convictions of others whatever they may be. These conditions do not exist in Germany at the present time.'[26] As his words were reported around the world, he was condemned in the German press as newspapers vied to demonstrate their allegiance to the Nazi regime. 'Good News of Einstein – He Is Not Coming Back!' read the headline in the *Berliner Lokalanzeiger*. The article seethed at how 'this puffed up bit of vanity dared to sit in judgement on Germany without knowing what is going on here – matters that forever must remain incomprehensible to a man who was never German in our eyes and who declares himself to be a Jew and nothing but a Jew'.[27]

Einstein's comments left Planck in a quandary. On 19 March he wrote to Einstein of his 'profound distress' over 'all kinds of rumours which have emerged in this unquiet and difficult time about your public and private statements of a political nature'.[28] Planck complained that 'these reports make it exceedingly difficult for all those who esteem and revere you to stand up for you'. He blamed Einstein for making the difficult situation of his 'tribal companions and co-religionists' worse. When his ship docked at Antwerp in Belgium on 28 March, Einstein asked to be driven to the German embassy in Brussels. There he surrendered his passport, renounced his German citizenship for a second time, and handed over a letter of resignation from the Prussian Academy.

While he pondered what to do and where to go, Einstein and Elsa moved into a villa in the small resort of Le Coq-sur-Mer on the Belgian coast. As rumours circulated that Einstein's life might be at risk, the Belgian government assigned two guards to protect him. In Berlin, Planck was relieved when he learnt of Einstein's resignation. It was the only honourable way to sever ties with the Academy and 'at the same time save your friends from an immeasurable amount of grief and pain', he wrote to Einstein.[29] There were few prepared to stand up for him in the new Germany.

On 10 May 1933, swastika-clad students and academics carrying torches marched down Unter den Linden to the Opernplatz just across from Berlin University's main entrance and set fire to some 20,000 books plundered from the shelves of the city's libraries and bookstores. A crowd of 40,000 watched as the flames consumed the 'un-German' and 'Jewish-Bolshevik' works by the likes of Marx, Brecht, Freud, Zola, Proust, Kafka, and Einstein. It was a scene repeated in every major university town in the country, and men like Planck read the smoke signals and did little, if anything, to resist. The book-burning was just the beginning of the Nazi assault on 'degenerate' art and culture, but a far more significant event had already occurred for German Jews when anti-Semitism was effectively legalised.

The 'Law for the Restoration of the Career Civil Service', passed on 7 April, applied to some 2 million state employees. The law was designed to target the Nazis' political opponents, socialists, communists, and the Jews. Paragraph 3 contained the infamous 'Aryan clause': 'Civil servants not of Aryan origin are to retire.'[30] The law defined a non-Aryan as a person who had one parent or grandparent who was not Aryan. Sixty-two years after their emancipation in 1871, German Jews were once again the subject of legalised state discrimination. It was the springboard for the Nazi persecution of the Jews that followed.

Universities were state institutions, and soon more than a thousand academics, including 313 professors, were dismissed or resigned. Almost a quarter of the pre-1933 physics community was forced into exile, including half of all theorists. By 1936 more than 1,600 scholars had been ousted; a third of these were scientists, including twenty who had been or would be awarded the Nobel Prize: eleven in physics, four in chemistry, five in medicine.[31] Formally, the new law did not apply to those employed before the First World War, or who were veterans of that war, or anyone who had lost a father or son during the war. But as the Nazi purge of the civil service continued unabated and claimed an increasing number who were exempt, on 16 May 1933 Planck, as president of the Kaiser Wilhelm Society, went to see Hitler. He thought he could limit the damage being done to German science.

Incredibly, Planck told Hitler that 'there are different sorts of Jews, some valuable for mankind and others worthless', and that 'distinctions must be made'.[32] 'That's not right', said Hitler.[33] 'A Jew is a Jew; all Jews stick together like leeches. Wherever there is one Jew, other Jews of all sorts immediately gather.' His opening gambit having failed, Planck changed tack. The wholesale expulsion of Jewish scientists would be harmful to Germany's interests, argued Planck. Hitler flew into a rage at the very suggestion: 'Our national policies will not be revoked or modified, even for scientists.' 'If the dismissal of Jewish scientists means the annihilation of contemporary German science, then we shall do without science for a few years!'[34]

In November 1918, in the immediate aftermath of defeat, Planck had rallied the dispirited members of the Prussian Academy of Sciences: 'If the enemy has taken from our fatherland all defence and power, if severe domestic crises have broken in upon us and perhaps still more severe crises stand before us, there is one thing which no foreign or domestic enemy has yet taken from us: that is the position which German science occupies in the world.'[35] For Planck, who had lost his eldest son on the battlefield, all the sacrifices had to be worth something. As his disastrous meeting with Hitler came to an abrupt end, Planck knew that the Nazis were on the verge of achieving what no one else had: the destruction of German science.

Two weeks earlier, the Nazi physicist and Nobel laureate Johannes Stark had been appointed director of the Physikalisch-Technische Reichsanstalt, the Imperial Institute of Physics and Technology. Soon Stark wielded even greater power in the service of 'Aryan physics', as he was placed in charge of disbursing government research funds. From these positions of power he was determined to exact revenge. In 1922 he had stepped down from his professorship at the University of Würzburg to try his hand at business. Anti-Semitic, dogmatic and quarrelsome, Stark had alienated virtually everyone bar the like-minded fellow Nobel laureate and Nazi Philipp Lenard, the leading and long-time proponent of so-called 'Deutsch Physik'. When Stark wanted to return to academia after the failure of his business venture, no one who was in a position to do so was prepared to offer him a job. Already bitterly opposed to the 'Jewish physics' of Einstein and dismissive of modern theoretical physics, Stark was determined to have a say in all appointments to professorial chairs of physics and lobbied to have them occupied by supporters of 'German physics'.

Heisenberg had long wanted to be Sommerfeld's successor at Munich. In 1935 Stark called Heisenberg the 'spirit of Einstein's spirit' and launched a concerted campaign against him and theoretical physics. It culminated on 15 July 1937 with the publication of an article in the SS journal, *Das Schwarze Korps*, in which Heisenberg was branded a 'white Jew'. He spent the next year trying to remove the slur that, if it stuck, would place him in real danger of being isolated and dismissed. He turned to Heinrich Himmler, head of the SS, who happened to be a family acquaintance. Himmler exonerated Heisenberg, but blocked his appointment as Sommerfeld's successor. There was also a proviso that in future he should 'clearly separate for your audiences, in the acknowledgement of scientific research results, the personal and political characteristics of the researcher'.[36] Heisenberg duly obliged in separating the scientist from the science. There would be no more mention by him of Einstein's name in public.

The Göttingen physicists James Franck and Max Born were exempt as war veterans from the 'Aryan clause'. But neither man chose to exercise his right, believing that to do so was tantamount to collusion with the Nazis. Franck was condemned by no fewer than 42 of his colleagues when he submitted his letter of resignation, for fuelling anti-German propaganda by stating that 'we Germans of Jewish descent are being treated as aliens and enemies of the Fatherland'.[37] Born, who had no intention of resigning, discovered his name on a list of suspended civil servants published in the local newspaper. 'All I had built up in Göttingen, during twelve years hard work, was shattered', he wrote later.[38] 'It seemed to me like the end of the world.' He shuddered at the thought of 'standing in front of students who, for whatever reason, have thrown me out, or living among colleagues who were able to live with this so easily'.[39]

Suspended but not yet sacked, Born had never felt particularly Jewish, he confessed to Einstein. But now he was 'extremely conscious of it, not only because we are considered to be so, but because oppression and injustice provoke me to anger and resistance'.[40] Born hoped to settle in England, 'for the English seem to be accepting the refugees most nobly and generously'.[41] His wish was granted when he was offered a three-year lectureship at Cambridge University. Believing that he might be depriving a deserving English physicist, Born accepted only after being reassured that the post had been created especially for him. He was one of the lucky few whose contributions to physics were internationally recognised, unlike the 'young ones' for whom Einstein said his 'heart aches'.[42] But even scientists of Born's calibre had to endure periods of deep uncertainty about their future. After his time in Cambridge was up, Born spent six months in Bangalore, India and was seriously considering a post in Moscow, when in 1936 he was offered the chair of natural philosophy at the University of Edinburgh.

Heisenberg had tried to convince Born that he was safe, since 'only the very least are affected by the law – you and Franck certainly not'. He hoped, like others, that things would eventually settle down and 'the political revolution could take place without any damage to Göttingen physics'.[43] But the damage was already done. It had taken the Nazis a matter of weeks to transform Göttingen, the cradle of quantum mechanics, from a great university to a second-rate institution. The Nazi minister of education asked David Hilbert, the most fêted mathematician in Göttingen, whether it was true 'that your Institute suffered so much from the departure of the Jews and their friends?' 'Suffered? No, it didn't suffer, Herr Minister', replied Hilbert. 'It just doesn't exist any more.'[44]

As news spread of what was happening in Germany, scientists and their professional bodies quickly swung into action to help colleagues fleeing Nazi oppression with money and jobs. Aid organisations supported by gifts and donations from individuals and private foundations were set up. In England, the Academic Assistance Council, with Rutherford as its president, was established in May 1933 as a 'clearing house' that found temporary posts and offered help for refugee scientists, artists, and writers. Many initially escaped to Switzerland, Holland or France and stayed only a short while before travelling on to Britain and the United States.

In Copenhagen, Bohr's institute became a staging post for many physicists. In December 1931, the Danish Academy of Sciences and Letters had chosen Bohr as the next occupant of the Aeresbolig, 'The House of Honour', a mansion built by the founder of the Carlsberg breweries. His new status as Denmark's leading citizen meant he enjoyed even more influence at home and abroad, which he exercised to help others. In 1933 he and his brother Harald helped set up 'The Danish Committee for Support of Intellectual Workers in Exile'. Through colleagues and former students, Bohr was able to get new posts established or have vacancies filled by refugees. It was Bohr who got James Franck to Copenhagen on a three-year visiting professorship in April 1934. After a year or so, Franck moved on to a tenured position in the United States, which, along with Sweden, was the final destination of many who arrived in Denmark. One man who did not have to worry about a job was Einstein.

In early September, as fears for his safety in Belgium grew, Einstein left for England. For the next month he kept a low profile, staying in a cottage on the Norfolk coast. Soon the tranquillity by the seaside was shattered when he learnt that Paul Ehrenfest, in a fit of despair while estranged from his wife, had committed suicide. It happened during a visit to an Amsterdam hospital to see his sixteen-year-old son Vassily, who suffered from Down's syndrome. Einstein was shocked at the news that Ehrenfest had also shot Vassily. Remarkably, the boy survived but was blinded in one eye.

Although deeply upset at Ehrenfest's suicide, Einstein's thoughts soon turned to the speech he had agreed to give at a fundraising rally highlighting the plight of refugees. The meeting, chaired by Rutherford, took place on 3 October at the Royal Albert Hall. A public eager to get a glimpse of the great man meant that there was not even standing room on the night. Einstein succeeded in addressing the audience of 10,000 in his heavily accented English without once mentioning Germany by name, at the request of the organisers. For the Refugee Assistance Council believed that 'the issue raised at the moment is not a Jewish one alone; many who have suffered or are threatened had no Jewish connection'.[45] Four days later, on the evening 7 October, Einstein left for America. Due to spend the next five months at the Institute for Advanced Study, he never returned to Europe.

As he was being driven from New York to Princeton, Einstein was handed a letter from Abraham Flexner. The institute's director was asking him not to attend any public events and to exercise discretion for own his safety. The reason Flexner gave was the danger posed to Einstein by the 'bands of irresponsible Nazis' to be found in America.[46] Yet his real concern was the damage that Einstein's public statements might inflict on the reputation of his fledgling institute, and therefore on the donations it relied on. Within a matter of weeks, Einstein found Flexner's restrictions and increasing interference suffocating. Once he even gave his new address as 'Concentration camp, Princeton'.[47]

Einstein wrote to the trustees of the institute to complain of Flexner's behaviour, and asked them to guarantee him 'security for undisturbed and dignified work, in such a way that there is no interference at every step of a kind that no self-respecting person can tolerate'.[48] If they could not, then he would have to 'discuss with you ways and means of severing my relations with your Institute in a dignified manner'.[49] Einstein gained the right to do as he pleased, but at a price. He would never have any real influence in the running of the institute. When he backed Schrödinger for a post at the institute, it effectively ruled the Austrian out of the running.

Schrödinger did not have to leave Berlin, but did so as a matter of principle. He had been in exile at Magdalen College, Oxford University less than a week when, on 9 November 1933, he received some unexpected news. The president of the college, George Gordon, informed Schrödinger that *The Times* had called to say that he would be among the winners of the Nobel Prize that year. 'I think you may believe it. The *Times* do not say a thing unless they really know', said Gordon proudly.[50] 'As for me, I was truly astonished, for I thought you had the prize.'

Schrödinger and Dirac were each awarded a half share of the 1933 Nobel Prize, with the deferred prize of 1932 going to Heisenberg alone. Dirac's first reaction was to refuse it because he did not want the publicity. He accepted after Rutherford convinced him that refusing it would generate even greater publicity. While Dirac toyed with the idea of rejecting the prize, Born was deeply hurt at being ignored by the Swedish Academy.

'I have a bad conscience regarding Schrödinger, Dirac, and Born', Heisenberg wrote to Bohr.[51] 'Schrödinger and Dirac both deserved an entire prize at least as much as I do, and I would have gladly shared with Born, since we have worked together.' Earlier he replied to a letter of congratulations from Born: 'The fact that I am to receive the Nobel Prize alone, for work done in Göttingen in collaboration – you, Jordan and I – this fact depresses me and I hardly know what to write to you.'[52] 'That Heisenberg's matrices bear his name is not altogether justified, as in those days he actually had no idea what a matrix was', Born complained to Einstein two decades later.[53] 'It was he who reaped all the rewards of our work together, such as the Nobel Prize and that sort of thing.' He admitted that 'for the last twenty years I have not been able to rid myself of a certain sense of injustice'. Born was finally awarded the Nobel in 1954 for 'his fundamental work in quantum mechanics and especially for his statistical interpretation of the wave function'.

After the difficult start, by the end of November 1933 Princeton was beginning to appeal to Einstein. 'Princeton is a wonderful little spot, a quaint and ceremonious village of puny demigods on stilts', he wrote to Queen Elizabeth of Belgium. 'Yet, by ignoring certain special conventions, I have been able to create for myself an atmosphere conducive to study and free from distractions.'[54] In April 1934 Einstein made public that he would be staying in Princeton indefinitely. The 'bird of passage' had found a place to nest for the rest of his life.

Einstein had always been an outsider, even in physics, beginning with his days in the Patent Office. Yet he had led the way for so long and so often. He hoped to do so again as he came up with a new challenge for Bohr and the Copenhagen interpretation.

Chapter 13

QUANTUM REALITY

'Princeton is a madhouse' and 'Einstein is completely cuckoo', wrote Robert Oppenheimer.[1] It was January 1935 and America's leading home-grown theoretical physicist was 31. Twelve years later, after directing the building of the atomic bomb, he would return to the Institute for Advanced Study to take charge of the 'madhouse' and its 'solipsistic luminaries shining in separate and helpless desolation'.[2] Einstein accepted that his critical attitude towards quantum mechanics ensured that 'here in Princeton I am considered an old fool'.[3]

It was a sentiment widely shared by the younger generation of physicists who, having been weaned on the theory, agreed with Paul Dirac's assessment that quantum mechanics explained 'most of physics and all of chemistry'.[4] That a few old men were fighting about the meaning of the theory was, for them, neither here nor there, given its enormous practical success. By the end of the 1920s, as one problem after another in atomic physics was solved, attention shifted from the atom to the nucleus. During the early 1930s, the discovery of the neutron by James Chadwick in Cambridge, and the work of Enrico Fermi and his team in Rome on the reactions induced by the impact of neutrons on nuclei, opened up the new frontier of nuclear physics.[5] In 1932 John Cockcroft and Ernest Walton, Chadwick's colleagues in Rutherford's Cavendish Laboratory, constructed the first particle accelerator and used it to split an atom by breaking apart its nucleus.

Einstein might have moved from Berlin to Princeton, but physics was moving on without him. He knew as much, but felt he had earned the right to pursue the physics that interested him. When he arrived at the institute in October 1933, Einstein was shown to his new office and asked what equipment he needed. 'A desk or table, a chair, paper and pencils', he replied.[6] 'Oh yes, and a large wastebasket, so I can throw away all my mistakes.' And there were plenty, but Einstein was never disheartened as he sought his holy grail – a unified field theory.

Just as Maxwell had unified electricity, magnetism and light into a single all-encompassing theoretical structure in the nineteenth century, Einstein hoped to unify electromagnetism and general relativity. For him such a unification was the next step, as logical as it was inevitable. It was in 1925 that he undertook the first of his many attempts at constructing such a theory that ended up in the wastebasket. After the discovery of quantum mechanics, Einstein believed that a unified field theory would yield this new physics as a by-product.

In the years following Solvay 1930, there was little direct contact between Bohr and Einstein. A valuable channel of communication ceased with Paul Ehrenfest's suicide in September 1933. In a moving tribute, Einstein wrote of his friend's inner struggle to understand quantum mechanics and 'the increasing difficulty of adaptation to new thoughts which always confronts the man past fifty. I do not know how many readers of these lines will be capable of fully grasping that tragedy.'[7]

There were many who read Einstein's words and mistook them as a lament at his own plight. Now in his mid-fifties, he knew he was regarded as a relic from a bygone age, refusing, or unable, to live with quantum mechanics. But he also knew what separated him and Schrödinger from most of their colleagues: 'Almost all the other fellows do not look from the facts to the theory but from the theory to the facts; they cannot extricate themselves from a once accepted conceptual net, but only flop about in it in a grotesque way.'[8]

In spite of these mutual misgivings, there were always young physicists eager to work with Einstein. One was Nathan Rosen, a 25-year-old

New Yorker who arrived from MIT in 1934 to serve as his assistant. A few months before Rosen, the 39-year-old Russian-born Boris Podolsky had joined the institute. He had first met Einstein at Caltech in 1931 and they had collaborated on a paper. Einstein had an idea for another paper. It would mark a new phase in his debate with Bohr, as it unleashed a fresh assault on the Copenhagen interpretation.

At Solvay 1927 and 1930, Einstein attempted to circumvent the uncertainty principle to show that quantum mechanics was inconsistent and therefore incomplete. Bohr, aided by Heisenberg and Pauli, had successfully dismantled each thought experiment and defended the Copenhagen interpretation. Afterwards, Einstein accepted that although quantum mechanics was logically consistent it was not the definitive theory that Bohr claimed. Einstein knew he needed a new strategy to demonstrate that quantum mechanics is incomplete, that it does not fully capture physical reality. To this end he developed his most enduring thought experiment.

For several weeks early in 1935, Einstein met Podolsky and Rosen in his office to thrash out his idea. Podolsky was assigned the task of writing the resulting paper, while Rosen did most of the necessary mathematical calculations. Einstein, as Rosen recalled later, 'contributed the general point of view and its implications'.[9] Only four pages long, the Einstein-Podolsky-Rosen paper, or the EPR paper as it became known, was completed and mailed by the end of March. 'Can Quantum Mechanical Description of Physical Reality Be Considered Complete?', with its missing 'the', was published on 15 May in the American journal *Physical Review*.[10] The EPR answer to the question posed was a defiant 'No!'. Even before it appeared in print, Einstein's name ensured that the EPR paper generated the kind of publicity nobody wanted.

On Saturday, 4 May 1935, the *New York Times* carried an article on page eleven under the attention-grabbing headline 'Einstein Attacks Quantum Theory': 'Professor Einstein will attack science's important theory of quantum mechanics, a theory of which he was a sort of grandfather. He concluded that while it is "correct" it is not "complete".' Three days later, the *New York Times* carried a statement from a clearly disgruntled Einstein.

Although no stranger to talking to the press, he pointed out that: 'It is my invariable practice to discuss scientific matters only in the appropriate forum and I deprecate advance publication of any announcement in regard to such matters in the secular press.'[11]

In the published paper, Einstein, Podolsky and Rosen started by differentiating between reality as it is and the physicist's understanding of it: 'Any serious consideration of a physical theory must take into account the distinction between the objective reality, which is independent of any theory, and the physical concepts with which the theory operates. These concepts are intended to correspond with the objective reality, and by means of these concepts we picture this reality to ourselves.'[12] In gauging the success of any particular physical theory, EPR argued that two questions had to be answered with an unequivocal 'Yes': Is the theory correct? Is the description given by the theory complete?

'The correctness of the theory is judged by the degree of agreement between the conclusions of the theory and human experience', said EPR. It was a statement that every physicist would accept when 'experience' in physics takes the form of experiment and measurement. To date there had been no conflict between the experiments performed in the laboratory and the theoretical predictions of quantum mechanics. It appeared to be a correct theory. Yet for Einstein it was not enough for a theory to be correct, in agreement with experiments; it also had to be complete.

Whatever the meaning of the term 'complete', EPR imposed a necessary condition for the completeness of a physical theory: *every element of the physical reality must have a counterpart in the physical theory.*[13] This completeness criterion required EPR to define a so-called 'element of reality' if they were to carry through their argument.

Einstein did not want to get stuck in the philosophical quicksand, which had swallowed so many, of trying to define 'reality'. In the past, none had emerged unscathed from an attempt to pinpoint what constituted reality. Astutely avoiding a 'comprehensive definition of reality' as 'unnecessary' for their purpose, EPR adopted what they deemed to be a 'sufficient' and 'reasonable' criterion for designating an 'element of reality': '*If, without in*

any way disturbing a system, we can predict with certainty (i.e. with prob-
ability equal to unity) the value of a physical quantity, then there exists an
element of physical reality corresponding to this physical quantity.[14]

Einstein wanted to disprove Bohr's claim that quantum mechanics was
a complete, fundamental theory of nature by demonstrating that there
existed objective 'elements of reality' which the theory did not capture.
Einstein had shifted the focus of the debate with Bohr and his supporters
away from the internal consistency of quantum mechanics to the nature of
reality and the role of theory.

EPR asserted that for a theory to be complete there had to be one-to-
one correspondence between an element of the theory and an element of
reality. A sufficient condition for the reality of a physical quantity, such as
momentum, is the possibility of predicting it with certainty without disturbing the system. If there existed an element of physical reality that was
unaccounted for by the theory, then the theory was incomplete. The situation would be akin to a person finding a book in a library and when trying
to check it out, being told by the librarian that according to the catalogue
there was no record of the library having the book. With the book bearing
all the necessary markings indicating that it was indeed a part of the collection, the only possible explanation would be that the library's catalogue
was incomplete.

According to the uncertainty principle, a measurement that yields
an exact value for the momentum of a microphysical object or system
excludes even the possibility of simultaneously measuring its position. The
question that Einstein wanted to answer was: Does the inability to measure
its exact position directly mean that the electron does not have a definite
position? The Copenhagen interpretation answered that in the absence
of a measurement to determine its position, the electron has no position.
EPR set out to demonstrate that there are elements of physical reality, such
as an electron having a definite position, that quantum mechanics cannot
accommodate – and therefore, it is incomplete.

EPR attempted to clinch their argument with a thought experiment.
Two particles, A and B, interact briefly and then move off in opposite

directions. The uncertainty principle forbids the exact measurement, at any given instant, of both the position and the momentum of either particle. However, it does allow an exact and simultaneous measurement of the total momentum of the two particles, A and B, and the relative distance between them.

The key to the EPR thought experiment is to leave particle B undisturbed by avoiding any direct observation of it. Even if A and B are light years apart, nothing within the mathematical structure of quantum mechanics prohibits a measurement of the momentum of A yielding information about the exact momentum of B without B being disturbed in the process. When the momentum of particle A is measured exactly, it indirectly but simultaneously allows, via the law of conservation of momentum, an exact determination of the momentum of B. Therefore, according to the EPR criterion of reality, the momentum of B must be an element of physical reality. Similarly, by measuring the exact position of A, it is possible, because the physical distance separating A and B is known, to deduce the position of B without directly measuring it. Hence, EPR argue, it too must be an element of physical reality. EPR appeared to have contrived a means to establish with certainty the exact values of *either* the momentum *or* the position of B due to measurements performed on particle A, without the slightest possibility of particle B being physically disturbed.

Given their reality criterion, EPR argued that they had thus proved that both the momentum and position of particle B are 'elements of reality', that B can have simultaneously exact values of position and momentum. Since quantum mechanics via the uncertainty principle rules out any possibility of a particle simultaneously possessing both these properties, these 'elements of reality' have no counterparts in the theory.[15] Therefore the quantum mechanical description of physical reality, EPR conclude, is incomplete.

Einstein's thought experiment was not designed to simultaneously measure the position and momentum of particle B. He accepted that it was impossible to measure either of these properties of a particle directly without causing an irreducible physical disturbance. Instead, the two-particle thought experiment was constructed to show that such properties could have a definite simultaneous existence, that both the position and the momentum of a particle are 'elements of reality'. If these properties of particle B can be determined without B being observed (measured), then these properties of B must exist as elements of physical reality independently of being observed (measured). Particle B has a position that is real and a momentum that is real.

EPR were aware of the possible counter-argument that 'two or more physical quantities can be regarded as simultaneous elements of reality *only when they can be simultaneously measured or predicted*'.[16] This, however, made the reality of the momentum and position of particle B dependent upon the process of measurement carried out on particle A, which could be light years away and which does not disturb particle B in any way. 'No reasonable definition of reality could be expected to permit this', said EPR.[17]

Central to the EPR argument was Einstein's assumption of locality – that some mysterious, instantaneous action-at-a-distance does not exist. Locality ruled out the possibility of an event in a certain region of space instantaneously, faster-than-light, influencing another event elsewhere. For Einstein, the speed of light was nature's unbreakable limit on how fast anything could travel from one place to another. For the discoverer of relativity it was inconceivable for a measurement on particle A to affect instantaneously, at a distance, the independent elements of physical reality possessed by particle B.

As soon as the EPR paper appeared, the alarm was raised among the leading quantum pioneers throughout Europe. 'Einstein has once again made a public statement about quantum mechanics, and even in the issue of *Physical Review* of May 15 (together with Podolsky and Rosen, not good company by the way)', wrote a furious Pauli in Zurich to Heisenberg in Leipzig.[18] 'As is well known,' he continued, 'that is a disaster whenever it happens.' Pauli nevertheless conceded, as only he could, 'that if a student in one of his earlier semesters had raised such objections, I would have considered him quite intelligent and promising'.[19]

With the zeal of a quantum missionary, Pauli urged Heisenberg to publish an immediate rebuttal to prevent any confusion or wavering among fellow physicists in the wake of Einstein's latest challenge. Pauli admitted that he had considered, for 'educational' reasons, 'squandering paper and ink in order to formulate those facts demanded by quantum theory which cause Einstein particular intellectual difficulties'.[20] In the end it was Heisenberg who drafted a reply to the EPR paper and sent Pauli a copy. But Heisenberg withheld the publication of his paper, as Bohr had already taken up arms in defence of the Copenhagen interpretation.

The EPR 'onslaught came down upon us as a bolt from the blue', recalled Léon Rosenfeld, who was in Copenhagen at the time.[21] 'Its effect on Bohr was remarkable.' Immediately abandoning everything else, Bohr was convinced that a thorough examination of the EPR thought experiment would reveal where Einstein had gone wrong. He would show them 'the right way to speak about it'.[22] Excitedly, Bohr started dictating to Rosenfeld the draft of a reply. But soon he began to hesitate. 'No, this won't do, we must try all over again', Bohr mumbled to himself. 'So it went on for a while, with growing wonder at the unexpected subtlety of the [EPR] argument', recalled Rosenfeld. 'Now and then, he would turn to me and ask: "What can they mean? Do you understand it?"'[23] After a while, an increasingly agitated Bohr realised that the argument Einstein had deployed was both ingenious and subtle. A refutation of the EPR paper would be harder than he first thought, and he announced that he 'must sleep on it'.[24] The next day he was calmer. 'They do it smartly,' he told Rosenfeld, 'but what counts is to do it right.'[25] For the next six weeks, day and night, Bohr worked on nothing else.

Even before he had finished his reply to EPR, Bohr wrote a letter on 29 June for publication in the journal *Nature*. Entitled 'Quantum Mechanics and Physical Reality', it briefly spelled out his counter-attack.[26] Once again, the *New York Times* smelt a story. 'Bohr and Einstein at Odds/ They Begin a Controversy Concerning the Fundamental Nature of Reality' were the headlines of the article that appeared on 28 July. 'The Einstein-Bohr controversy has just begun this week in the current issue of *Nature*, the British scientific publication,' the paper told its readers, 'with a preliminary challenge by Professor Bohr to Professor Einstein and with a promise by Professor Bohr that "a fuller development of this argument will be given in an article to be published shortly in the *Physical Review*".'

Bohr had deliberately chosen the same forum as Einstein, and his six-page response, received on 13 July, was also entitled 'Can Quantum-Mechanical Description of Physical Reality Be Considered Complete?'[27] Published on 15 October, Bohr's answer was an emphatic 'Yes'. However, unable to identify any error in the EPR argument, Bohr was reduced to arguing that Einstein's evidence for quantum mechanics being incomplete was not strong enough to bear the weight of such a claim. Using a debating tactic with a long and illustrious history, Bohr began his defence of the Copenhagen interpretation by simply rejecting the major component of Einstein's case for incompleteness: the criterion of physical reality. Bohr believed that he had identified a weakness in the EPR definition: the need to conduct a measurement 'without in any way disturbing a system'.[28]

Bohr hoped to exploit what he described as an 'essential ambiguity when it is applied to quantum phenomena' of the reality criterion, as he publicly retreated from the position that an act of measurement resulted in an unavoidable physical disturbance. He had relied on disturbance to undermine Einstein's previous thought experiments by demonstrating that it was impossible to know simultaneously the exact momentum and position of a particle because the act of measuring one caused an uncontrollable disturbance that ruled out an exact measurement of the other. Bohr knew perfectly well that EPR did not seek to challenge Heisenberg's uncertainty principle, since their thought experiment was not designed to simultaneously measure the position and momentum of a particle.

Bohr acknowledged as much when he wrote that in the EPR thought experiment 'there is no question of a mechanical disturbance of the system under investigation'.[29] It was a significant public concession, one he had made in private a few years earlier as he, Heisenberg, Hendrik Kramers and Oskar Klein sat around the fire at his country cottage in Tisvilde. 'Isn't it odd,' said Klein, 'that Einstein should have such great difficulties in accepting the role of chance in atomic physics?'[30] It is because 'we cannot make observations without disturbing the phenomena', said Heisenberg; 'the quantum effects we introduce with our observation automatically introduce a degree of uncertainty into the phenomenon to be observed.'[31] 'This Einstein refuses to accept, although he knows the facts perfectly well.' 'I don't entirely agree with you', Bohr told Heisenberg.[32] 'In any case,' he continued, 'I find all such assertions as "observation introduces uncertainty into the phenomenon" inaccurate and misleading. Nature has taught us that the word "phenomenon" cannot be applied to atomic processes unless we also specify what experimental arrangement or what observational instruments are involved. If a particular experimental set up has been defined and a particular observation follows, then we can admittedly speak of a phenomenon, but not of its disturbance by observation.'[33] Yet before, during, and after the Solvay conferences, an act of measurement disturbing the observed object peppered Bohr's writings and was central to his dismantling of Einstein's thought experiments.

Feeling the pressure from Einstein's continued probing of the Copenhagen interpretation, Bohr abandoned his previous reliance on 'disturbance' because he knew that it implied that an electron, for example, existed in a state that could be disturbed. Instead, Bohr now emphasised that any microphysical object being measured and the apparatus doing the measuring formed an indivisible whole – the 'phenomenon'. There simply was no room for a physical disturbance due to an act of measurement. This was why Bohr believed the EPR reality criterion was ambiguous.

Alas, Bohr's response to EPR was less than clear. Years later, in 1949, he admitted to a certain 'inefficiency of expression' when he re-read his paper. Bohr tried to clarify that the 'essential ambiguity' he had alluded to in his EPR rejoinder lay in referring to 'physical attributes of objects when dealing with phenomena where no sharp distinction can be made between the behaviour of the objects themselves and their interaction with the measuring instruments'.[34]

Bohr did not object to EPR predicting the results of possible measurements of particle B based on knowledge acquired by measuring particle A. Once the momentum of particle A is measured, it is possible to predict accurately the result of a similar measurement of the momentum of particle B as outlined by EPR. However, Bohr argued that that does not mean that momentum is an independent element of B's reality. Only when an 'actual' momentum measurement is carried out on B can it be said to possess momentum. A particle's momentum becomes 'real' only when it interacts with a device designed to measure its momentum. A particle does not exist in some unknown but 'real' state prior to an act of measurement. In the absence of such a measurement to determine either the position or the momentum of a particle, Bohr argued that it was meaningless to assert that it actually possessed either.

For Bohr, the role of the measuring apparatus was pivotal in defining EPR's elements of reality. Thus, once a physicist sets up the equipment to measure the exact position of particle A, from which the position of particle B can be calculated with certainty, it excludes the possibility of measuring the momentum of A and hence deducing the momentum of B.

If, as Bohr conceded to EPR, there is no direct physical disturbance of particle B, then its 'elements of physical reality', he argued, must be defined by the nature of the measuring device and the measurement made on A.

For EPR, if the momentum of B is an element of reality, then a momentum measurement on particle A cannot affect B. It merely allows the calculation of the momentum that particle B has independently of any measurement. EPR's reality criterion assumes that if particles A and B exert no physical force on each other, then whatever happens to one cannot 'disturb' the other. However, according to Bohr, since A and B had once interacted before travelling apart, they were forever entwined as parts of a single system and could not be treated individually as two separate particles. Hence, subjecting A to a momentum measurement was practically the same as performing a direct measurement on B, leading instantly to it having a well-defined momentum.

Bohr agreed that there was no 'mechanical' disturbance of particle B due to an observation of particle A. Like EPR, he too excluded the possibility of any physical force, a push or pull, acting at a distance. However, if the reality of the position or momentum of particle B is determined by measurements performed on particle A, then there appears to be some instantaneous 'influence' at a distance. This violates locality, that what happens to A cannot instantaneously affect B, and separability, that A and B exist independently of each other. Both concepts lay at the heart of the EPR argument and Einstein's view of an observer-independent reality. However, Bohr maintained that a measurement of particle A somehow instantaneously 'influences' particle B.[35] He did not expand on the nature of this mysterious '*influence on the very conditions which define the possible types of predictions regarding the further behaviour of the system*'.[36] Bohr concluded that since 'these conditions constitute an inherent element of the description of any phenomenon to which the term "physical reality" can be properly attached, we see that the argumentation of the mentioned authors does not justify their conclusion that quantum-mechanical description is essentially incomplete'.[37]

Einstein mocked Bohr's 'voodoo forces' and 'spooky interactions'. 'It seems hard to look into the cards of the Almighty', he wrote later.[38] 'But I won't for one minute believe that he throws dice or uses "telepathic" devices (as he is being credited with by the present quantum theory).' He told Born that 'physics should represent reality in time and space, free from spooky action at a distance'.[39]

The EPR paper expressed Einstein's view that the Copenhagen interpretation of quantum theory and the existence of an objective reality were incompatible. He was right and Bohr knew it. 'There is no quantum world. There is only an abstract quantum mechanical description', argued Bohr.[40] According to the Copenhagen interpretation, particles do not have an independent reality, they do not possess properties when they are not being observed. It was a view that was later concisely summarised by the American physicist John Archibald Wheeler: no elementary phenomenon is a real phenomenon until it is an observed phenomenon. A year before

EPR, Pascual Jordan took the Copenhagen rejection of an observer-independent reality to its logical conclusion: 'We ourselves produce the results of measurement.'[41]

'Now we have to start all over again,' said Paul Dirac, 'because Einstein proved that it does not work.'[42] He initially believed that Einstein had delivered a fatal blow against quantum mechanics. But soon, like most physicists, Dirac accepted that Bohr had once more emerged victorious from a battle with Einstein. Quantum mechanics had long proved its worth, and few were interested in examining Bohr's reply to the EPR argument too closely, for it was obscure even by his own standards.

Shortly after the EPR paper appeared in print, Einstein received a letter from Schrödinger: 'I was very happy that in the paper just published in P.R. you have evidently caught dogmatic q.m. by the coat-tails.'[43] After offering an analysis of some of the finer points of the EPR paper, Schrödinger explained his own reservation concerning the theory he had done so much to create: 'My interpretation is that we do not have a q.m. that is consistent with relativity theory, i.e. with a finite transmission speed of all influences. We have only the analogy of the old absolute mechanics ... The separation process is not at all encompassed by the orthodox scheme.'[44] As Bohr struggled to formulate his response, Schrödinger believed that the central role of separability and locality in the EPR argument meant that quantum mechanics was not a complete description of reality.

In his letter Schrödinger used the term 'verschränkung', later translated into English as 'entanglement', to describe the correlations between two particles that interact and then separate, as in the EPR experiment. He accepted, like Bohr, that having interacted, instead of two one-particle systems, there was just a single two-particle system and therefore any changes to one particle would affect the other, despite the distance that separated them. 'Any "entanglement of predictions" that takes place can obviously only go back to the fact that the two bodies at some earlier time formed in a true sense *one* system, that is were interacting, and have left behind *traces* on each other', he wrote in a famous paper published later in the year.[45] 'If two separated bodies, each by itself known maximally, enter a situation in which they influence each other, and separate again, then there occurs regularly that which I have just called *entanglement* of our knowledge of the two bodies.'[46]

Although he did not share Einstein's intellectual and emotional commitment to locality, Schrödinger was not prepared to reject it. He put forward an argument for undoing the entanglement. Any measurement on either separated part A or B of an entangled two-particle state breaks the entanglement and both are once more independent of each other. 'Measurements on separated systems,' he concluded, 'cannot directly influence each other – that would be magic.'

Schrödinger must have been surprised when he read the letter, dated 17 June, that arrived from Einstein. 'From the point of view of principles,' he wrote, 'I absolutely do not believe in a statistical basis for physics in the sense of quantum mechanics, despite the singular success of the formalism of which I am well aware.'[47] This Schrödinger already knew, but Einstein declared: 'This epistemology-soaked orgy ought to come to an end.' Even as he wrote the words, Einstein knew how he sounded: 'No doubt, however, you smile at me and think that, after all, many a young heretic turns into an old fanatic, and many a young revolutionary becomes an old reactionary.'

Their letters had crossed in the post. Two days after having written his, Einstein received Schrödinger's on the EPR paper and replied immediately. 'What I really intended has not come across very well,' Einstein explained, 'on the contrary the main point was, so to speak, buried by erudition.'[48] The EPR paper written by Podolsky lacked the clarity and style that characterised Einstein's published work in German. He was unhappy that the fundamental role of separability, that the state of one object cannot depend upon the kind of measurement made on another spatially separated object, had been obscured in the paper. Einstein wanted the separation principle to be an explicit feature of the EPR argument and not as it appeared, on the last page, as some sort of afterthought. He wanted to draw out the incompatibility of separability and the completeness of quantum mechanics. Both could not be true.

'The actual difficulty lies in the fact that physics is a kind of metaphysics', he told Schrödinger; 'physics describes reality; we know it only through its physical description.'[49] Physics was nothing less than a 'description of reality', but that description, Einstein wrote, 'can be "complete" or "incomplete"'. He attempted to illustrate what he meant by asking Schrödinger to imagine two closed boxes, one of which contains a ball. Opening the lid of a box and looking inside is 'making an observation'. Prior to looking inside the first box, the probability that it contains the ball is ½, in other words there is a 50 per cent chance that the ball is inside the box. After the box is opened, there is either a probability of 1 (the ball is in the box) or 0 (the ball is not in the box). But, says Einstein, in reality the ball was always in one of the two boxes. So, he asks, is the statement 'The probability is ½ that the ball is in the first box' a complete description of reality? If no, then a complete description would be 'The ball is (or is not) in the first box'. If before the box is open is deemed to be a complete description, then such a description would be 'The ball is not in *one* of the two boxes'. The ball's existence in a definite box occurs only when one of the boxes is opened. 'In this way arises the statistical character of the world of experience or its empirical systems of laws', concluded Einstein. So he poses the question, is the state *before* the box is opened completely described by the probability ½?

To decide, Einstein brought in the 'separation principle' – the second box and its contents is independent of anything that happens to the first box. Therefore, according to him, the answer is no. Assigning the probability of ½ that the first box contains the ball is an incomplete description of reality. It was Bohr's violation of Einstein's separation principle that resulted in the 'spooky action at a distance' in the EPR thought experiment.

On 8 August 1935, Einstein followed up his ball-in-the-box with a more explosive scenario to demonstrate to Schrödinger the incompleteness of quantum mechanics because the theory could only offer probabilities where there was certainty. He asked Schrödinger to consider a keg of unstable gunpowder that spontaneously combusts at some time during the next year. At the beginning the wave function describes a well-defined state – a keg of unexploded gunpowder. But after a year the wave function 'describes a sort of blend of not-yet and of already-exploded systems'.[50] 'Through no art of interpretation can this wave-function be turned into an adequate description of a real state of affairs,' Einstein told Schrödinger, '[for] in reality there is just no intermediary between exploded and not-exploded.'[51] Either the keg had exploded or it had not. It was, said Einstein, a 'crude macroscopic example' that exhibited the same 'difficulties' as encountered in the EPR thought experiment.

The flurry of letters he exchanged with Einstein between June and August 1935 had inspired Schrödinger to scrutinise the Copenhagen interpretation. The fruit of this dialogue was a three-part essay published between 29 November and 13 December. Schrödinger said he did not know whether to call 'The Present Situation in Quantum Mechanics' a 'report' or a 'general confession'. Either way, it contained a single paragraph about the fate of a cat that was to have a lasting impact:

'A cat is penned up in a steel chamber, along with the following diabolical device (which must be secured against direct interference by the cat): in a Geiger counter there is a tiny bit of radioactive substance, so small, that *perhaps* in the course of one hour one of the atoms decays, but also, with equal probability, perhaps none; if it happens, the counter tube discharges and through a relay releases a hammer which shatters a small flask of hydrocyanic acid. If one has left this entire system to itself for an hour, one would say that the cat still lives *if* meanwhile no atom has decayed. The first atomic decay would have poisoned it. The wave function of the entire system would express this by having in it the living and the dead cat (pardon the expression) mixed or smeared out in equal parts.'[52]

According to Schrödinger and common sense, the cat is either dead or alive, depending on whether or not there has been a radioactive decay. But according to Bohr and his followers, the realm of the subatomic is an *Alice in Wonderland* sort of place: because only an act of observation can decide if there has been a decay or not, it is only this observation that determines whether the cat is dead or alive. Until then the cat is consigned to quantum purgatory, a superposition of states in which it is neither dead nor alive.

Although he chided Schrödinger for choosing to publish in a German journal while there remained German scientists prepared to tolerate the Nazi regime, Einstein was delighted. The cat shows, he told Schrödinger, 'that we agree completely with respect to the character of the present theory'. A wave function that contains a living and a dead cat 'cannot be considered to describe a real state'.[53] Years later, in 1950, Einstein inadvertently blew up the cat, as he forgot that it was he who devised the exploding gunpowder keg. Writing to Schrödinger about 'contemporary physicists', he could not conceal his dismay at their insistence 'that the quantum theory provides a description of reality, and even a *complete* description'.[54] Such an interpretation, Einstein told Schrödinger, was 'refuted most elegantly by your system of radioactive atom + Geiger counter + amplifier + charge of gunpowder + cat in a box, in which the wave function of the system contains the cat both alive and blown to bits'.[55]

Schrödinger's famous feline thought experiment also highlighted the difficulty of where to draw the line between the measuring apparatus, which is part of the macro world of the everyday, and the object being measured, which is part of the micro world of the quantum. For Bohr, there was no sharp 'cut' between the classical and quantum worlds. To explain his point about the unbreakable bond between observer and observed, Bohr offered the example of a blind man with a cane. Where, he asked, was the break between the blind man and the unseen world in which he lived? The blind man is inseparable from his cane, argued Bohr; it is an extension of him, as he uses it to get information about the world around him. Does the world start at the tip of the blind man's cane? No, said Bohr. Through the tip of his cane the blind man's sense of touch reaches into the world, and the two are inextricably bound together. Bohr suggested that the same applies when an experimenter attempts to measure some property of a microphysical particle. The observer and the observed are entwined in an intimate embrace through the act of measurement such that it is impossible to say where one begins and the other ends.

Nevertheless, the Copenhagen view assigns a privileged position to the observer, be it human or a mechanical device, in the construction of reality.

But all matter is made up of atoms and therefore subject to the laws of quantum mechanics, so how can the observer or measuring apparatus have a privileged position? This is the measurement problem. The Copenhagen interpretation's assumption of the prior existence of the classical world of the macroscopic measuring device appears circular and paradoxical.

Einstein and Schrödinger believed it to be a glaring indication of the incompleteness of quantum mechanics as a total world-view, and Schrödinger tried to highlight it with his cat-in-a-box. Measurement in the Copenhagen interpretation remains an unexplained process, since there is nothing in the mathematics of quantum mechanics that specifies how or when the wave function collapses. Bohr 'solved' the problem by simply declaring that measurements can indeed be made, but never offered an explanation of how.

Schrödinger met Bohr while in England in March 1936 and reported the encounter to Einstein: 'Recently in London spent a few hours with Niels Bohr, who in his kind, courteous way repeatedly said that he found it "appalling", even found it "high treason" that people like Laue and I, but in particular someone like you, should want to strike a blow against quantum mechanics with the known paradoxical situation, which is so necessarily contained in the way of things, so supported by experiment. It is as if we are trying to force nature to accept our preconceived conception of "reality". He speaks with the deep inner conviction of an extraordinarily intelligent man, so that it is difficult for one to remain unmoved in one's position.' Yet Einstein and Schrödinger both remained steadfast in their opposition to the Copenhagen interpretation.[56]

In August 1935, two months before the EPR paper was published, Einstein finally bought a house. There was nothing to distinguish 112 Mercer Street from its neighbours, but because of its owner it became one of the most famous addresses in the world. It was conveniently located within walking distance of his office at the Institute for Advanced Study, although he preferred to work in his study at home. Located on the first floor, a large table covered with the usual paraphernalia of the scholar dominated the centre of the study. On the walls there were portraits of Faraday and Maxwell, later joined by one of Gandhi.

The small clapboard house with its green shutters was also home to Elsa's younger daughter Margot, and Helen Dukas. All too soon the domestic tranquillity was shattered as Elsa was diagnosed with heart disease. As her condition worsened, Einstein became 'miserable and depressed', Elsa wrote to a friend.[57] She was pleasantly surprised: 'I never thought he was so attached to me. That, too, helps.'[58] She died aged 60 on 20 December 1936. With two women to look after him, Einstein quickly came to terms with his loss.

'I am settling down splendidly here', he wrote to Born.[59] 'I hibernate like a bear in its cave, and really feel more at home than ever before in all my varied existence.' He explained that this 'bearishness has been accentuated still further by the death of my mate, who was more attached to human beings than I'. Born found Einstein's almost casual announcement of Elsa's death 'rather strange' but unsurprising. 'For all his kindness, sociability, and love of humanity,' Born said later, 'he was nevertheless totally detached from his environment and the human beings included in it.'[60] Almost. There was one person to whom Einstein was deeply attached, his sister Maja. She came to live with him in 1939 after Mussolini's racial laws forced her to leave Italy, and stayed until her death in 1951.

After Elsa's death, Einstein established a routine that as the years passed varied less and less. Breakfast between 9 and 10 was followed by a walk to the institute. After working until 1pm he would return home for lunch and a nap. Afterwards he would work in his study until dinner between 6.30 and 7pm. If not entertaining guests, he would return to work until he went to bed between 11 and 12. He rarely went to the theatre or to a concert, and unlike Bohr, hardly ever watched a movie. He was, Einstein said in 1936, 'living in the kind of solitude that is painful in one's youth but in one's more mature years is delicious'.[61]

In early February 1937, Bohr arrived in Princeton, together with his wife and their son Hans, for a week-long stay as part of a six-month world tour. It was the first opportunity that Einstein and Bohr had had to meet face-to-face since the publication of the EPR paper. Could Bohr finally convince Einstein to accept the Copenhagen interpretation? 'The discussion on quantum mechanics was not at all heated', recalled Valentin Bargmann, who later served as one of Einstein's assistants.[62] 'But to the outside observer, Einstein and Bohr were talking past each other.' Any meaningful discussion, he believed, required 'days and days'. Alas, during the encounter he witnessed, 'So many things were left unsaid'.[63]

What was left unsaid between them each man already knew. Their debate about the interpretation of quantum mechanics came down to a philosophical belief about the status of reality. Did it exist? Bohr believed that quantum mechanics was a complete fundamental theory of nature, and he built his philosophical worldview on top of it. It led him to declare: 'There is no quantum world. There is only an abstract quantum mechanical description. It is wrong to think that the task of physics is to find out how nature is. Physics concerns what we can say about nature.'[64] Einstein, on the other hand, chose the alternative approach. He based his assessment of quantum mechanics on his unshakeable belief in the existence of a causal, observer-independent reality. Consequently he could never accept the Copenhagen interpretation. 'What we call science,' Einstein argued, 'has the sole purpose of determining what *is*.'[65]

For Bohr the theory came first, then the philosophical position, the interpretation constructed to make sense of what the theory says about reality. Einstein knew that it was dangerous to build a philosophical worldview on the foundations of any scientific theory. If the theory is found wanting in the light of new experimental evidence, then the philosophical position it supports collapses with it. 'It is basic for physics that one assumes a real world existing independently from any act of perception', said Einstein. 'But this we do not *know*.'[66]

Einstein was a philosophical realist and knew that such a position could not be justified. It was a 'belief' concerning reality that was not susceptible to proof. While that may be so, for Einstein 'it is existence and reality that one wishes to comprehend'.[67] 'I have no better expression than "religious" for confidence in the rational nature of reality insofar as it is accessible to human reason', he wrote to Maurice Solovine. 'Wherever this feeling is absent, science degenerates into uninspired empiricism.'[68]

Heisenberg understood that Einstein, and Schrödinger, wanted 'to return to the reality concept of classical physics or, to use a more general philosophic term, to the ontology of materialism'.[69] The belief in an 'objective real world whose smallest parts exist objectively in the same sense as stones or trees exist, independently of whether or not we observe them', was for Heisenberg a throw-back to 'simplistic materialist views that prevailed in the natural sciences of the nineteenth century'.[70] Heisenberg was only partly right when he identified that Einstein and Schrödinger wanted 'to change the philosophy without changing the physics'.[71] Einstein accepted that quantum mechanics was the best theory available, but it was 'an incomplete representation of real things, although it is the only one which can be built out of the fundamental concepts of force and material points (quantum corrections to classical mechanics)'.[72]

Einstein was desperately seeking to change the physics as well; for he was not the conservative relic many thought. He was convinced that the concepts of classical physics would have to be replaced by new ones. Since the macroscopic world is described by classical physics and its concepts, Bohr agued that even to seek to go beyond them was a waste of time. He had developed his framework of complementarity in order to save classical concepts. For Bohr there was no underlying physical reality that exists independently of measuring equipment, and that meant, as Heisenberg pointed out, 'we cannot escape the paradox of quantum theory, namely, the necessity of using the classical concepts'.[73] It is the Bohr-Heisenberg call to retain classical concepts that Einstein called a 'tranquilizing philosophy'.[74]

Einstein never abandoned the ontology of classical physics, an observer-independent reality, but he was prepared to make a decisive break with classical physics. The view of reality endorsed by the Copenhagen interpretation was all the evidence he needed of the necessity to do so. He wanted a revolution more radical than the one offered by quantum mechanics. It was hardly surprising that Einstein and Bohr left so much unsaid.

In January 1939, Bohr returned to Princeton and stayed for four months as a visiting professor at the institute. Although the two men still enjoyed a warm, friendly relationship, their ongoing dispute over quantum reality had inevitably led to a cooling. 'Einstein was only a shadow of himself', recalled Rosenfeld, who had accompanied Bohr to America.[75] They did meet, usually at formal receptions, but they no longer talked about the physics that mattered so much to them. During Bohr's stay Einstein gave only one lecture, on his search for a unified field theory. With Bohr in the audience, he expressed the hope that quantum physics would be derivable from such a theory. But Einstein had already made it known that he would rather not discuss the issue further. 'Bohr was profoundly unhappy about this', said Rosenfeld.[76] With Einstein unwilling to talk about quantum physics, Bohr found that there were plenty of others in Princeton eager to discuss the latest developments in nuclear physics, given the ominous events in Europe that would lead once again to a world at war.

'No matter how deeply one immerses oneself in work,' Einstein wrote to Queen Elizabeth of Belgium, 'a haunting feeling of inescapable tragedy persists.'[77] The letter was dated 9 January 1939, two days before Bohr sailed for America and brought with him the news of a discovery that others had made: the splitting apart of a large nucleus into smaller nuclei, with an accompanying release of energy – nuclear fission. It was during the voyage that Bohr realised it was the uranium-235 isotope that undergoes nuclear fission when it is bombarded by slow-moving neutrons, and not uranium-238. At the age of 53, it was Bohr's last major contribution to physics. With Einstein unwilling to debate the nature of quantum reality, Bohr concentrated on working out the details of nuclear fission with the American John Wheeler from Princeton University.

After Bohr returned to Europe, Einstein sent a letter, dated 2 August, to President Roosevelt urging him to examine the feasibility of developing an atomic bomb, given that Germany had stopped the sale of uranium ore from mines it now controlled in Czechoslovakia. Roosevelt replied in October, thanking Einstein for his letter and informing him that he had set up a committee to investigate the issues raised. In the meantime, on 1 September 1939, Germany attacked Poland.

Still a pacifist, Einstein was prepared to compromise until Hitler and the Nazis were defeated. In a second letter, dated 7 March 1940, he urged Roosevelt that more needed to be done: 'Since the outbreak of the war, interest in uranium has intensified in Germany. I have now learned that research there is carried out in great secrecy.'[78] Unknown to Einstein, the man in charge of the German atomic bomb programme was Werner Heisenberg. Once again, the letter failed to solicit much of a response. Bohr's discovery that it was uranium-235 that underwent fission was far more important to the creation of the atom bomb than anything achieved by Einstein's two letters to Roosevelt. The American government did not seriously begin thinking about developing an atomic bomb, codenamed the Manhattan Project, until October 1941.

Even though Einstein had become an American citizen in 1940, the authorities considered him a security risk because of his political views. He was never asked to work on the atomic bomb. Bohr was. On 22 December 1943 he stopped off at Princeton on his way to Los Alamos in New Mexico, where the bomb was being built. He had dinner with Einstein and Wolfgang Pauli, who had joined the Institute for Advanced Study in 1940. Much had happened since the last time Bohr met Einstein.

In April 1940, German forces had occupied Denmark. Bohr chose to remain in Copenhagen, hoping that his international reputation would provide some semblance of protection to others at his institute. And it did until August 1943, when the illusion of Danish self-rule was finally shattered as the Nazis declared martial law after the government rejected a demand that a state of emergency be declared and acts of sabotage be punishable by death. Then on 28 September, Hitler ordered the deportation of Denmark's 8,000 Jews. A sympathetic German official informed two Danish politicians that the round-up was to begin at 9pm on 1 October. As word quickly spread of the Nazi plan, almost every Jew disappeared, hidden in the homes of fellow Danes or finding sanctuary in churches, or disguised as patients in hospitals. The Nazis managed to round up fewer than 300 Jews. Bohr, whose mother had been Jewish, managed to escape to Sweden with his family. From there he flew to Scotland in a British bomber, almost dying from a lack of oxygen because he was travelling in the bomb-bay and had an ill-fitting oxygen mask. After meeting British politicians he soon travelled to America, where after his fleeting visit to Princeton he worked on the atomic bomb under the alias 'Nicholas Baker'.

After the war, Bohr returned to his institute in Copenhagen, and Einstein said he felt 'no friendship for any real German'.[79] Yet he had abiding sympathy for Planck, who outlived all four children from his first marriage. The death of his youngest son was the bitterest of all the blows Planck endured in his long life. Erwin, an undersecretary of state in the Reich Chancellery before the Nazis came to power, was a suspect in an attempt to assassinate Hitler in July 1944. He was arrested and tortured by the Gestapo and found guilty of complicity in the assassination plot. At one point there was a glimmer of hope as Planck set, in his words, 'Heaven and Hell in motion' to have the death penalty commuted to a prison sentence.[80] Then, without warning, Erwin was hanged in Berlin in February 1945. Planck had been denied the opportunity to see his son one last time: 'He was a precious part of my being. He was my sunshine, my pride, my hope. No words can describe what I have lost with him.'[81]

When he heard the news that Planck had died, aged 89, following a stroke on 4 October 1947, Einstein wrote to his widow of the 'beautiful and fruitful time' he had been privileged to spend with him. As he offered his condolence, Einstein recalled that the 'hours which I was permitted to spend at your house, and the many conversations which I conducted face to face with that wonderful man, will remain among my most beautiful recollections for the rest of my life'.[82] It was something, he reassured her, which could not 'be altered by the fact that a tragic fate tore us apart'.

After the war, Bohr was made a permanent non-resident member of the Institute for Advanced Study and could come and stay whenever he wanted to. His first trip in September 1946 was brief, as he came to take part in the bicentennial celebrations of the founding of Princeton University. Then in 1948 he arrived in February and stayed until June. This time Einstein was willing to talk physics. Abraham Pais, a young Dutch physicist who helped Bohr during his visit, later described the occasion when the Dane came bursting into his office 'in a state of angry despair', saying, 'I am sick of myself'.[83] When Pais asked what was wrong, Bohr replied that he had been to see Einstein and they had got into an argument about the meaning of quantum mechanics.

The renewal of their friendship was signalled by the fact that Einstein let Bohr use his office. One day Bohr was dictating a draft of a paper in honour of Einstein's 70th birthday to Pais. Stuck on what to say next, Bohr stood looking out of the window, every now and then muttering Einstein's name aloud. At that moment Einstein tiptoed into the office. His doctor had banned him from buying any tobacco, but had said nothing about stealing it. Pais later recounted what happened next: 'Always on tiptoes, he made a beeline for Bohr's tobacco pot, which stood on the table at which I was sitting. Bohr, unaware, was standing at the window, muttering, "Einstein … Einstein …" I was at a loss what to do, especially because I had at that moment not the faintest idea of what Einstein was up to. Then Bohr, with a firm "Einstein", turned around. There they were, face to face, as if Bohr had summoned him forth. It is an understatement to say that for a moment Bohr was speechless. I myself, who had seen it coming, had distinctly felt uncanny for a moment, so I could well understand Bohr's own reaction. A moment later the spell was broken when Einstein explained his mission. Soon we were all bursting with laughter.'[84]

There were other visits to Princeton, but Bohr never managed to get Einstein to change his mind on quantum mechanics. Nor did Heisenberg, who saw him only once after the war during a lecture tour of the United States that overlapped with Bohr's last visit in 1954. Einstein invited Heisenberg to his home and, over coffee and cakes, they chatted for most of the afternoon. 'Of politics we said nothing', recalled Heisenberg.[85] 'Einstein's whole interest focused on the interpretation of quantum theory, which continued to disturb him, just as it had done in Brussels twenty-five years before.' Einstein remained resolute. '"I don't like your kind of physics", he said.'[86]

'The necessity of conceiving of nature as an objective reality is said to be superannuated prejudice while the quantum theoreticians are vaunted', Einstein had once written to his old friend Maurice Solovine.[87] 'Men are even more susceptible to suggestion than horses, and each period is dominated by a mood, with the result that most men fail to see the tyrant who rules over them.'

When Chaim Weizmann, the first president of Israel, died in November 1952, the prime minister David Ben-Gurion felt compelled to offer Einstein the presidency. 'I am deeply moved by the offer from our state of Israel, and at once saddened and ashamed because I cannot accept it', said Einstein.[88] He highlighted the fact that he lacked 'both a natural aptitude and the experience to deal properly with people and to exercise official functions'. 'For these reasons alone,' he explained, 'I should be unsuited to fulfil the duties of high office, even if advancing age was not making increasing inroads on my strength.'

Ever since the summer of 1950 when doctors discovered that his aortic aneurysm, a bulge in the aorta, was getting larger, Einstein knew he was living on borrowed time. He wrote his will and made it clear that he wanted to be cremated after a private funeral. He lived to celebrate his 76th birthday, and one of his last acts was to sign a declaration written by the philosopher Bertrand Russell calling for nuclear disarmament. Einstein wrote to Bohr asking him to sign it. 'Don't frown like that! This has nothing to do with our old controversy on physics, but rather concerns a matter on which we are in complete agreement.'[89] On 13 April 1955, Einstein experienced severe chest pains, and two days later he was taken to hospital. 'I want to go when *I* want', he said, refusing surgery. 'It is tasteless to prolong life artificially; I have done my share, it is time to go.'[90]

As fate would have it, his step-daughter Margot was staying in the same hospital. She saw Einstein twice and they chatted for a few hours. Hans Albert, who had arrived in America with his family in 1937, rushed from Berkeley in California to his father's bedside. For a while Einstein seemed better and asked for his notes, unable to abandon his search for a unified field theory even at the end. Shortly after 1am on 18 April, the aneurysm burst. After saying a few words in German that the night nurse could not understand, Einstein died. Later that day he was cremated, but not before his brain was removed and his ashes scattered at an undisclosed location. 'If everyone lived a life like mine there would be no need for novels', Einstein once wrote to his sister. The year was 1899 and he was twenty.[91]

'Except for the fact that he was the greatest physicist since Newton,' said Banesh Hoffmann, one of Einstein's Princeton assistants, 'one might almost say that he was not so much a scientist as an artist of science.'[92] Bohr paid his own heartfelt tribute. He recognised Einstein's achievements to be 'as rich and fruitful as any in the whole history of our culture', and said that 'mankind will always be indebted to Einstein for the removal of the obstacles to our outlook which were involved in the primitive notions of absolute space and time. He gave us a world picture with a unity and harmony surpassing the boldest dreams of the past.'[93]

The Einstein-Bohr debate did not end with Einstein's death. Bohr would argue as if his old quantum foe were still alive: 'I can still see Einstein's smile, both knowing, humane and friendly.'[94] Often his first thought when thinking about some fundamental issue in physics was to wonder what Einstein would have said about it. On Saturday, 17 November 1962, Bohr gave the last of five interviews concerning his role in the development of quantum physics. After lunch on Sunday, Bohr went to take his usual nap. When he called out, his wife Margrethe rushed to the bedroom and found him unconscious. Bohr, aged 77, had suffered a fatal heart attack. The last drawing on the blackboard in his study, made the night before as he replayed the argument over once more, was of Einstein's light box.

PART IV

DOES GOD PLAY DICE?

'I want to know how God created this world. I am not interested in this or that phenomenon, in the spectrum of this or that element. I want to know His thoughts, the rest are details.'
—ALBERT EINSTEIN

Chapter 14

FOR WHOM BELL'S THEOREM TOLLS

'You believe in the God who plays dice, and I in complete law and order in a world which objectively exists, and which I, in a wildly speculative way, am trying to capture', Einstein wrote to Born in 1944.[1] 'I firmly *believe*, but I hope that someone will discover a more real-istic way, or rather a more tangible basis than it has been my lot to find. Even the great initial success of quantum theory does not make me believe in the fundamental dice game, although I am well aware that our younger colleagues interpret this as a consequence of senility. No doubt the day will come when we shall see whose instinctive attitude was the correct one.' Twenty years passed before a discovery brought that day of judgement closer.

In 1964 the radio astronomers Arno Penzias and Robert Woodrow detected the echo of the big bang; the evolutionary biologist Bill Hamilton published his theory of the genetic evolution of social behaviour; and the theoretical physicist Murray Gell-Mann predicted the existence of a new family of fundamental particles called quarks. These were just three of the landmark scientific breakthroughs that year. Yet according to the physicist and historian of science Henry Stapp, none rivalled Bell's theorem, 'the most profound discovery of science'.[2] It was ignored.

Most physicists were too busy using quantum mechanics as it continued to notch up one success after another to be bothered about the subtleties of the arguments between Einstein and Bohr over its meaning and interpreta-tion. It was little wonder they failed to recognise that a 34-year-old Irish physicist, John Stewart Bell, had discovered what Einstein and Bohr could not: a mathematical theorem that could decide between their two oppos-ing philosophical worldviews. For Bohr there was 'no quantum world', only 'an abstract quantum mechanical description'.[3] Einstein believed in a reality independent of perception. The debate between Einstein and Bohr was as much about the kind of physics that was acceptable as a meaning-ful theoretical description of reality as it was about the nature of reality itself.

Einstein was convinced that Bohr and the supporters of the Copenhagen interpretation were playing a 'risky game' with reality.[4] John Bell was sympathetic to Einstein's position, but part of the inspiration behind his ground-breaking theorem lay in the work done in the early 1950s by an American physicist forced into exile.

David Bohm was a talented PhD student of Robert Oppenheimer's at the University of California at Berkeley. Born in Wilkes-Barre, Pennsylvania in December 1917, Bohm was prevented from joining the top-secret research facility in Los Alamos, New Mexico to work on the development of the atomic bomb in 1943 after Oppenheimer was appointed its director. The authorities cited Bohm's many relatives in Europe, nineteen of whom were to die in Nazi concentration camps, as the reason they considered him to be a security risk. In truth, having been questioned by US army intel-ligence, and attempting to secure his position as the scientific leader of the Manhattan Project, Oppenheimer had named Bohm as a possible member of the American Communist party.

Four years later, in 1947, the self-confessed 'shatterer of worlds' took charge of the 'madhouse', as Oppenheimer once called the Institute for Advanced Study in Princeton.[5] Maybe in an attempt to atone for his earlier naming of Bohm, of which his protégé was unaware, Oppenheimer helped him obtain an assistant professorship at Princeton University. Amid the anti-Communist paranoia sweeping the United States after the Second World War, Oppenheimer was soon under suspicion because of his earlier left-wing political views. Having watched him closely for some years, the FBI had compiled a large dossier on the man who knew America's atomic secrets.

In an attempt to smear Oppenheimer, some of his friends and colleagues were investigated by the House Un-American Activities Committee and forced to appear before it. In 1948 Bohm, who had joined the American Communist party in 1942 but left after only nine months, invoked the Fifth Amendment that protected him against self-incrimination. Within a year he was subpoenaed to appear before a grand jury, and once again pleaded the Fifth. In November 1949 Bohm was arrested, charged with contempt of court and briefly imprisoned before being released on bail. Princeton University, worried about losing wealthy donors, suspended him. Although he was acquitted when his case came to trial in June 1950, the university chose to pay off the remaining year of Bohm's contract, provided he did not set foot on campus. Bohm was blacklisted and unable to find another academic post in the United States, and Einstein seriously considered appointing him as his research assistant. Oppenheimer opposed the idea and was among those who advised his former student to leave the country. In October 1951, Bohm left for Brazil and the University of São Paulo.

He had been in Brazil only a matter of weeks when the American embassy, fearing that his final destination might be the Soviet Union, confiscated Bohm's passport and reissued it as valid only for travel to the United States. Worried that his South American exile would cut him off from the international physics community, Bohm acquired Brazilian nationality to circumvent the travel ban imposed by the Americans. Back in the United States, Oppenheimer faced a hearing. Pressure on him intensified the moment it emerged that Klaus Fuchs, a physicist he had selected to work on the atomic bomb, was a Soviet spy. Einstein advised Oppenheimer to turn up, tell the committee they were fools, and return home. He did no such thing, but another hearing in the spring of 1954 revoked Oppenheimer's security clearance.

Bohm left Brazil in 1955 and spent two years at the Technion Institute in Haifa, Israel before moving to England. After four years at Bristol University, in 1961 Bohm settled once and for all in London after being appointed professor of theoretical physics at Birkbeck College. During his troubled time in Princeton, Bohm had largely devoted himself to studying the structure and interpretation of quantum mechanics. In February 1951 he published *Quantum Theory*, one of the first textbooks to examine in some detail the interpretation of the theory and the EPR thought experiment.

Einstein, Podolsky and Rosen had conjured up an imaginary experiment that involved a pair of correlated particles, A and B, so far apart that it should be impossible for them to physically interact with one another. EPR argued that a measurement carried out on particle A could not physically disturb particle B. Since any measurement is performed on only one of the particles, EPR believed they could cut off Bohr's counter-attack – an act of measurement causes a 'physical disturbance'. Since the properties of the two particles are correlated, they argued that by measuring a property of particle A, such as its position, it is possible to know the corresponding property of B without disturbing it. EPR's aim was to demonstrate that particle B possessed the property independently of being measured, and since this was something that quantum mechanics failed to describe, it was therefore incomplete. Bohr countered, never so succinctly, that the pair of particles were entangled and formed a single system no matter how far apart they were. Therefore, if you measured one, then you also measured the other.

'If their [EPR] contention could be proved,' wrote Bohm, 'then one would be led to search for a more complete theory, perhaps containing something like hidden variables, in terms of which the present quantum theory would be a limiting case.'[6] But he concluded 'that quantum theory is inconsistent with the assumption of hidden causal variables'.[7] Bohm looked at quantum theory from the prevailing Copenhagen viewpoint.

However, in the process of writing his book he became dissatisfied with Bohr's interpretation, even as he agreed with the dismissal by others of the EPR argument as 'unjustified, and based on assumptions concerning the nature of matter which implicitly contradict the quantum theory at the outset'.[8]

It was the subtlety of the EPR thought experiment, and what he came to regard as the reasonable assumptions on which it was constructed, that led Bohm to question the Copenhagen interpretation. It was a brave step for a young physicist whose contemporaries were busy using quantum theory to make their reputations rather than risking career suicide by raking over the embers of a dying fire. But Bohm was already a marked man after his appearance before the House Un-American Activities Committee, and, suspended by Princeton, he had little left to lose.

Bohm presented Einstein with a copy of *Quantum Theory* and discussed his reservations with Princeton's most famous resident. Encouraged to examine the Copenhagen interpretation more closely, Bohm produced two papers that appeared in January 1952. In the first of these he publicly thanked Einstein 'for several interesting and stimulating discussions'.[9] By then Bohm was in Brazil, but the papers had been written and sent to the *Physical Review* in July 1951, just four months after the publication of his book. Bohm appeared to have had a Paul-like conversion on the road not to Damascus, but Copenhagen.

In his papers Bohm outlined an alternative interpretation of quantum theory and argued that 'the mere possibility of such an interpretation proves that it is not necessary for us to give up a precise, rational, and objective description of individual systems at a quantum level of accuracy'.[10] Reproducing the predictions of quantum mechanics, it was a mathematically more sophisticated and coherent version of Louis de Broglie's pilot wave model, which the French prince had abandoned after it was severely criticised at the 1927 Solvay conference.

Whereas the wave function in quantum mechanics is an abstract wave of probability, in the pilot wave theory it is a real, physical wave that guides particles. Just as an ocean current carries along a swimmer or a ship, the pilot wave produces a current that is responsible for the motion of a particle. The particle has a well-defined trajectory determined by the precise values of position and velocity that it possesses at any given time but which the uncertainty principle 'hides' by preventing an experimenter from measuring them.

On reading Bohm's two papers, Bell said that he 'saw the impossible done'.[11] Like almost everyone else, he thought that Bohm's alternative to the Copenhagen interpretation had been ruled out as impossible. He asked why no one had told him about the pilot wave theory: 'Why is the pilot wave picture ignored in textbooks? Should it not be taught, not as the only way, but as an antidote to the prevailing complacency? To show that vagueness, subjectivity, and indeterminism, are not forced on us by experimental facts, but by deliberate theoretical choice?'[12] A part of the answer was the legendary Hungarian-born mathematician John von Neumann.

The eldest of three brothers, the Jewish banker's son was a mathematical prodigy. When his first paper was published at eighteen, von Neumann was a student at Budapest University but spent most of his time in Germany at the universities of Berlin and Göttingen, returning only to take his exams. In 1923 he enrolled at the ETH in Zurich to study chemical engineering after his father insisted that he have something more practical to fall back on than mathematics. After graduating from the ETH and gaining a doctorate from Budapest in double-quick time, von Neumann became at 23 the youngest-ever *privatdozent* appointed by Berlin University in 1927. Three years later he began teaching at Princeton University and in 1933 joined Einstein as a professor at the Institute for Advanced Study, remaining there for the rest of his life.

A year earlier, in 1932, the then 28-year-old von Neumann wrote a book that became the quantum physicist's bible, *Mathematical Foundations of Quantum Mechanics*.[13] In it he asked whether quantum mechanics could be reformulated as a deterministic theory by the introduction of hidden variables, which, unlike ordinary variables, are inaccessible to measurement and therefore not subject to the restrictions imposed by the uncertainty principle. Von Neumann argued that 'the present system of quantum mechanics would have to be objectively false in order that another description of the elementary processes than the statistical one may be possible'.[14] In other words, the answer was 'No', and he offered a mathematical proof that outlawed the 'hidden variables' approach that Bohm would adopt twenty years later.

It was an approach with a history. Ever since the seventeenth century, men like Robert Boyle had studied the various properties of gases as their pressure, volume and temperature were varied, and had discovered the gas laws. Boyle found the law describing the relationship between the volume of a gas and its pressure. He established that if a certain quantity of a gas was kept at a fixed temperature and its pressure was doubled, its volume was halved. If the pressure was increased threefold, then its volume was reduced to a third. At constant temperature, the volume of a gas is inversely proportional to the pressure.

The correct physical explanation of the gas laws had to wait until Ludwig Boltzmann and James Clerk Maxwell developed the kinetic theory of gases in the nineteenth century. 'So many of the properties of matter, especially when in gaseous form, can be deduced from the hypothesis that their minute parts are in rapid motion, the velocity increasing with temperature,' wrote Maxwell in 1860, 'that the precise nature of this motion becomes the subject of rational curiosity'.[15] It led him to conclude that 'the relations between pressure, temperature, and density in a perfect gas can be explained by supposing the particles to move with uniform velocity in straight lines, striking against the sides of the containing vessel and thus producing pressure'.[16] Molecules in a continual state of motion, haphazardly colliding into one another and the walls of the container holding the gas, produced the relationships between pressure, temperature and volume expressed in the gas laws. Molecules could be regarded as the unobserved microscopic 'hidden variable' that explained the observed macroscopic properties of gases.

Einstein's explanation of Brownian motion in 1905 is an example where the 'hidden variable' is the molecules of the fluid in which the pollen grains are suspended. The reason behind the erratic movement of the grains that had so perplexed everyone was suddenly clear after Einstein pointed out that it was due to the bombardment by invisible, but very real, molecules.

The appeal of hidden variables in quantum mechanics had its roots in Einstein's claim that the theory is incomplete. Maybe that incompleteness was due to the failure to capture the existence of an underlying layer of reality. This untapped seam in the form of hidden variables – possibly hidden particles, forces, or something completely new – would restore an independent, objective reality. Phenomena that at one level appear probabilistic would with the help of hidden variables be revealed as deterministic, and particles would possess a definite velocity and position at all times.

As von Neumann was acknowledged as one of the great mathematicians of the day, most physicists simply accepted, without bothering to check, that he had proscribed hidden variables when it came to quantum mechanics. For them the mere mention of 'von Neumann' and 'proof' was enough. However, von Neumann admitted that there remained the possibility, though small, that quantum mechanics might be wrong. 'In spite of the fact that quantum mechanics agrees well with experiment, and that it has opened up for us a qualitatively new side of the world, one can never say of the theory that it has been proved by experience, but only that it is the best known summarization of experience',[17] he wrote. Yet despite these words of caution, von Neumann's proof was held to be sacrosanct. Virtually everyone misinterpreted it as proving that no theory of hidden variables could reproduce the same experimental results as quantum mechanics.

When he analysed von Neumann's argument, Bohm believed that it was wrong but could not clearly pinpoint the weakness. Nevertheless, encouraged by his discussions with Einstein, Bohm attempted to construct the hidden variables theory that was deemed to be impossible. It would be Bell who demonstrated that one of the assumptions used by von Neumann was unwarranted, and therefore that his 'impossibility' proof was incorrect.

Born in July 1928 in Belfast, John Stewart Bell was descended from a family of carpenters, blacksmiths, farm workers, labourers and horse dealers.

'My parents were poor but honest', he once said.[18] 'Both of them came from large families of eight or nine that were traditional of the working class people of Ireland at that time.' With a father who was in and out of work, Bell's childhood was far removed from the comfortable middle-class upbringing of the quantum pioneers. Nevertheless, before he reached his teens, the bookish Bell had earned the nickname 'The Prof', even before he told his family that he wanted to become a scientist.

There was an older sister and two younger brothers, and though their mother believed that a good education was the route to future prosperity for her children, John was the only one who went on to secondary school aged eleven. It was not a lack of ability that denied his siblings the same opportunity, only a shortage of money for a family always struggling to make ends meet. Luckily the family came into a small sum of money that enabled Bell to enrol at the Belfast Technical High School. Not as prestigious as some of the other schools in the city, it offered a curriculum that combined the academic and the practical that suited him. In 1944, aged sixteen, Bell gained the qualifications necessary to study at Queen's University in his home town.

With seventeen the minimum age for admission and his parents unable to finance his university studies, Bell looked for work and fortuitously found it as an assistant technician in the laboratory of the physics department at Queen's University. Before long, the two senior physicists recognised Bell's abilities and allowed him to attend the first-year lectures whenever his duties permitted. His enthusiasm and obvious talent were rewarded with a small scholarship, and this, together with the money he was able to set aside, meant that he returned after his year as a technician as a fully-fledged physics student. With the sacrifices that he and his parents had made, Bell was focused and driven. He proved to be an exceptional student and in 1948 obtained a degree in experimental physics. A year later he gained another in mathematical physics.

Bell admitted that he 'had a very bad conscience about having lived off my parents for so long, and thought I should get a job'.[19] With his two degrees and glowing references, he went to England to work for the

United Kingdom Atomic Energy Research Establishment. In 1954 Bell married a fellow physicist, Mary Ross. In 1960, having gained a PhD from Birmingham University, he and his wife moved to CERN, the Conseil Européen pour la Recherche Nucléaire, near Geneva, Switzerland. For a man who would make his name as a quantum theorist, Bell's job was designing particle accelerators. He was proud to call himself a quantum engineer.

Bell first came across von Neumann's proof in 1949, his last year as a student in Belfast, when he read Max Born's new book, *Natural Philosophy of Cause and Chance*. 'I was very impressed that somebody – von Neumann – had actually proved that you couldn't interpret quantum mechanics as some sort of statistical mechanics', he later recalled.[20] But Bell did not read von Neumann's book as it was written in German and he did not know the language. Instead he accepted Born's word for the soundness of von Neumann's proof. According to Born, von Neumann had put quantum mechanics on an axiomatic basis by deriving it from a few postulates of a 'very plausible and general character', such that the 'formalism of quantum mechanics is uniquely determined by these axioms'.[21] In particular, Born said, it meant that 'no concealed parameters can be introduced with the help of which the indeterministic description could be transformed into a deterministic one'.[22] Implicitly, Born was arguing in favour of the Copenhagen interpretation, because 'if a future theory should be deterministic, it cannot be a modification of the present one but must be essentially different'.[23] Born's message was that quantum mechanics is complete, therefore it cannot be modified.

It was 1955 before von Neumann's book was published in English, but by then Bell had read Bohm's papers on hidden variables. 'I saw that von Neumann must have been just wrong', he said later.[24] Yet Pauli and Heisenberg branded Bohm's hidden variables alternative as 'metaphysical' and 'ideological'.[25] The ready acceptance of von Neumann's impossibility proof proved only one thing to Bell, a 'lack of imagination'.[26] Nevertheless, it had allowed Bohr and the advocates of the Copenhagen interpretation to consolidate their position even while some of them suspected that von Neumann might be wrong. Even though he later dismissed Bohm's work, Pauli in his published lectures on wave mechanics wrote that 'no proof of the impossibility of extending [i.e. completing quantum theory by hidden variables] has been given'.[27]

For 25 years, hidden variable theories had been ruled impossible by the authority of von Neumann. However, if such a theory could be constructed to yield the same predictions as quantum mechanics, then there would be no reason for physicists to simply accept the Copenhagen interpretation. When Bohm demonstrated that such an alternative was possible, the Copenhagen interpretation was so well entrenched as the *only* interpretation of quantum mechanics that he was either ignored or attacked. Einstein, who had initially encouraged him, dismissed Bohm's hidden variables as 'too cheap'.[28]

'I think he was looking for a much more profound rediscovery of quantum phenomena', Bell said as he tried to understand Einstein's reaction.[29] 'The idea that you could just add a few variables and the whole thing would remain unchanged apart from the interpretation, which was a kind of trivial addition to ordinary quantum mechanics, must have been a disappointment to him.' Bell was convinced that Einstein wanted to see some grand new principle emerge on a par with the conservation of energy. Instead, what Bohm offered Einstein was an interpretation that was 'non-local', requiring the instantaneous transmission of so-called 'quantum mechanical forces'. There were other horrors lurking in Bohm's alternative. 'For example,' clarified Bell, 'the trajectories that were assigned to the elementary particles were instantaneously changed when anyone moved a magnet anywhere in the universe.'[30]

It was in 1964, during a year-long sabbatical from CERN and his day job designing particle accelerators, that Bell found the time to enter the Einstein-Bohr debate. Bell decided to find out if non-locality was a peculiar feature of Bohm's model or if it was a characteristic of any hidden variable theory that aimed to reproduce the results of quantum mechanics. 'I knew, of course, that the Einstein-Podolsky-Rosen setup was the critical one, because it led to distant correlations', he explained. 'They ended their paper by stating that if you somehow completed the quantum mechanical description, non-locality would only be apparent. The underlying theory would be local.'[31]

Bell started out trying to preserve locality by attempting to construct a 'local' hidden variable theory in which if one event caused another, then there had to be enough time between the two to allow a signal travelling at the speed of light to pass between them. 'Everything I tried didn't work', he said later.[32] 'I began to feel that it very likely couldn't be done.' In his attempt to eliminate what Einstein decried as 'spooky actions at a distance', non-local influences that were transmitted instantly between one place and another, Bell derived his celebrated theorem.[33]

He began by looking at a version of the EPR thought experiment first devised by Bohm in 1951 that was simpler than the original. Whereas Einstein, Podolsky and Rosen had used two properties of a particle, position and momentum, Bohm used only one, quantum spin. First proposed in 1925 by the young Dutch physicists George Uhlenbeck and Samuel Goudsmit, the quantum spin of a particle had no analogue in classical physics. An electron had just two possible spin states, 'spin-up' and 'spin-down'. Bohm's adaptation of EPR involved a spin-zero particle that disintegrates and in the process produces two electrons, A and B. Since their combined spin must remain zero, one electron must have spin-up and the other spin-down.[34] Flying off in opposite directions until they are far enough apart to rule out any physical interaction between them, the quantum spin of each electron is measured at exactly the same time by a spin detector. Bell was interested in the correlations that could exist between the results of these simultaneous measurements carried out on pairs of such electrons.

The quantum spin of an electron can be measured independently in any one of three directions at right angles to each other, labelled x, y, and z.[35] These directions are just the normal three dimensions of the everyday world in which everything moves – left and right (x-direction), up and down (y-direction), and back and forth (z-direction). When the spin of electron A is measured along the x-direction by a spin-detector placed in its path, it will be either 'spin-up' or 'spin-down'. The odds are 50-50, the same as those for flipping a coin to see whether it lands heads or tails. In both cases, whether it is one or the other is pure chance. But as with flipping a coin repeatedly, if the experiment is done again and again, then electron A will be found to have spin-up in half the measurements and spin-down in the rest.

Unlike two coins that are flipped at the same time, each of which can be heads or tails, as soon as the spin of electron A is measured as spin-up, then a simultaneous measurement of the spin of electron B along the same direction will reveal it to be spin-down. There is a perfect correlation between the results of the two spin measurements. Bell later attempted to demonstrate that there was nothing strange about the nature of these correlations: 'The philosopher in the street, who has not suffered a course in quantum mechanics, is quite unimpressed by Einstein-Podolsky-Rosen correlations. He can point to many examples of similar correlations in everyday life. The case of Bertlemann's socks is often cited. Dr Bertlemann likes to wear two socks of different colours. Which colour he will have on a given foot on a given day is quite unpredictable. But when you see that the first sock is pink you can be already sure that the second sock will not be pink. Observation of the first, and experience of Bertlemann, gives immediate information about the second. There is no accounting for tastes, but apart from that there is no mystery here. And is not the EPR business the same?'[36] As with the colour of Bertlemann's socks, given that the spin of the parent particle is zero, it is no surprise that once the spin of electron A along any direction is measured as spin-up, the spin of electron B in the same direction is confirmed as spin-down.

According to Bohr, until a measurement is made, neither electron A nor electron B has a pre-existing spin in any direction. 'It is as if we had come to deny the reality of Bertlemann's socks,' said Bell, 'or at least of their colours, when not looked at.'[37] Instead, before they are observed, the electrons exist in a ghostly superposition of states so that they are spin-up and spin-down at the same time. Since the two electrons are entangled, the information concerning their spin states is given by a wave function similar to ψ = (A spin-up and B spin-down)+(A spin-down and B spin-up). Electron A has no x-component of spin until a measurement to determine it causes the wave function of the system, A and B, to collapse, and then it is either spin-up or spin-down. At that very moment, its entangled partner B acquires the opposite spin in the same direction, even if it is on the other side of the universe. Bohr's Copenhagen interpretation is non-local.

Einstein would explain the correlations by arguing that both electrons possess definite values of quantum spin in each of the three directions x, y, and z whether they are measured or not. For Einstein, said Bell, 'these correlations simply showed that the quantum theorists had been hasty in dismissing the reality of the microscopic world'.[38] Since the pre-existing spin states of the electron pair cannot be accommodated by quantum mechanics, this led Einstein to conclude that the theory was incomplete. He did not dispute the correctness of the theory, only that it was not a complete picture of physical reality at the quantum level.

Einstein believed in 'local realism': that a particle cannot be instantly influenced by a distant event and that its properties exist independently of any measurement. Unfortunately, Bohm's clever reworking of the original EPR experiment could not distinguish between the positions of Einstein and Bohr. Both men could account for the results of such an experiment. Bell's stroke of genius was to discover a way out of the impasse by changing the relative orientation of the two spin detectors.

If the spin detectors measuring electrons A and B are aligned so that they are parallel, then there is a 100 per cent correlation between the two sets of measurements – whenever spin-up is measured by one detector, spin-down is recorded by the other and vice versa. If one of the detectors is rotated slightly, then they are no longer perfectly aligned. Now if the spin states of many pairs of entangled electrons are measured, when A is found to be spin-up, the corresponding measurement of its partner B will sometimes also be spin-up. Increasing the angle of orientation between the detectors results in a reduction in the degree of correlation. If the detectors are at 90 degrees to each other and the experiment is once again repeated many times, when A is measured along the x-direction as spin-up, only in half of these instances will B be detected as spin-down. If the detectors are orientated at 180 degrees to one another, then the pair of electrons will be completely anti-correlated. If A's spin state is measured as spin-up, then B's will also be spin-up.

Although a thought experiment, it was possible to calculate the exact degree of spin correlation for a given orientation of the detectors predicted by quantum mechanics. However, it was not possible to do a similar calculation using an archetypal hidden variables theory that preserved locality. The only thing that such a theory would predict was a less than perfect match between spin states of A and B. This was not enough to decide between quantum mechanics and a local hidden variables theory.

Bell knew that any actual experiment that found spin correlations in line with the predictions of quantum mechanics could easily be disputed. After all, it was possible that in the future someone might develop a hidden variables theory that also exactly predicted the spin correlations for different orientations of the detectors. Bell then made an astonishing discovery. It was possible to decide between the predictions of quantum mechanics and *any* local hidden variables theory by measuring the correlations of pairs of electrons for a given setting of the spin detectors and then repeating the experiment with a different orientation.

This enabled Bell to calculate the total correlation for both sets of orientations in terms of the individual results predicted by *any* local hidden variables theory. Since in any such theory the outcome of a measurement at one detector cannot be affected by what is measured at the other, it is possible to distinguish between hidden variables and quantum mechanics.

Bell was able to calculate the limits on the degree of spin correlation between pairs of entangled electrons in a Bohm-modified EPR experiment. He found that in the ethereal realm of the quantum there is a greater level of correlation if quantum mechanics reigns supreme than in any world that depends on hidden variables and locality. Bell's theorem said that no local hidden variables theory could reproduce the same set of correlations as quantum mechanics. Any local hidden variables theory would lead to spin correlations that generated numbers, called the correlation coefficients, between −2 and +2. However, for certain orientations of the spin detectors, quantum mechanics predicted correlation coefficients that lay outside of the range known as 'Bell's inequality' that ran from −2 to +2.[39]

Although Bell, with his red hair and pointed beard, was difficult to miss, his extraordinary theorem was ignored. This was hardly surprising, since in 1964 the journal to get noticed in was the *Physical Review*, published by the American Physical Society. The problem for Bell was that the *Physical Review* charged, and it was your university that usually paid the bill once your paper was accepted. As a guest at Stanford University in California at the time, Bell did not want to abuse the hospitality he had been shown by asking the university to pay. Instead, his six-page paper, 'On the Einstein Podolsky Rosen Paradox', was published in the third issue of *Physics*, a little-read, short-lived journal that actually paid its contributors.[40]

In fact this was the second paper that Bell wrote during his sabbatical year. The first reconsidered the verdict of von Neumann and others that 'quantum mechanics does not permit a hidden variable interpretation'.[41] Unfortunately, mis-filed by the *Review of Modern Physics*, with a letter from the editor going astray causing a further delay, the paper was not published until July 1966. It was, wrote Bell, aimed at those 'who believe that "the question concerning the existence of such hidden variables received an early and rather decisive answer in the form of von Neumann's proof on the mathematical impossibility of such variables in quantum theory"'.[42] He went on to show, once and for all, that von Neumann had been wrong.

A scientific theory that does not agree with experimental facts will either be modified or discarded. Quantum mechanics, however, had passed every test it had been subjected to. There was no conflict between theory and experiment. For the vast majority of Bell's colleagues, young and old alike, the dispute between Einstein and Bohr over the correct interpretation of quantum mechanics was more philosophy than physics. They shared Pauli's view, expressed in a letter to Born in 1954, that 'one should no more rack one's brain about the problem of whether something one cannot know anything about exists all the same, than about the ancient question of how many angels are able to sit on the point of a needle'.[43] To Pauli it seemed 'that Einstein's questions are ultimately always of this kind' in his critique of the Copenhagen interpretation.[44]

Bell's theorem changed that. It allowed the local reality advocated by Einstein, that the quantum world exists independently of observation and that physical effects cannot be transmitted faster than the speed of light, to be tested against Bohr's Copenhagen interpretation. Bell had brought the Einstein-Bohr debate into a new arena, experimental philosophy. If Bell's inequality held, then Einstein's contention that quantum mechanics was incomplete would be right. However, should the inequality be violated, then Bohr would emerge the victor. No more thought experiments; it was Einstein vs. Bohr in the laboratory.

It was Bell who first challenged the experimentalists to put his inequality to the test when he wrote in 1964 that 'it requires little imagination to envisage the measurements involved actually being made'.[45] But like Gustav Kirchhoff and his imaginary blackbody a century earlier, it is easier for a theorist to 'envisage' an experiment than for his colleagues to realise it in practice. Five years passed before Bell received a letter in 1969 from a young physicist at Berkeley in California. John Clauser, then 26, explained that he and others had devised an experiment to test the inequality.

Two years earlier, Clauser had been a doctoral student at New York's Columbia University when he first came across Bell's inequality. Convinced that it was worth testing, Clauser went to see his professor and was bluntly told that 'no decent experimentalist would ever go to the effort of actually trying to measure it'.[46] It was a reaction in keeping with the near 'universal acceptance of quantum theory and its Copenhagen interpretation as gospel', Clauser wrote later, 'along with a total unwillingness to even mildly question the theory's foundations'.[47] Nevertheless, by the summer of 1969 Clauser had devised an experiment with the help of Michael Horne, Abner Shimony and Richard Holt. It required the quartet to fine-tune Bell's inequality so that it could be tested in a real laboratory rather than in the imaginary laboratory of the mind equipped with perfect instruments.

Clauser's search for a postdoctoral position took him to the University of California at Berkeley, where he had to settle for a job doing radio astronomy. Luckily, when Clauser explained to his new boss the experiment he really wanted to perform, he was allowed to devote half of his time to it. Clauser found a willing graduate student, Stuart Freedman, to help. Instead of electrons, Clauser and Freedman used pairs of correlated photons in their experiment. The switch was possible because photons have a property called polarisation that for the purposes of the test played the role of quantum spin. Although a simplification, a photon can be regarded as being polarised either 'up' or 'down'. Just like electrons and spin, if the polarisation of one photon along the x-direction is measured as 'up', then the other will be measured as 'down', since the combined polarisations of both photons must be zero.

The reason for employing photons rather than electrons is that they are easier to produce in the laboratory, especially since the experiment would involve numerous pairs of particles being measured. It was 1972 before Clauser and Freedman were ready to put Bell's inequality to the test. They heated calcium atoms until they acquired enough energy for an electron to jump from the ground state to a higher energy level. As the electron fell back down to the ground state, it did so in two stages and emitted a pair of entangled photons, one green and the other blue. The photons were sent in opposite directions until detectors simultaneously measured their polarisations. The two detectors were initially oriented at 22.5 degrees relative to each other for the first set of measurements, and then realigned at 67.5 degrees for the second set. Clauser and Freedman found, after 200 hours of measurements, that the level of photon correlations violated Bell's inequality.

It was a result in favour of Bohr's non-local Copenhagen interpretation of quantum mechanics with its 'spooky action at a distance', and against the local reality backed by Einstein. But there were serious reservations as to the validity of the outcome. Between 1972 and 1977 different teams of experimenters conducted nine separate tests of Bell's inequality. It was violated in only seven.[48] Given these mixed results, there were misgivings concerning the accuracy of the experiments. One problem was the inefficiency of the detectors that resulted in only a small fraction of the total number of pairs generated being measured. No one knew precisely what effect this had on the level of correlations. There were other loopholes that needed to be closed before it could be conclusively shown for whom Bell's theorem tolled.

As Clauser and others were busy planning and executing their experiments, a French physics graduate was doing voluntary work in Africa and spending his spare time reading up on quantum mechanics. It was while working his way through an influential French textbook on the subject that Alain Aspect first became fascinated by the EPR thought experiment. After reading Bell's seminal papers, he began thinking about subjecting Bell's inequality to a rigorous test. In 1974, after three years in Cameroon, Aspect returned to France.

The 27-year-old set about making his African dream come true in a basement laboratory at the Institut d'Optique Théoretique et Appliquée, Université Paris-Sud in Orsay. 'Do you have a permanent position?' Bell asked, when Aspect went to see him in Geneva.[49] Aspect explained that he was just a graduate student aiming for a doctorate. 'You must be a very courageous graduate student', replied Bell.[50] He was concerned that the young Frenchman could be damaging his future prospects by attempting to conduct such a difficult experiment.

It took longer than he imagined at the outset, but in 1981 and 1982 Aspect and his collaborators used the latest technological innovations, including lasers and computers, to perform not one but three delicate experiments to test Bell's inequality. Like Clauser, Aspect measured the correlation of the polarisation of entangled pairs of photons moving in opposite directions after being simultaneously emitted from individual calcium atoms. However, the rate at which photon pairs were created and measured was many times higher. His experiments revealed, said Aspect, 'the strongest violation of Bell's inequalities ever achieved, and excellent agreement with quantum mechanics'.[51]

Bell was one of the examiners when Aspect received his doctorate in 1983, but some doubts remained concerning the results. Since the nature of quantum reality hung in the balance, every possible loophole, however improbable, had to be considered. For example, the possibility that the detectors might somehow be signalling to each other was later eliminated by the random switching of their orientation while the photons were in mid-flight. Although it fell short of being the definitive experiment, further refinements and other investigations in the years since have led to Aspect's original results being confirmed. Although no experiment has been conducted in which every possible loophole is closed, most physicists accept that Bell's inequality has been violated.

Bell derived the inequality from just two assumptions. First, there exists an observer-independent reality. This translates into a particle having a well-defined property such as spin before it is measured. Second, locality is preserved. There is no faster-than-light influence, so that what happens here cannot possibly instantaneously affect what happens way over there. Aspect's results mean that one of these two assumptions has to be given up, but which one? Bell was prepared to give up locality. 'One wants to be able to take a realistic view of the world, to talk about the world as if it is really there, even when it is not being observed', he said.[52]

Bell, who died in October 1990 at the age of 62 from a brain haemorrhage, was convinced that 'quantum theory is only a temporary expedient' that would eventually be replaced by a better theory.[53] Nevertheless, he conceded that experiments had shown that 'Einstein's world view is not tenable'.[54] Bell's theorem tolled for Einstein and local reality.

Chapter 15

THE QUANTUM DEMON

'I thought a hundred times as much about the quantum problems as I have about general relativity theory', Einstein once admitted.[1] Bohr's rejection of the existence of an objective reality as he tried to understand what quantum mechanics was telling him about the atomic world was a sure sign for Einstein that the theory contained, at best, only a part of the whole truth. The Dane insisted that there is no quantum reality beyond what is revealed by an experiment, an act of observation. 'To believe this is logically possible without contradiction,' Einstein conceded, 'but it is so very contrary to my scientific instinct that I cannot forgo the search for a more complete conception.'[2] He continued to 'believe in the possibility of giving a model of reality which shall represent events themselves and not merely the probability of their occurrence'.[3] Yet, in the end, he failed to refute Bohr's Copenhagen interpretation. 'About relativity he spoke with detachment, about quantum theory with passion', recalled Abraham Pais, who had known Einstein in Princeton.[4] 'The quantum was his demon.'

———————

'I think I can safely say that nobody understands quantum mechanics', said the celebrated American Nobel laureate Richard Feynman in 1965, ten years after Einstein's death.[5] With the Copenhagen interpretation as firmly established as the quantum orthodoxy as any papal edict issued from Rome, most physicists simply followed Feynman's advice. 'Do not keep asking yourself, if you can possibly avoid it, "but how can it be like that?"' he warned.[6] 'Nobody knows how it can be like that.' Einstein never thought it was like that, but what would he have thought of Bell's theorem and the experiments showing that it tolled for him?

At the core of Einstein's physics was his unshakeable belief in a reality that exists 'out there' independently of whether or not it is observed. 'Does the moon exist only when you look at it?' he asked Abraham Pais in an attempt to highlight the absurdity of thinking otherwise.[7] The reality that Einstein envisaged had locality and was governed by causal laws that it was the job of the physicist to discover. 'If one abandons the assumption that what exists in different parts of space has its own independent, real existence,' he told Max Born in 1948, 'then I simply cannot see what it is that physics is meant to describe.'[8] Einstein believed in a realism, causality, and locality. Which, if any, would he have been prepared to sacrifice?

'God does not play dice', said Einstein memorably and often.[9] Just like any modern-day advertising copywriter, he knew the value of an unforgettable tagline. It was his snappy denunciation of the Copenhagen interpretation and not a cornerstone of his scientific worldview. This was not always clear, even to someone like Born who knew him for almost half a century. It was Pauli who eventually explained to Born what really lay at the heart of Einstein's opposition to quantum mechanics.

During Pauli's two-month stay in Princeton in 1954, Einstein gave him a draft of a paper written by Born that touched on determinism. Pauli read it and wrote to his old boss that 'Einstein does not consider the concept of "determinism" to be as fundamental as it is frequently held to be.'[10] It was something that Einstein told him 'emphatically many times' over the years.[11] 'Einstein's point of departure is "realistic" rather than "deterministic",' explained Pauli, 'which means that his philosophical prejudice is a different one.'[12] By 'realistic' Pauli meant that Einstein assumed that electrons, for example, have pre-existing properties prior to any act of measurement. He accused Born of having 'erected some dummy Einstein for yourself, which you then knocked down with great pomp'.[13] Surprisingly, Born, given their long friendship, had never fully grasped that what really troubled Einstein was not dice-playing, but the Copenhagen interpretation's 'renunciation of the representation of a reality thought of as independent of observation'.[14]

One possible reason for the misunderstanding may be that Einstein first said that God 'is not playing at dice' in December 1926 when he tried to convey to Born his unease at the role of probability and chance in quantum mechanics and the rejection of causality and determinism.[15] Pauli, however, understood that Einstein's objections went far beyond the theory being expressed in the language of probability. 'In particular it seems to me misleading to bring the concept of determinism into the dispute with Einstein', he warned Born.[16]

At the heart of the problem,' wrote Einstein in 1950 of quantum mechanics, 'is not so much the question of causality but the question of realism.'[17] For years he had hoped that he 'may yet work out the quantum puzzle without having to renounce the representation of reality'.[18] For the man who discovered relativity, that reality had to be local, with no place for faster-than-light influences. The violation of Bell's inequality meant that if he wanted a quantum world that existed independently of observers, then Einstein would have had to give up locality.

Bell theorem cannot decide whether quantum mechanics is complete or not, but only between it and any local hidden variables theory. If quantum mechanics is correct – and Einstein believed it was, since it had passed every experimental test in his day – then Bell's theorem implied that any hidden variables theory that replicated its results had to be non-local. Bohr would have regarded, as others do, the results of Alain Aspect's experiments as support for the Copenhagen interpretation. Einstein would probably have accepted the validity of the results testing Bell's inequality without attempting to save local reality through one of the loopholes in these experiments that remained to be closed. However, there was another way out that Einstein might have accepted, even though some have said that it violates the spirit of relativity – the no signalling theorem.

It was discovered that it is impossible to exploit non-locality and quantum entanglement to communicate useful information instantaneously from one place to another, since any measurement of one particle of an entangled pair produces a completely random result. After performing such a measurement, an experimenter learns nothing more than the probabilities of the outcome of a possible measurement on the other entangled particle conducted at a distant location by a colleague. Reality may be non-local, allowing faster-than-light influences between entangled pairs of particles in separate locations, but it is benign, with no 'spooky communication at a distance'.

Whereas Aspect's team and others who tested Bell's inequality ruled out either locality or an objective reality but allowed a non-local reality, in 2006 a group from the universities of Vienna and Gdansk became the first to put non-locality *and* realism to the test. The experiment was inspired by the work of the British physicist Sir Anthony Leggett. In 1973 and not yet knighted, Leggett had the idea of amending Bell's theorem by assuming the existence of instantaneous influences passing between entangled particles. In 2003, the year he won the Nobel Prize for his work on the quantum properties of liquid helium, Leggett published a new inequality that pitted non-local hidden variable theories against quantum mechanics.

The Austrian-Polish group led by Markus Aspelmeyer and Anton Zeilinger measured previously untested correlations between pairs of entangled photons. They found that the correlations violated Leggett's inequality, just as quantum mechanics predicted. When the results were published in the journal *Nature*, in April 2007, Alain Aspect pointed out that the philosophical 'conclusion one draws is more a question of taste than logic'.[19] The violation of Leggett's inequality implies only that realism and a certain type of non-locality are incompatible; it did not rule out *all* possible non-local models.

Einstein never proposed a hidden variables theory, even though he seemed to implicitly advocate such an approach in 1935 at the end of the EPR paper: 'While we have thus shown that the wave function does not provide a complete description of the physical reality, we left open the question of whether or not such a description exists. We believe, however, that such a theory is possible.'[20] And as late as 1949, in a reply to those who had contributed to a collection of papers to mark his 70th birthday, Einstein wrote: 'I am, in fact, firmly convinced that the essentially statistical character of contemporary quantum theory is solely to be ascribed to the fact that this [theory] operates with an incomplete description of physical systems.'[21]

The introduction of hidden variables to 'complete' quantum mechanics seemed to be in accordance with Einstein's view that the theory is 'incomplete', but by the beginning of the 1950s he was no longer sympathetic to any such attempt to complete it. By 1954 he was adamant that 'it is not possible to get rid of the statistical character of the present quantum theory by merely adding something to the latter, without changing the fundamental concepts about the whole structure'.[22] He was convinced that something more radical was required than a return to the concepts of classical physics at the sub-quantum level. If quantum mechanics is incomplete, only a part of the whole truth, then there must be a complete theory waiting to be discovered.

Einstein believed that this was the elusive unified field theory that he spent the last 25 years of his life searching for – the marriage of general relativity with electromagnetism. It would be the complete theory that would contain within it quantum mechanics. 'What God has put asunder, let no man join together', was Pauli's caustic judgement on Einstein's dream of unification.[23] Although at the time most physicists ridiculed Einstein as out of touch, the search for such a theory would become the holy grail of physics as the discoveries of the weak nuclear force responsible for radioactivity and the strong nuclear force that held the nucleus together brought the number of forces that physicists had to contend with to four.

When it came to quantum mechanics there were those, like Werner Heisenberg, who simply accused Einstein of being 'unable to change his attitude' after a career spent probing the 'objective world of physical processes which runs its course in space and time, independent of us, according to firm laws'.[24] It was hardly surprising, Heisenberg implied, that Einstein found it impossible to accept a theory asserting that, on the atomic scale, 'this objective world of time and space did not even exist'.[25] Born believed that Einstein 'could no longer take in certain new ideas in physics which contradicted his own firmly held philosophical convictions'.[26] While acknowledging that his old friend had been 'a pioneer in the struggle for conquering the wilderness of quantum phenomena', the fact that 'he kept himself aloof and sceptical' about quantum mechanics, Born lamented, was a 'tragedy', as Einstein 'gropes his way in loneliness, and for us who miss our leader and standard-bearer'.[27]

As Einstein's influence waned, Bohr's grew. With missionaries like Heisenberg and Pauli spreading the message among their own flocks, the Copenhagen interpretation became synonymous with quantum mechanics. When he was a student in the 1960s, John Clauser was often told that Einstein and Schrödinger 'had become senile' and their opinions on matters quantum could not be trusted.[28] 'This gossip was repeated to me by a large number of well-known physicists from many different prestigious institutions', he recalled years after becoming the first to test Bell's inequality in 1972. In stark contrast, Bohr was deemed to possess almost supernatural powers of reasoning and intuition. Some have even suggested that while others needed to perform calculations, Bohr did not.[29] Clauser recalled that during his student days 'open inquiry into the wonders and peculiarities of quantum mechanics' that went beyond the Copenhagen interpretation was 'virtually prohibited by the existence of various religious stigmas and social pressures, that taken together, amounted to an evangelical crusade against such thinking'.[30] But there were unbelievers prepared to challenge the Copenhagen orthodoxy. One of them was Hugh Everett III.

When Einstein died in April 1955, Everett was 24 and studying for his master's degree at Princeton University. Two years later he obtained a PhD with a thesis entitled 'On the Foundations of Quantum Mechanics' in which he demonstrated that it was possible to treat each and every possible outcome of a quantum experiment as actually existing in a real world. According to Everett, for Schrödinger's cat trapped in its box this would mean that the moment the box was opened the universe would divide, leaving one universe in which the cat was dead and another in which it was still alive.

Everett called his interpretation the 'relative state formulation of quantum mechanics' and showed that his assumption that all quantum possibilities exist led to the same quantum mechanical predictions for the results of experiments as the Copenhagen interpretation.

Everett published his alternative in July 1957 with an accompanying note from his supervisor, the distinguished Princeton physicist John Wheeler. It was his very first paper and it went virtually unnoticed for more than a decade. By then disillusioned by the lack of interest, Everett had already left academia and was working for the Pentagon, applying game theory to strategic war planning.

'There is no question that there is an unseen world', the American film director Woody Allen once said. 'The problem is how far is it from midtown and how late is it open?'[31] Unlike Allen, most physicists balked at the implications of accepting an infinite number of co-existing parallel alternative realities in which every conceivable outcome of every possible experimental result is realised. Sadly, Everett, who died of a heart attack aged 51 in 1982, did not live to see the 'many worlds interpretation', as it became known, taken seriously by quantum cosmologists as they struggled to explain the mystery of how the universe came into being. The many worlds interpretation allowed them to circumvent a problem to which the Copenhagen interpretation had no answer – what act of observation could possibly bring about the collapse of the wave function of the entire universe?

The Copenhagen interpretation requires an observer outside the universe to observe it, but since there is none – leaving God aside – the universe should never come into existence but remain forever in a superposition of many possibilities. This is the long-standing measurement problem writ large. Schrödinger's equation that describes quantum reality as a superposition of possibilities, and attaches a range of probabilities to each possibility, does not include the act of measurement. There are no observers in the mathematics of quantum mechanics. The theory says nothing about the collapse of the wave function, the sudden and discontinuous change of the state of a quantum system upon observation or measurement, when the possible becomes the actual. In Everett's many worlds interpretation there was no need for an observation or measurement to collapse the wave function, since each and every quantum possibility coexists as an actual reality in an array of parallel universes.

'This problem of getting the interpretation proved to be rather more difficult than just working out the equations', said Paul Dirac 50 years after the 1927 Solvay conference.[32] The American Nobel laureate Murray Gell-Mann believes part of the reason was that 'Niels Bohr brain-washed a whole generation of physicists into believing that the problem had been solved'.[33] A poll conducted in July 1999 during a conference on quantum physics held at Cambridge University revealed the answers of a new generation to the vexed question of interpretation.[34] Of the 90 physicists polled, only four voted for the Copenhagen interpretation, but 30 favoured the modern version of Everett's many worlds.[35] Significantly, 50 ticked the box labelled 'none of the above or undecided'.

The unresolved conceptual difficulties, such as the measurement problem and the inability to say exactly where the quantum world ends and the classical world of the everyday begins, have led to an increasing number of physicists willing to look for something deeper than quantum mechanics. 'A theory that yields "maybe" as an answer,' says the Dutch Nobel Prize-winning theorist Gerard 't Hooft, 'should be recognized as an inaccurate theory.'[36] He believes the universe is deterministic, and is in search of a more fundamental theory that would account for all the strange, counter-intuitive features of quantum mechanics. Others like Nicolas Gisin, a leading experimenter exploring entanglement, 'have no problem thinking that quantum theory is incomplete'.[37]

The emergence of other interpretations and the claim to completeness of quantum mechanics being in serious doubt have led to a reconsideration of the long-standing verdict against Einstein in his long-running debate with Bohr. 'Can it really be true that Einstein, in any significant sense, was as profoundly "wrong" as the followers of Bohr might maintain?' asks the British mathematician and physicist Sir Roger Penrose. 'I do not believe so. I would, myself, side strongly with Einstein in his belief in a submicroscopic reality, and with his conviction that present-day quantum mechanics is fundamentally incomplete.'[38]

Although he never managed to deliver a decisive blow in his encounters with Bohr, Einstein's challenge was sustained and thought-provoking. It encouraged men like Bohm, Bell and Everett to probe and evaluate Bohr's Copenhagen interpretation when it was all-prevailing and few distinguished theory from interpretation. The Einstein-Bohr debate about the nature of reality was the inspiration behind Bell's theorem. The testing of Bell's inequality directly or indirectly helped spawn new areas of research including quantum cryptography, quantum information theory, and quantum computing. Among the most remarkable of these new fields is quantum teleportation, which exploits the phenomena of entanglement. Although it appears to belong to the realm of science fiction, in 1997 not one but two teams of physicists succeeded in teleporting a particle. The particle was not physically transported, but its quantum state was transferred to a second particle located elsewhere, thereby effectively teleporting the initial particle from one place to another.

After having been marginalised during the last 30 years of his life because of his criticism of the Copenhagen interpretation and his attempts to slay his quantum demon, Einstein has been vindicated, in part. Einstein versus Bohr had little to do with the equations and numbers generated by the mathematics of quantum mechanics. What does quantum mechanics mean? What does it say about the nature of reality? It was their answers to these types of questions that separated the two men. Einstein never put forward an interpretation of his own, because he was not trying to shape his philosophy to fit a physical theory. Instead he used his belief in an observer-independent reality to assess quantum mechanics and found the theory wanting.

In December 1900, classical physics had a place for everything and almost everything in its place. Then Max Planck stumbled across the quantum, and physicists are still struggling to come to terms with it. Fifty long years of 'conscious brooding', said Einstein, had not brought him any closer to understanding the quantum.[39] He kept trying to the end, taking solace in the words of the German playwright and philosopher Gotthold Lessing: 'The aspiration to truth is more precious than its assured possession.'[40]

TIMELINE

1858	23 April: Max Planck is born in Kiel, Germany.
1871	30 August: Ernest Rutherford is born in Spring Grove, New Zealand.
1879	14 March: Albert Einstein is born in Ulm, Germany.
1882	11 December: Max Born is born in Breslau, Silesia, Germany.
1885	7 October: Niels Bohr is born in Copenhagen, Denmark.
1887	12 August: Erwin Schrödinger is born in Vienna, Austria.
1892	15 August: Louis de Broglie is born in Dieppe, France.
1893	**February:** Wilhelm Wien discovers the displacement law for blackbody radiation.
1895	**November:** Wilhelm Röntgen discovers X-rays.
1896	**March:** Henri Becquerel discovers that uranium compounds emit previously unknown radiation that he calls 'uranic rays'. **June:** Wien publishes a distribution law for blackbody radiation that is in agreement with the available data.
1897	**April:** J.J. Thomson announces the discovery of the electron.
1900	25 April: Wolfgang Pauli is born in Vienna, Austria. **July:** Einstein graduates from the Federal Polytechnikum in Zurich. **September:** The breakdown of Wien's distribution law is confirmed beyond any doubt in the far infrared part of the blackbody spectrum. **October:** Planck announces his blackbody radiation law at a meeting in Berlin of the German Physical Society. **14 December:** Planck presents the derivation of his blackbody radiation law in a lecture to the German Physical Society. The introduction of the quantum of energy is barely noticed. At best, it is deemed to be a theorist's sleight of hand to be eliminated later.

1901	**5 December:** Werner Heisenberg is born in Würzburg, Germany.
1902	**June:** Einstein begins work as an 'Expert Class III' at the Patent Office in Bern, Switzerland.
	8 August: Paul Dirac is born in Bristol, England.
1905	**June:** Einstein's paper on the existence of light-quanta and the photoelectric effect is published in *Annalen der Physik*.
	July: Einstein's paper explaining Brownian motion is published in *Annalen der Physik*.
	September: Einstein's paper 'On the Electrodynamics of Moving Bodies', outlining his special theory of relativity, is published in *Annalen der Physik*.
1906	**January:** Einstein receives his PhD from Zurich University at the third attempt with a thesis entitled 'A New Determination of Molecular Dimensions'.
	April: Einstein is promoted to 'Expert Class II' at the Patent Office in Bern.
	September: Ludwig Boltzmann commits suicide while on holiday near Trieste, Italy.
	December: Einstein's paper on the quantum theory of specific heat is published in *Annalen der Physik*.
1907	**May:** Rutherford takes up the post of professor and head of physics at Manchester University.
1908	**February:** Einstein becomes *privatdozent* at Bern University.
1909	**May:** Einstein is appointed extraordinary professor of theoretical physics at Zurich University, effective the following October.
	September: Einstein delivers the keynote lecture at the annual meeting of the Gesellschaft Deutscher Naturforscher und Ärzte, held that year in Salzburg, Austria. Einstein says that 'the next stage in the development of theoretical physics will bring us a theory of light that may be conceived of as a sort of fusion of the wave and of the emission theory of light'.
	December: Bohr receives his Master's degree from Copenhagen University.

1911	**January:** Einstein is appointed to full professorship at the German University in Prague. The appointment begins in April 1911.
	March: Rutherford announces the discovery of the atomic nucleus at a meeting in Manchester, England.
	May: Bohr receives his doctorate from Copenhagen University with a thesis on the electron theory of metals.
	September: Bohr arrives at Cambridge University to begin postgraduate work with J.J. Thomson.
	30 October–4 November: The first Solvay conference is held in Brussels. Einstein, Planck, Marie Curie and Rutherford are among the invited participants.
1912	**January:** Einstein is appointed professor of theoretical physics at the Eidgenossische Technische Hochschule (ETH) in Zurich, the new name for the Federal Polytechnikum where he was a student.
	March: Bohr transfers from Cambridge to Rutherford's laboratory at Manchester University.
	September: Bohr is appointed *privatdozent* and assistant to the professor of physics at Copenhagen University.
1913	**February:** Bohr hears about Balmer's formula for the spectral lines of hydrogen for the first time, a vital clue as he develops the quantum model of the atom.
	July: The first in a trilogy of papers by Bohr on the quantum theory of the hydrogen atom is published in the *Philosophical Magazine*. Planck and Walther Nernst travel to Zurich to entice Einstein to Berlin. He accepts their offer.
	September: Bohr presents his new theory of the quantum atom at the British Association for the Advancement of Science (BAAS) conference in Birmingham, England.
1914	**April:** The Franck-Hertz experiment confirms Bohr's concept of quantum jumps and atomic energy levels. They bombard mercury vapour with electrons and measure the frequencies of the emitted radiation, which corresponds to the transitions between different energy levels. Einstein arrives in Berlin to take up professorships at the Prussian Academy of Sciences and Berlin University.

1914 *(cont.)*	**August:** The First World War begins.
	October: Bohr returns to work at Manchester University. Planck and Röntgen are among the signatories of the *Manifesto of the Ninety-Three*, asserting that Germany bears no responsibility for the war, has not violated Belgian neutrality, and committed no atrocities.
1915	**November:** Einstein completes his general theory of relativity.
1916	**January:** Arnold Sommerfeld proposes a theory to explain the fine structure of the spectral lines in hydrogen and introduces a second quantum number as he replaces Bohr's circular orbits with elliptical orbits.
	May: Bohr is appointed professor of theoretical physics at Copenhagen University.
	July: Einstein returns to work on quantum theory and discovers the phenomena of spontaneous and induced emission of a photon from an atom. Sommerfeld adds the magnetic quantum number to Bohr's original atomic model.
1918	**September:** Pauli leaves Vienna to study at Munich University with Arnold Sommerfeld.
	November: The First World War ends.
1919	**November:** Planck is awarded the 1918 Nobel Prize for physics. At a joint meeting of the Royal Society and the Royal Astronomical Society in London, an official announcement is made that Einstein's prediction that light is deflected by a gravitational field was confirmed by measurements made by two British expeditions during a solar eclipse in May. Einstein becomes a global celebrity overnight.
1920	**March:** Sommerfeld introduces a fourth quantum number.
	April: Bohr visits Berlin and meets Planck and Einstein for the first time.
	August: A public rally at the Berlin Philharmonic Hall against relativity theory. An angry Einstein replies to his critics in a newspaper article. He visits Bohr in Copenhagen for the first time.
	October: Heisenberg enrols to study physics at Munich University and meets fellow student Wolfgang Pauli.

1921   **March:** With Bohr as its founder and director, the Institute for Theoretical Physics in Copenhagen is officially opened.

April: Born arrives in Göttingen from Frankfurt as professor and director of the institute of theoretical physics, determined to make it the equal of Sommerfeld's institute in Munich.

October: After obtaining his doctorate from Munich University, Pauli becomes Born's assistant in Göttingen.

1922   **April:** Preferring city life to that in a small, provincial university town, Pauli leaves Göttingen to take up an assistant's position at Hamburg University.

June: Bohr gives a series of celebrated lectures in Göttingen on atomic theory and the periodic table. At this 'Bohr Festspiele', Heisenberg and Pauli meet the Dane for the first time. Bohr is deeply impressed by both young men.

October: Heisenberg begins a six-months' sojourn in Göttingen with Born. Pauli arrives in Copenhagen to be Bohr's assistant until September 1923.

November: Einstein is awarded the 1921 Nobel Prize and Bohr the prize for 1922.

1923   **May:** Arthur Compton's comprehensive report concerning his discovery of the scattering of X-ray photons by atomic electrons is published. The 'Compton effect', as it became known, is taken as irrefutable evidence in support of Einstein's 1905 light-quanta hypothesis.

July: Einstein's second visit to see Bohr in Copenhagen. Heisenberg just manages to obtain his doctorate from Munich University after poorly answering questions on experimental physics during his oral examination.

September: De Broglie links waves with electrons as he extends wave-particle duality to incorporate matter.

October: Heisenberg becomes Born's assistant in Göttingen. Pauli returns to Hamburg after a year-long stay in Copenhagen.

1924 **February:** Bohr, Hendrik Kramers and John Slater propose that in atomic processes energy is only conserved statistically, in an attempt to counter Einstein's light-quanta hypothesis. The BKS idea is experimentally disproved in April–May 1925.

March: Heisenberg pays his first visit to Bohr in Copenhagen.

September: Heisenberg leaves Göttingen to work at Bohr's institute until May 1925.

November: De Broglie successfully defends his doctoral thesis extending wave-particle duality to matter. Sent a copy of the thesis by de Broglie's supervisor, Einstein had earlier given it his nod of approval.

1925 **January:** Pauli discovers the exclusion principle.

June: Heisenberg goes to the small island of Helgoland in the North Sea to recover from a severe bout of hay fever. During his stay he takes the all-important first steps towards matrix mechanics, his version of the much sought-after theory of quantum mechanics.

September: Heisenberg's first ground-breaking paper on matrix mechanics, 'On a Quantum-Theoretical Reinterpretation of Kinematics and Mechanical Relations', is published in the *Zeitschrift für Physik*.

October: Samuel Goudsmit and George Uhlenbeck propose the concept of quantum spin.

November: Pauli applies matrix mechanics to the hydrogen atom. A veritable tour de force, it is published in March 1926.

December: While enjoying a secret rendezvous with a former lover in the Alpine ski resort of Arosa, Schrödinger constructs what will become his celebrated wave equation.

1926 **January:** Back in Zurich, Schrödinger applies his wave equation to the hydrogen atom and finds that it reproduces the series of energy levels of the Bohr-Sommerfeld hydrogen atom.

February: The three-man paper written by Heisenberg, Born and Pascual Jordan offering a detailed account of the mathematical structure of matrix mechanics is published after being submitted to the *Zeitschrift für Physik* in November 1925.

1926
(cont.)
March: Schrödinger's first paper on wave mechanics is published in the *Annalen der Physik* after being submitted in January. Another five papers follow in quick succession. Schrödinger and others prove that wave mechanics and matrix mechanics are mathematically equivalent. They are two forms of the same theory – quantum mechanics.

April: Heisenberg delivers a two-hour lecture on matrix mechanics attended by Einstein and Planck. Afterwards Einstein invites the young turk back to his apartment where the two of them discuss, Heisenberg recalled later, 'the philosophical background of my recent work'.

May: Heisenberg is appointed Bohr's assistant and lecturer at Copenhagen University. As Bohr recovers from a severe case of flu, Heisenberg begins using wave mechanics to account for the spectral lines of helium.

June: Dirac receives his PhD from Cambridge University with a thesis entitled 'Quantum Mechanics'.

July: Born puts forward the probability interpretation of the wave function. Schrödinger delivers a lecture in Munich and during the question-and-answer session, Heisenberg complains about the shortcomings of wave mechanics.

September: Dirac goes to Copenhagen and during his stay develops transformation theory, which shows that Schrödinger's wave mechanics and Heisenberg's matrix mechanics are special cases of a more general formulation of quantum mechanics.

October: Schrödinger visits Copenhagen. He, Bohr and Heisenberg fail to reach any sort of accord over the physical interpretation of either matrix or wave mechanics.

1927
January: Clinton Davisson and Lester Germer obtain conclusive evidence that wave-particle duality also applies to matter as they succeed in diffracting electrons.

February: After months of trying, tempers fray as Bohr and Heisenberg are no closer to developing a coherent physical interpretation of quantum mechanics. Bohr leaves on a month-long skiing holiday in Norway. In Bohr's absence, Heisenberg discovers the uncertainty principle.

1927 *(cont.)*	**May:** The uncertainty principle is published after arguments between Heisenberg and Bohr over its interpretation.
	September: The Volta conference at Lake Como, Italy. Bohr presents his principle of complementarity and the central elements of what later became known as the Copenhagen interpretation of quantum mechanics. Born, Heisenberg and Pauli are among those present, but not Schrödinger or Einstein.
	October: At the fifth Solvay conference in Brussels, the Einstein-Bohr debate begins over the foundations of quantum mechanics and the nature of reality. Schrödinger succeeds Planck as professor of theoretical physics at Berlin University. Compton is awarded the Nobel Prize for the discovery of the 'Compton effect'. Heisenberg, aged only 25, is appointed to a professorship at Leipzig University.
	November: George Thomson, son of J.J. Thomson, the discoverer of the electron, reports the successful diffraction of electrons employing a different technique than Davisson and Germer.
1928	**January:** Pauli is appointed professor of theoretical physics at the ETH in Zurich.
	February: Heisenberg delivers his inaugural lecture as professor of theoretical physics at Leipzig University.
1929	**October:** De Broglie receives the Nobel Prize for the discovery of the wave nature of the electron.
1930	**October:** The sixth Solvay conference in Brussels, the second round of the Einstein-Bohr debate as Bohr refutes Einstein's 'clock-in-the-box' thought experiment challenging the consistency of the Copenhagen interpretation.
1931	**December:** The Danish Academy of Sciences and Letters selects Bohr as the next occupant of the Aeresbolig, 'The House of Honour', a mansion built by the founder of the Carlsberg breweries.
1932	John von Neumann's book *The Mathematical Foundations of Quantum Mechanics* is published in German. It contains his famous 'impossibility proof' – no hidden variables theory can reproduce the predictions of quantum mechanics. Dirac is elected Lucasian Professor of Mathematics at Cambridge University – a post once held by Isaac Newton.

1933   **January:** The Nazis seize power in Germany. Luckily, Einstein is in America as a visiting professor at the California Institute of Technology.

March: Einstein publicly declares that he will not return to Germany. He resigns from the Prussian Academy of Sciences as soon as he arrives in Belgium and severs all links with official German institutions.

April: The Nazis introduce the 'Law for the Restoration of the Career Civil Service', designed to target political opponents, socialists, communists, and the Jews. Paragraph 3 contains the infamous 'Aryan clause': 'Civil servants not of Aryan origin are to retire.' By 1936 more than 1,600 scholars would be ousted, a third of them scientists, including twenty who had been or would be awarded the Nobel Prize.

May: 20,000 books are burned in Berlin, with similar bonfires of 'un-German' works throughout the country. Although unaffected by Nazi regulations, unlike Born and many other colleagues, Schrödinger leaves Germany for Oxford. Heisenberg stays. The Academic Assistance Council, with Rutherford as its president, is set up in England to help refugee scientists, artists and writers.

September: As fears over his safety increase, Einstein leaves Belgium for England. Paul Ehrenfest commits suicide.

October: Einstein arrives in Princeton, New Jersey for a scheduled visit. Intending to stay for only a few months at the Institute for Advanced Study (IAS), Einstein never returns to Europe.

November: Heisenberg receives the deferred 1932 Nobel Prize, while Dirac and Schrödinger share the prize for 1933.

1935   **May:** The Einstein, Podolsky and Rosen (EPR) paper, 'Can Quantum Mechanical Description of Physical Reality Be Considered Complete?', is published in the *Physical Review*.

October: Bohr's reply to EPR is published in the *Physical Review*.

1936   **March:** Schrödinger and Bohr meet in London. Bohr says that it's 'appalling' and 'high treason' that Schrödinger and Einstein want to strike a blow against quantum mechanics.

October: Born takes up a post as professor of natural philosophy at Edinburgh University after spending nearly three years at Cambridge and a few months in Bangalore, India. He stayed until his retirement in 1953.

1937 **February:** Bohr arrives in Princeton for a week-long stay as part of a world tour. Einstein and Bohr discuss the interpretation of quantum mechanics face-to-face for the first time since the publication of the EPR paper, but talk past each other as many things are left unsaid.

July: Heisenberg is branded a 'white Jew' in an SS journal for teaching 'Jewish' physics such as Einstein's theory of relativity.

October: Rutherford dies aged 66 in Cambridge after surgery for a strangulated hernia.

1939 **January:** Bohr arrives at the IAS as a visiting professor for the entire semester. Einstein avoids any discussions with Bohr, and during the next four months they meet only once at reception.

August: Einstein signs a letter to President Roosevelt raising the possibility of making an atomic bomb and the danger of the Germans constructing such a weapon.

September: The Second World War begins.

October: Schrödinger arrives in Dublin after stints at the universities of Graz and Ghent. He remained in Dublin as senior professor at the Institute for Advanced Studies until 1956 when he returned to Vienna.

1940 **March:** Einstein sends a second letter to President Roosevelt concerning the atomic bomb.

August: Pauli leaves war-torn Europe and joins Einstein at the Institute for Advanced Study in Princeton. He remained there until 1946 when he returned to Zurich and the ETH.

1941 **October:** Heisenberg visits Bohr in Copenhagen. Denmark had been occupied by German forces since April 1940.

1943 **September:** Bohr and his family escape to Sweden.

December: Bohr visits Princeton to have dinner with Einstein and Pauli before heading to Los Alamos in New Mexico to work on the atomic bomb. It was the first meeting between Einstein and Bohr since the Dane's visit in January 1939.

1945 **May:** Germany surrenders. Heisenberg is arrested by Allied forces.

August: Atomic bombs are dropped on Hiroshima and then Nagasaki. Bohr returns to Copenhagen.

November: Pauli is awarded the Nobel Prize for the discovery of the exclusion principle.

1946	**July:** Heisenberg is appointed director of the Kaiser Wilhelm Institute for Physics in Göttingen, later renamed the Max Planck Institute.
1947	**October:** Planck dies in Göttingen aged 89.
1948	**February:** Bohr arrives at the IAS as a visiting professor until June. Relations with Einstein are more cordial than during previous visits as both men continue to disagree over the interpretation of quantum mechanics. In Princeton, Bohr writes an account of the debate with Einstein at the Solvay conferences of 1927 and 1930 as his contribution to a volume of papers to celebrate Einstein's 70th birthday in March 1949.
1950	**February:** Bohr is at the IAS until May.
1951	**February:** David Bohm publishes his book *Quantum Theory*. It contains a novel and simplified version of the EPR thought experiment.
1952	**January:** Two papers by Bohm are published in which he does what von Neumann said was impossible: he offers a hidden variables interpretation of quantum mechanics.
1954	**September:** Bohr is at the IAS until December.
	October: Bitterly disappointed at being overlooked when Heisenberg was honoured in 1932, Born is finally awarded the Nobel Prize for 'his fundamental work in quantum mechanics and especially for his statistical interpretation of the wave function'.
1955	**April:** Einstein dies in Princeton aged 76. After a simple ceremony, his ashes are scattered at an undisclosed location.
1957	**July:** Hugh Everett III puts forward the 'relative state' formulation of quantum mechanics, later known as the many worlds interpretation.
1958	**December:** Pauli dies in Zurich aged 58.
1961	**January:** Schrödinger dies in Vienna aged 73.
1962	**November:** Bohr dies in Copenhagen aged 77.
1964	**November:** John Bell's discovery that any hidden variables theory whose predictions agree with those of quantum mechanics must be non-local is published in a little-read journal. Known as Bell's inequality, it derives limits on the degree of correlation of the quantum spins of entangled pairs of particles that have to be satisfied by any local hidden variables theory.

1966	**July:** Bell shows conclusively that von Neumann's proof ruling out hidden variables theories, published in 1932 in his book *The Mathematical Foundations of Quantum Mechanics*, is flawed. Bell had submitted his paper to the journal *Review of Modern Physics* at the end of 1964, but an unfortunate series of mishaps delayed its publication.
1970	**January:** Born dies in Göttingen aged 87.
1972	**April:** John Clauser and Stuart Freedman at the University of California, Berkeley, having conducted the first test of Bell's inequality, report that it is violated – any local hidden variables cannot reproduce the predictions of quantum mechanics. However, there are doubts about the accuracy of their results.
1976	**February:** Heisenberg dies in Munich aged 75.
1982	After years of preliminary work, Alain Aspect and his collaborators at the Institut d'Optique Théoretique et Appliquée, Université Paris-Sud, subject Bell's inequality to the most rigorous test then possible. Their results show that the inequality is violated. Although certain loopholes remain to be closed, most physicists, including Bell, accept the results.
1984	**October:** Dirac dies in Tallahassee, Florida aged 82.
1987	**March:** De Broglie dies in France aged 94.
1997	**December:** A team at the University of Innsbruck led by Anton Zeilinger reports that it has succeeded in transferring the quantum state of a particle from one place to another – in effect, teleporting it. An integral part of the process is the phenomenon of quantum entanglement. A group at Rome University, under the leadership of Francesco DeMartini, also successfully carries out quantum teleportation.
2003	**October:** Anthony Leggett publishes a Bell-type inequality derived on the basis that reality is non-local.
2007	**April:** An Austrian-Polish team led by Markus Aspelmeyer and Anton Zeilinger announce that measurements of previously untested correlations between pairs of entangled photons show that Leggett's inequality is violated. The experiment rules out only a subset of possible non-local hidden variables theories.
20??	A quantum theory of gravity? A Theory of Everything? A theory beyond the quantum?

GLOSSARY

Terms in *italics* have an entry in the glossary.

Alkali elements   Elements such as lithium, sodium and potassium in group one of the *periodic table* that share the same chemical properties.

Alpha decay   A process of radioactive decay in which the *nucleus* of an *atom* emits an *alpha particle*.

Alpha particle   A subatomic particle consisting of two *protons* and two *neutrons* bound together. Emitted during *alpha decay*, it is identical to the *nucleus* of a helium *atom*.

Amplitude   The maximum displacement of a *wave* or an oscillation that is equal to half the distance from the top of the wave (or oscillation) to the bottom. In *quantum mechanics*, the amplitude of a process is a number that is linked to the probability of that process occurring.

Angular momentum   A property of a rotating object akin to the *momentum* of an object moving in a straight line. The angular momentum of an object depends on its mass, its size, and the speed with which it is spinning. One object orbiting another also possesses angular momentum that depends on its mass, the radius of its orbit, and its velocity. In the atomic realm, angular momentum is *quantised*. It can change only by amounts that are whole-number multiples of *Planck's constant* divided by 2π.

Atom   The smallest component of an element consisting of a positively-charged *nucleus* surrounded by a bound system of negatively-charged *electrons*. Since an atom is neutral, the number of positively-charged *protons* in the nucleus is equal to the number of electrons.

Atomic number (Z)   The number of *protons* in the *nucleus* of an *atom*. Every element has a unique atomic number. Hydrogen, with a single proton making up its nucleus and one electron orbiting it, has an atomic number of 1. Uranium, with 92 protons and 92 electrons, has an atomic number of 92.

Balmer series   The set of emission or absorption lines in the spectrum of hydrogen caused by the transitions of its *electron* between the second and higher *energy levels*.

Bell's inequality   A mathematical condition derived by John Bell in 1964 concerning the degree of correlation of the quantum spins of entangled pairs of particles that has to be satisfied by any local *hidden variables* theory.

Bell's theorem   A mathematical proof discovered by John Bell in 1964 that any *hidden variables* theory whose predictions agree with those of quantum mechanics must be non-local. See *non-locality*.

Beta particle   A fast moving *electron* ejected from the *nucleus* of a radioactive element due to the interconversion of *protons* and *neutrons*. Faster and more penetrating than *alpha particles*, it can be stopped by a thin sheet of metal.

Blackbody   A hypothetical, idealised body that absorbs and emits all *electromagnetic radiation* that strikes it. It can be approximated in the laboratory as a heated box with a pinhole in one of its walls.

Blackbody radiation   *Electromagnetic radiation* emitted by a *blackbody*.

Brownian motion   The erratic motion of pollen grains suspended in a fluid first observed, in 1827, by Robert Brown. In 1905 Einstein explained that Brownian motion was due to the random buffeting of the pollen grains by the molecules of the fluid.

Causality   Every cause has an effect.

Classical mechanics   The name given to the physics that originates from Newton's three laws of motion. Also called Newtonian mechanics, in which the properties of particles such as position and *momentum* are, in principle, simultaneously measurable with unlimited accuracy.

Classical physics   The description applied to all non-quantum physics such as *electromagnetism* and *thermodynamics*. Although Einstein's *general theory of relativity* is regarded by physicists as 'modern' twentieth-century physics, it is nevertheless a 'classical' theory.

Cloud chamber   A device invented by C.T.R. Wilson around 1911 that enables the detection of particles by observing their tracks through a chamber containing saturated vapour.

Collapse of the wave function   According to the *Copenhagen interpretation*, until it is observed or measured, a microphysical object like an *electron* does not exist anywhere. Between one measurement and the next it has no existence outside the abstract possibilities of the *wave function*. It is only when an observation or measurement is made that one of the 'possible' states of the electron becomes its 'actual' state and the probabilities of all the other possibilities become zero. This sudden, discontinuous change in the wave function due to an act of measurement is called the 'collapse of the wave function'.

Commutativity   Two variables A and B are said to commute if A×B=B×A. For example, if A and B are the numbers 5 and 4, then 5×4=4×5. Multiplication of numbers is commutative, since the order in which they are multiplied makes no difference. If A and B are *matrices*, then A×B does not necessarily equal B×A. When this happens, A and B are said to be non-commutative.

Complementarity   A principle advocated by Niels Bohr that the wave and particle aspects of light and matter are complementary but exclusive. This dual nature of light and matter is like the two sides of the same coin that can display either face, but not both simultaneously. For example, an experiment can be devised to reveal either the wave properties of light or its particle nature, but not both at the same time.

Complex number   A number written in the form a+ib, where a and b are ordinary real numbers familiar from arithmetic. i is the square root of −1, so that $(\sqrt{-1})^2 = -1$, and b is called the 'imaginary' part of the complex number.

Compton effect   The *scattering* of *photons* by atomic *electrons* discovered by the American physicist Arthur H. Compton in 1923.

Conjugate variables   A pair of *dynamical variables* such as position and *momentum*, or *energy* and time, that are related to one another through the *uncertainty principle*, are called conjugate variables or conjugate pairs.

Conservation law   A law which states that some physical quantity, such as *momentum* or *energy*, is conserved in all physical processes.

Conservation of energy The principle that *energy* cannot be created or destroyed, but can only be converted from one form to another. For example, when an apple falls from a tree, its *potential energy* is converted into *kinetic energy*.

Copenhagen interpretation An interpretation of *quantum mechanics*, whose principal architect Niels Bohr was based in Copenhagen. Over the years there were differences of opinion between Bohr and other leading advocates of the Copenhagen interpretation such as Werner Heisenberg. However, all agreed on its central tenets: Bohr's *correspondence principle*, Heisenberg's *uncertainty principle*, Born's *probability interpretation* of the *wave function*, Bohr's principle of *complementarity*, and the *collapse of the wave function*. There is no quantum reality beyond what is revealed by an act of measurement or observation. Hence it is meaningless to say, for example, that an *electron* exists somewhere independent of an actual observation. Bohr and his supporters maintained that quantum mechanics was a complete theory, a claim challenged by Einstein.

Correspondence principle A guiding principle advocated by Niels Bohr in which the laws and equations of quantum physics reduce to those of *classical physics* under conditions where the impact of *Planck's constant* is negligible.

De Broglie wavelength The *wavelength* λ of a particle is related to the *momentum* p of the particle by the relationship $\lambda=h/p$, where h is *Planck's constant*.

Degrees of freedom A system is said to have n degrees of freedom if it requires n coordinates to specify each state of the system. Each degree of freedom represents an independent way in which a body can move or a system can change. An object in the everyday world has three degrees of freedom corresponding to the three directions in which it can move – up and down, back and forth, and side to side.

Determinism In *classical mechanics*, if the positions and momenta of all the particles in the universe at some instant of time were known, and if all the forces between those particles were also known, then the subsequent state of the universe could in principle be determined. In *quantum mechanics* it is impossible to specify simultaneously the position and momentum of any particle at any instant. The theory therefore leads to an indeterministic view of the universe, one in which its future cannot be determined in principle. Nor can a particle's.

Diffraction   The spreading out of waves when they pass a sharp edge or through an aperture, such as water waves entering a harbour through a gap in the wall.

Dynamical variables   Quantities used to characterise the state of a particle such as position, *momentum*, *potential energy*, and *kinetic energy*.

Electromagnetic radiation   *Electromagnetic waves* differ in the amount of energy they transfer, called electromagnetic radiation. Lower-frequency waves like radio waves emit less electromagnetic radiation than higher-frequency waves such as gamma rays. Electromagnetic waves and electromagnetic radiation are often used interchangeably. See *electromagnetic waves* and *radiation*.

Electromagnetic spectrum   The entire range of *electromagnetic waves*: radio waves, *infrared radiation*, visible light, *ultraviolet radiation*, *X-rays*, and *gamma rays*.

Electromagnetic waves   Generated by oscillating electric charges, they differ in *wavelength* and *frequency*, but all electromagnetic waves have the same speed in empty space, approximately 300,000 kilometres per second. This is the speed of light, and it was the experimental confirmation that light was an electromagnetic wave.

Electromagnetism   Electricity and magnetism were regarded as two distinct phenomena described by their own sets of equations until the latter half of the nineteenth century. Following the experimental work of men like Michael Faraday, James Clerk Maxwell succeeded in developing a theory that unified electricity and magnetism into electromagnetism and described its behaviour in a set of four equations.

Electron   An elementary particle with a negative electric charge that, unlike the *proton* and the *neutron*, is not composed of more fundamental components.

Electron volt (eV) A unit of energy used in atomic, nuclear and particle physics that is about ten-billionth-billionths of a joule (1.6×10^{-19} joules).

Energy A physical quantity that can exist in different forms, such as *kinetic energy*, *potential energy*, chemical energy, thermal energy, and radiant energy.

Energy levels The discrete set of allowed internal energy states of an *atom* corresponding to the different *quantum* energy states of the atom itself.

Entanglement A quantum phenomenon in which two or more particles remain inexorably linked no matter how far apart they are.

Entropy In the nineteenth century, Rudolf Clausius defined entropy as the amount of heat in or out of a body, or a system, divided by the temperature at which the transfer takes place. Entropy is the measure of the disorder of a system; the higher the entropy, the greater the disorder. No physical process that would lead to a decrease in the entropy of an isolated system can occur in nature.

Ether A hypothetical, invisible medium that was believed to fill all of space and through which light and all other *electromagnetic waves* were thought to travel.

Exclusion principle No two *electrons* can occupy the same quantum state, i.e. have the same set of four *quantum numbers*.

Fine structure The splitting of an *energy level* or *spectral line* into several distinct components.

Frequency (ν) The number of complete cycles executed by a vibrating or an oscillating system in one second. The frequency of a wave is the number of complete *wavelengths* that pass a fixed point in one second. The unit of measurement is the hertz (Hz) and is equal to one cycle or wavelength per second.

Gamma rays Extremely short-*wavelength electromagnetic radiation*. It is the most penetrating of the three types of radiation emitted by radioactive substances.

Ground state The lowest energy state that an atom can possess. All other atomic states are called excited states. The lowest energy state of a hydrogen atom corresponds to its electron occupying the lowest energy level. If it occupies any other energy level, the hydrogen atom is in an excited state.

Harmonic oscillator   A vibrating or oscillating system whose *frequency* of vibration or oscillation does not depend on the *amplitude*.

Hidden variables   An interpretation of *quantum mechanics* based on the belief that the theory is incomplete and that there is an underlying layer of reality that contains additional information about the quantum world. This extra information is in the form of the hidden variables, unseen but real physical quantities. The identification of these hidden variables would lead to exact predictions for the outcomes of measurements and not just probabilities of obtaining certain results. Its adherents believe that it would restore a reality that exists independently of observation, denied by the *Copenhagen interpretation*.

Infrared radiation   *Electromagnetic radiation* with *wavelengths* longer than visible red light.

Interference   This is a characteristic phenomenon of wave motion in which two waves interact. Where two wave troughs or crests meet, they coalesce to produce a new, bigger trough or crest; this is known as constructive interference. But where a trough meets a crest or vice versa, they cancel each other out, a process called destructive interference.

Isotopes   Different forms of the same element that have the same number of *protons* in the *nucleus*, i.e. that share the same *atomic number*, but each having a different number of *neutrons*. For example, there are three forms of hydrogen with their nuclei containing zero, one, and two neutrons respectively. All three have similar chemical properties but different masses.

Joule   A unit of *energy* used in *classical physics*. A 100-watt light bulb converts 100 joules of electrical energy per second into heat and light.

Kinetic energy   *Energy* associated with the motion of an object. A stationary object, planet or particle has no kinetic energy.

Light   The human eye can detect only a small portion of all *electromagnetic waves*. These visible *wavelengths* of the *electromagnetic spectrum* range between 400nm (violet) and 700nm (red). White light is made up of red, orange, yellow, green, blue, indigo, and violet light. When a beam of white light is passed through a glass prism, these different strands of light are unpicked and form a rainbow band of colours called a continuum or continuous spectrum.

Light-quanta The name first used by Einstein in 1905 to describe particles of light, later renamed *photons.*

Locality The requirement that a cause and its effects occur at the same place, that there is no action at a distance. If an event A is the cause of another at B, there must be enough time between the two to allow a signal travelling at the speed of light from A to reach B. Any theory which has locality is called local. See *non-locality.*

Matrices Arrays of numbers (or other elements such as variables) with their own rules of algebra, matrices are extremely useful for expressing information about a physical system. An n×n square matrix has n columns and n rows.

Matrix mechanics A version of *quantum mechanics* discovered by Heisenberg in 1925 and then developed in conjunction with Max Born and Pascual Jordan.

Matter wave When a particle behaves as though it has a wave character, the wave representing it is called a matter wave or a de Broglie wave. See *de Broglie wavelength.*

Maxwell's equations A set of four equations derived by James Clerk Maxwell in 1864 that unified and described the disparate phenomena of electricity and magnetism as a single entity – *electromagnetism.*

Momentum (p) A physical property of an object that is equivalent to its mass times its *velocity.*

Nanometre (nm) One nanometre is equal to one billionth of a metre.

Neutron An uncharged particle that is similar in mass to a *proton.*

Non-locality An influence is allowed to pass between two systems or particles instantaneously, exceeding the limit set by the speed of light, so that a cause at one place can produce an immediate effect at some distant location. Any theory that allows non-locality is called non-local. See *locality.*

Nucleus The positively-charged mass at the heart of an *atom.* Initially believed to be made up only of *protons,* but later found to include *neutrons.*

It contains virtually the entire mass of an atom but occupies a tiny fraction of its volume. Discovered in 1911 by Ernest Rutherford and his co-workers at Manchester University.

Observable   Any *dynamical variable* of a system or object that can, in principle, be measured. For example, the position, *momentum*, and *kinetic energy* of an *electron* are all observables.

Period   The time it takes for a single *wavelength* to pass a fixed point, and also the time required to complete one cycle of an oscillation or vibration. The period is inversely proportional to the *frequency* of a wave, vibration, or oscillation.

Periodic table   The arrangement of the elements according to their *atomic number* into rows and columns that displays their recurring chemical properties.

Photoelectric effect   The emission of *electrons* from a given metal surface when *electromagnetic radiation* above a certain minimum *frequency* (*wavelength*) strikes it.

Photon   The quantum of light characterised by the energy $E=h\nu$ and momentum $p=h/\lambda$ where ν and λ are the *frequency* and *wavelength* of the radiation. The name was introduced in 1926 by the American chemist Gilbert Lewis. See *light-quanta*.

Planck's constant (h)   A fundamental constant of nature with a value of 6.626×10^{-34} joule-seconds that lies at the heart of quantum physics. Because Planck's constant is not zero, it is responsible for chopping up, quantising, energy and other physical quantities in the atomic realm.

Potential energy   The *energy* that an object or system has by virtue of its position or state. For example, the height of an object above the earth's surface determines its gravitational potential energy.

Probability interpretation   The interpretation suggested by Max Born that the *wave function* allowed only the probability of finding a particle at a particular location to be calculated. It is part and parcel of the idea that *quantum mechanics* can generate only the relative probabilities of obtaining certain results from the measurement of an *observable* and cannot predict which specific result will be obtained on a given occasion.

Proton   A particle contained in the *nucleus* of an *atom* that carries a positive charge equal and opposite to that on an *electron* and that has a mass some 2,000 times that of the electron's.

Quantised   Any physical quantity that can only have certain discrete values is quantised. An atom has only certain discrete energy levels and its energy is therefore quantised. The spin of an electron is quantised since it can only be either $+\frac{1}{2}$ (spin up) or $-\frac{1}{2}$ (spin down).

Quantum   A term introduced by Max Planck in 1900 to describe the indivisible packets of *energy* that an oscillator could emit or absorb in his model as he tried to derive an equation that reproduced the distribution of *blackbody radiation*. A quantum of energy (E) comes in various sizes determined by $E=h\nu$, where h is *Planck's constant* and ν is the *frequency* of the *radiation*. 'Quantum', more properly '*quantised*', can be applied to any physical property of a microphysical system or object that is discontinuous, that can change only by discrete units.

Quantum jump   Also known as a quantum leap, it is the transition of an *electron* between two *energy levels* inside an *atom* or molecule due to the emission or absorption of a *photon*.

Quantum mechanics   The theory of physics of the atomic and subatomic realm that replaced the ad hoc mixture of *classical mechanics* and quantum ideas that emerged between 1900 and 1925. Although dissimilar, Heisenberg's *matrix mechanics* and Schrödinger's *wave mechanics* are two mathematically equivalent representations of quantum mechanics.

Quantum number   Numbers that specify *quantised* physical quantities such as *energy*, *quantum spin* or *angular momentum*. For example, the quantised energy levels of a hydrogen atom are denoted by a set of numbers beginning with n=1 for the *ground state*, where n is the principal quantum number.

Quantum spin   A fundamental property of particles with no direct counterpart in *classical physics*. Any picturesque comparison of a 'spinning' *electron* to a spinning top is merely a poor aid that fails to capture the essence of this *quantum* concept. The quantum spin of a particle cannot be explained in terms of classical rotation since it can only have certain values that are equal to either a whole number or half a whole number multiplied by *Planck's constant* h divided by 2π ($\hbar$, a quantity called h-bar). Quantum spin is said to be either up (clockwise) or down (anti-clockwise) with respect to the direction of measurement.

Radiation The emission of energy or particles. Examples include *electromagnetic radiation*, thermal radiation and *radioactivity*.

Radioactivity When an unstable atomic *nucleus* spontaneously disintegrates to acquire a more stable configuration by emitting *alpha*, *beta* or *gamma* radiation, the process is called radioactivity or radioactive decay.

Realism The philosophical worldview which maintains that there exists a reality 'out there' independent of an observer. For a realist, the moon exists when no one is looking at it.

Relativity, general Einstein's theory of gravitation in which the gravitational force is explained as a distortion of space-time.

Relativity, special Einstein's 1905 theory of space-time in which the speed of light remains the same for all observers however fast they are moving. It is called 'special' because it does not describe objects that are accelerating, or gravity.

Scattering The deflection of one particle by another.

Schrödinger's cat A thought experiment devised by Erwin Schrödinger in which, according to the rules of *quantum mechanics*, a cat exists in a superposition of alive and dead states until it is observed.

Schrödinger's equation The fundamental equation of the *wave mechanics* version of *quantum mechanics* that governs the behaviour of a particle or the evolution of a physical system by encoding how its wave function varies with time.

$$-\frac{\hbar^2}{2m}\nabla^2\psi + V\psi = i\hbar\frac{\partial\psi}{\partial t}$$

where m is the mass of the particle, ∇^2 is a mathematical entity called the 'del-squared operator' which is responsible for tracking how the wave function ψ changes from place to place, V captures the forces acting on the particle, i is the square root of -1, $\partial\psi/\partial t$ describes how the wave function ψ changes in time, and ħ is Planck's constant h divided by 2π and is pronounced 'h-bar'. There is another form of the equation that gives a snapshot in time and is called the time-independent Schrödinger equation.

Spectral energy distribution of blackbody radiation   At any given temperature, it is the intensity of *electromagnetic radiation* emitted by a *blackbody* at each *wavelength* (or *frequency*). Also known simply as the blackbody spectrum.

Spectral lines   The pattern of coloured lines of light on a black background is called an emission spectrum. A series of black lines on a coloured background is called an absorption spectrum. Each element has a unique set of both emission and absorption spectral lines produced respectively by the emission and absorption of *photons* as *electrons* within the *atoms* of the element jump between different *energy levels*.

Spectroscopy   The area of physics concerned with analysing and studying absorption and emission spectra.

Spontaneous emission   The spontaneous emission of a *photon* as an *atom* makes the transition from an excited state to a lower energy state.

Stark effect   The splitting of *spectral lines* when atoms are placed in an electric field.

Stimulated emission   When an incident *photon* is not absorbed by an excited *atom*, but 'stimulates' it to emit a second photon of the same *frequency*.

Superposition   A *quantum* state composed of two or more other states. Such a state has certain probabilities for exhibiting the properties of the states out of which it is composed. See *Schrödinger's cat*.

Thermodynamics   Commonly described as the physics of the transformation of heat into and from other forms of *energy*.

Thermodynamics, the first law   The internal *energy* of an isolated system is a constant. Or equivalently, energy cannot be created or destroyed – the principle of the *conservation of energy*.

Thermodynamics, the second law   Heat does not flow spontaneously from cold to hot objects. Or equivalently, since there are different formulations of the law, the *entropy* of a closed system cannot decrease.

Thought experiment An idealised, imaginary experiment conceived as a means to test the consistency or limits of a physical theory or concept.

Ultraviolet catastrophe *Classical physics* distributes an infinite amount of *energy* among the high frequencies of *blackbody radiation*. This so-called ultraviolet catastrophe predicted by classical theory does not occur in nature.

Ultraviolet light *Electromagnetic radiation* with wavelengths shorter than those of visible violet light.

Uncertainty principle The principle discovered by Werner Heisenberg in 1927 that it is not possible to measure simultaneously certain pairs of *observables* – such as position and *momentum*, *energy* and time – with a degree of accuracy that exceeds a limit expressed in terms of *Planck's constant* h.

Velocity The speed of an object in a given direction.

Wave function (ψ) A mathematical function associated with the wave properties of a system or particle. The wave function represents everything that can be known about the state of a physical system or particle in *quantum mechanics*. For example, using the wave function of the hydrogen atom it is possible to calculate the probability of finding its *electron* at a certain point around the *nucleus*. See *probability interpretation* and *Schrödinger's equation*.

Wave mechanics A version of *quantum mechanics* developed in 1926 by Erwin Schrödinger.

Wave packet A *superposition* of many different waves that cancel each other out everywhere except within a small confined region of space, allowing the representation of a particle.

Wave-particle duality *Electrons* and *photons*, matter and *radiation*, may behave either like waves or like particles depending upon the experiment performed.

Wavelength (λ) The distance between two successive peaks or troughs of a wave. The wavelength of *electromagnetic radiation* determines which part of the *electromagnetic spectrum* it belongs to.

Wien's displacement law   Wilhelm Wien discovered in 1893 that as the temperature of a *blackbody* increases, the *wavelength* at which it emits the greatest intensity of *radiation* shifts to ever-shorter wavelengths.

Wien's distribution law   A formula discovered by Wilhelm Wien in 1896 that described the distribution of *blackbody radiation* in accordance with the experimental data then available.

X-rays   The *radiation* discovered by Wilhelm Röntgen in 1895 for which he was awarded the first Nobel Prize for physics in 1901. X-rays were later identified as *electromagnetic waves* of extremely short *wavelength*, emitted when very fast-moving *electrons* strike a target.

Zeeman effect   The splitting of *spectral lines* when *atoms* are placed in a magnetic field.

NOTES

PROLOGUE: THE MEETING OF MINDS

1 Pais (1982), p. 443.
2 Mehra (1975), quoted p. xvii.
3 Mehra (1975), quoted p. xvii.
4 Excluding the three professors (de Donder, Henriot and Piccard) from the Free University of Brussels invited as guests, Herzen representing the Solvay family, and Verschaffelt there in his capacity as the scientific secretary, then seventeen out of the 24 participants had already or would in due course receive a Nobel Prize. They were: Lorentz, 1902; Curie, 1903 (physics) and 1911 (chemistry); W.L. Bragg, 1915; Planck, 1918; Einstein, 1921; Bohr, 1922; Compton, 1927; Wilson, 1927; Richardson, 1928; de Broglie, 1929; Langmuir, 1932 (chemistry); Heisenberg, 1932; Dirac, 1933; Schrödinger, 1933; Pauli, 1945; Debye, 1936 (chemistry); and Born 1954. The seven who did not were Ehrenfest, Fowler, Brillouin, Knudsen, Kramers, Guye and Langevin.
5 Fine (1986), quoted p. 1. Letter from Einstein to D. Lipkin, 5 July 1952.
6 Snow (1969), p. 94.
7 Fölsing (1997), quoted p. 457.
8 Pais (1994), quoted p. 31.
9 Pais (1994), quoted p. 31.
10 Jungk (1960), quoted p. 20.
11 Gell-Mann (1981), p. 169.
12 Hiebert (1990), quoted p. 245.
13 Mahon (2003), quoted p. 149.
14 Mahon (2003), quoted p. 149.

CHAPTER 1: THE RELUCTANT REVOLUTIONARY

1 Planck (1949), pp. 33–4.
2 Hermann (1971), quoted p. 23. Letter from Planck to Robert Williams Wood, 7 October 1931.
3 Mendelssohn (1973), p. 118.
4 Heilbron (2000), quoted p. 5.
5 Mendelssohn (1973), p. 118.
6 Hermann (1971), quoted p. 23. Letter from Planck to Robert Williams Wood, 7 October 1931.
7 Heilbron (2000), quoted p. 3.
8 In the seventeenth century it was well known that passing a beam of sunlight through a prism resulted in the production of a spectrum of colours. It was believed that this rainbow of colours was the result of some sort of transformation of light as a result

of passing through the prism. Newton disagreed that somehow the prism adds colour and conducted two experiments. In the first he passed a beam of white light through a prism to produce the spectrum of colours and allowed a single colour to pass through a slit in a board and strike a second prism. Newton argued that if the colour had been the result of some change that light had undergone by passing through the first prism, passing it through a second would produce another change. Alas he found that, no matter which colour was selected as he repeated the experiment, passing it through a second prism left the original colour unchanged. In his second experiment Newton succeeded in mixing light of different colours to create white light.

9   Herschel made his serendipitous discovery on 11 September 1800, but it was published the following year. The spectrum of light can be viewed horizontally and vertically, depending on the arrangement apparatus. The prefix 'infra' came from the Latin word meaning 'below', when the light spectrum was viewed as a vertical strip with violet at the top and red at the bottom.

10   The wavelengths of red light and its various shades lie between 610 and 700 nanometres (nm), where a nanometre is a billionth of a metre. Red light of 700nm has a frequency of 430 trillion oscillations per second. At the opposite end of the visible spectrum, violet light ranges over 450nm to 400nm with the shorter wavelength having a frequency of 750 trillion oscillations per second.

11   Kragh (1999), quoted p. 121.

12   Teichmann et al. (2002), quoted p. 341.

13   Kangro (1970), quoted p. 7.

14   Cline (1987), quoted p. 34.

15   In 1900, London had a population of approximately 7,488,000, Paris of 2,714,000, and Berlin of 1,889,000.

16   Large (2001), quoted p. 12.

17   Planck (1949), p. 15.

18   Planck (1949), p. 16.

19   Planck (1949), p. 15.

20   Planck (1949), p. 16.

21   Planck (1949), p. 16.

22   Heat is not a form of energy as is commonly assumed, but a process that transfers energy from A to B due a temperature difference.

23   Planck (1949), p. 14.

24   Planck (1949), p. 13.

25   Lord Kelvin had also formulated a version of the second law: it is impossible for an engine to convert heat into work with 100 per cent efficiency. It was equivalent to Clausius. Both were saying the same thing but in two different languages.

26   Planck (1949), p. 20.

27   Planck (1949), p. 19.

28   Heilbron (2000), quoted p. 10.

29   Heilbron (2000), quoted p. 10.

30   Planck (1949), p. 20.

31   Planck (1949), p. 21.

32 Jungnickel and McCormmach (1986), quoted p. 52, Vol. 2.

33 Otto Lummer and Ernst Pringsheim christened Wien's discovery 'the displacement law' (*Verschiebungsgesetz*) only in 1899.

34 Given the inverse relationship between frequency and wavelength, as the temperature increases so does the frequency of the radiation of maximum intensity.

35 When the wavelength is measured in micrometres and the temperature in degrees Kelvin, then the constant is 2900.

36 In 1898 the Berlin Physical Society (Berliner Physikalische Gesellschaft), founded in 1845, changed its name to the German Physical Society (Deutsche Physikalische Gesellschaft zu Berlin).

37 The infrared part of the spectrum can be subdivided into roughly four wavelength bands: the near infrared, near the visible spectrum (0.0007–0.003mm), the intermediate infrared (0.003–0.006mm), the far infrared (0.006–0.015mm) and the deep infrared (0.015–1mm).

38 Kangro (1976), quoted p. 168.

39 Planck (1949), pp. 34–5.

40 Jungnickel and McCormmach (1986), Vol. 2, quoted p. 257.

41 Mehra and Rechenberg (1982), Vol. 1, Pt. 1, quoted p. 41.

42 Jungnickel and McCormmach (1986), Vol. 2, quoted p. 258.

43 Kangro (1976), quoted p. 187.

44 Planck (1900a), p. 79.

45 Planck (1900a), p. 81.

46 Planck (1949), pp. 40–1.

47 Planck (1949), p. 41.

48 Planck (1949), p. 41.

49 Planck (1993), p. 106.

50 Mehra and Rechenberg (1982), Vol. 1, p. 50, footnote 64.

51 Hermann (1971), quoted p. 23. Letter from Planck to Robert Williams Wood, 7 October 1931.

52 Hermann (1971), quoted p. 23. Letter from Planck to Robert Williams Wood, 7 October 1931.

53 Hermann (1971), quoted p. 24. Letter from Planck to Robert Williams Wood, 7 October 1931.

54 Hermann (1971), quoted p. 23. Letter from Planck to Robert Williams Wood, 7 October 1931.

55 Heilbron (2000), quoted p. 14.

56 Planck (1949), p. 32.

57 Hermann (1971), quoted p. 16.

58 Planck (1900b), p. 84.

59 The numbers have been rounded up.

60 Planck (1900b), p. 82.

61 Born (1948), p. 170.

62 Planck was also pleased because he had devised a way of measuring length, time and mass using a new set of units that would be valid and easily reproducible anywhere

in the universe. It was a matter of convention and convenience that had led to the introduction of various measuring systems at different places and times in human history, the latest being the measurement of length in metres, time in seconds, and mass in kilograms. Using h and two other constants, the speed of light c and Newton's gravitational constant G, Planck calculated values of length, mass and time that were unique and could serve as the basis of a universal scale of measurement. Given the smallness of the values of h and G, it could not be used for practical everyday purposes, but it would be the scale of choice to communicate with an extraterrestrial culture.

63   Heilbron (2000), quoted p. 38.
64   Planck (1949), pp. 44–5.
65   James Franck, Archive for the History of Quantum Physics (AHQP) interview, 7 September 1962.
66   James Franck, AHQP interview, 7 September 1962.

CHAPTER 2: THE PATENT SLAVE

1   Hentschel and Grasshoff (2005), quoted p. 131.
2   Collected Papers of Albert Einstein (CPAE), Vol. 5, p. 20. Letter from Einstein to Conrad Habicht, 30 June–22 September 1905.
3   Fölsing (1997), quoted p. 101.
4   Hentschel and Grasshoff (2005), quoted p. 38.
5   Einstein (1949a), p. 45.
6   CPAE, Vol. 5, p. 20. Letter from Einstein to Conrad Habicht, 18 or 25 May 1905.
7   CPAE, Vol. 5, p. 20. Letter from Einstein to Conrad Habicht, 18 or 25 May 1905.
8   Brian (1996), quoted p. 61.
9   CPAE, Vol. 9, Doc. 366.
10   CPAE, Vol. 9, Doc. 366.
11   Calaprice (2005), quoted p. 18.
12   CPAE, Vol. 1, xx, M. Einstein.
13   Einstein (1949a), p. 5.
14   Einstein (1949a), p. 5.
15   Einstein (1949a), p. 5.
16   Einstein (1949a), p. 8.
17   Oktoberfest started in 1810 as a fair to celebrate the marriage between the Bavarian Crown Prince Ludwig (the future King Ludwig I) and Princess Thérèse on 17 October. The event was so popular that it has been repeated annually ever since. It begins not in October, but September. It lasts sixteen days and ends on the first Sunday in October.
18   CPAE, Vol. 1, p. 158.
19   Fölsing (1997), quoted p. 35.
20   With 6 being the highest mark, Einstein received the following marks: algebra 6, geometry 6, history 6, descriptive geometry 5, physics 5–6, Italian 5, chemistry 5, natural history 5, German 4–5, geography 4, artistic drawing 4, technical drawing 4, and French 3.

21 CPAE, Vol. 1, pp. 15–16.
22 Einstein (1949a), p. 17.
23 Einstein (1949a), p. 15.
24 Fölsing (1997), quoted pp. 52–3.
25 Overbye (2001), quoted p. 19.
26 CPAE, Vol. 1, p. 123. Letter from Einstein to Mileva Maric, 16 February 1898.
27 Cropper (2001), quoted p. 205.
28 Einstein (1949a), p. 17.
29 CPAE, Vol. 1, p. 162. Letter from Einstein to Mileva Maric, 4 April 1901.
30 CPAE, Vol. 1, pp. 164–5. Letter from Hermann Einstein to Wilhelm Ostwald, 13 April 1901.
31 CPAE, Vol. 1, pp. 164–5. Letter from Hermann Einstein to Wilhelm Ostwald, 13 April 1901.
32 CPAE, Vol. 1, p. 165. Letter from Einstein to Marcel Grossmann, 14 April 1901.
33 CPAE, Vol. 1, p. 177. Letter from Einstein to Jost Winteler, 8 July 1901.
34 The advert appeared in the *Bundesblatt* (Federal Gazette) of 11 December 1901. CPAE, Vol. 1, p. 88.
35 CPAE, Vol. 1, p. 189. Letter from Einstein to Mileva Maric, 28 December 1901.
36 Berchtold V, Duke of Zähringen, founded the city in 1191. According to legend, Berchtold went hunting nearby and named the city Bärn after his first kill, a bear (Bär in German).
37 CPAE, Vol. 1, p. 191. Letter from Einstein to Mileva Maric, 4 February 1902.
38 Pais (1982), quoted pp. 46–7.
39 Einstein (1993), p. 7.
40 CPAE, Vol. 5, p. 28.
41 Hentschel and Grasshoff (2005), quoted p. 37.
42 Fölsing (1997), quoted p. 103.
43 Fölsing (1997), quoted p. 103.
44 Highfield and Carter (1994), quoted p. 210.
45 See CPAE, Vol. 5, p. 7. Letter from Einstein to Michele Besso, 22 January 1903.
46 CPAE, Vol. 5, p. 20. Letter from Einstein to Conrad Habicht, 30 June–22 September 1905.
47 Hentschel and Grasshoff (2005), quoted p. 23.
48 CPAE, Vol. 1, p. 193. Letter from Einstein to Mileva Maric, 17 February 1902.
49 Fölsing (1997), quoted p. 101.
50 Fölsing (1997), quoted p. 104.
51 Fölsing (1997), quoted p. 102.
52 Born (1978), p. 167.
53 Einstein (1949a), p. 15.
54 Einstein (1949a), p. 17.
55 CPAE, Vol. 2, p. 97.
56 Einstein (1905a), p. 178.
57 Einstein (1905a), p. 183.

58   Einstein also used his quantum of light hypothesis to explain Stoke's law of photolu-
      minescence and the ionisation of gases by ultraviolet light.
59   Mulligan (1999), quoted p. 349.
60   Susskind (1995), quoted p. 116.
61   Pais (1982), quoted p. 357.
62   During his Nobel Lecture, entitled 'The Electron and the light-quanta from the
      experimental point of view', Millikan also said: 'After ten years of testing and chang-
      ing and learning and sometimes blundering, all efforts being directed from the first
      toward the accurate experimental measurement of the energies of emission of photo-
      electrons, now as a function of the temperature, now of wavelength, now of material,
      this work resulted, contrary to my own expectations, in the first direct experimental
      proof in 1914 of the exact validity, within narrow limits of experimental errors, of
      the Einstein equation, and the first direct photoelectric determination of Planck's
      constant h.'
63   CPAE, Vol. 5, pp. 25–6. Letter from Max Laue to Einstein, 2 June 1906.
64   CPAE, Vol. 5, pp. 337–8. Proposal for Einstein's Membership in the Prussian Academy
      of Sciences, dated 12 June 1913 and signed by Max Planck, Walther Nernst, Heinrich
      Rubens and Emil Warburg.
65   Park (1997), quoted p. 208. Written in English, *Opticks* was first published in 1704.
66   Park (1997), quoted p. 208.
67   Park (1997), quoted p. 211.
68   Robinson (2006), quoted p. 103.
69   Robinson (2006), quoted p. 122.
70   Robinson (2006), quoted p. 96.
71   In German: 'War es ein Gott der diese Zeichen schrieb?'
72   Baierlein (2001), p. 133.
73   Einstein (1905a), p. 178.
74   Einstein (1905a), p. 193.
75   CPAE, Vol. 5, p. 26. Letter from Max Laue to Einstein, 2 June 1906.
76   In 1906 Einstein published *On the Theory of Brownian Motion* in which he presented
      his theory in a more elegant and extended form.
77   CPAE, Vol. 5, p. 63. Letter from Jakob Laub to Einstein, 1 March 1908.
78   CPAE, Vol. 5, p. 120. Letter from Einstein to Jakob Laub, 19 May 1909.
79   CPAE, Vol. 5, p. 120. Letter from Einstein to Jakob Laub, 19 May 1909.
80   CPAE, Vol. 5, p. 120. Letter from Einstein to Jakob Laub, 19 May 1909.
81   CPAE, Vol. 5, p. 120. Letter from Einstein to Jakob Laub, 19 May 1909.
82   CPAE, Vol. 2, p. 563.
83   CPAE, Vol. 5, p. 140. Letter from Einstein to Michele Besso, 17 November 1909.
84   Jammer (1966), quoted p. 57.
85   CPAE, Vol. 5, p. 187. Letter from Einstein to Michele Besso, 13 May 1911.
86   CPAE, Vol. 5, p. 190. Letter and invitation to the Solvay Congress from Ernst Solvay
      to Einstein, 9 June 1911.
87   CPAE, Vol. 5, p. 192. Letter from Einstein to Walter Nernst, 20 June 1911.
88   Pais (1982), quoted p. 399.

89 CPAE, Vol. 5, p. 241. Letter from Einstein to Michele Besso, 26 December 1911.
90 Brian (2005), quoted p. 128.
91 CPAE, Vol. 5, p. 220. Letter from Einstein to Heinrich Zangger, 7 November 1911.

CHAPTER 3: THE GOLDEN DANE

1 Niels Bohr Collected Works (BCW), Vol. 1, p. 559. Letter from Bohr to Harald Bohr, 19 June 1912.
2 Pais (1991), quoted p. 47. Since 1946 it has housed Copenhagen University's museum of medical history.
3 Pais (1991), quoted p. 46.
4 Pais (1991), quoted p. 99.
5 Pais (1991), quoted p. 48.
6 A second university in Aarhus was founded only in 1928.
7 Pais (1991), quoted p. 44.
8 Pais (1991), quoted p. 108.
9 Moore (1966), quoted p. 28.
10 Rozental (1967), p. 15.
11 Pais (1989a), quoted p. 61.
12 Niels Bohr, AHQP interview, 2 November 1962.
13 Niels Bohr, AHQP interview, 2 November 1962.
14 Heilbron and Kuhn (1969), quoted p. 223. Letter from Bohr to Margrethe Nørland, 26 September 1911.
15 BCW, Vol. 1, p. 523. Letter from Bohr to Ellen Bohr, 2 October 1911.
16 Weinberg (2003), quoted p. 10.
17 Aston (1940), p. 9.
18 Pais (1991), quoted p. 120.
19 BCW, Vol. 1, p. 527. Letter from Bohr to Harald Bohr, 23 October 1911.
20 BCW, Vol. 1, p. 527. Letter from Bohr to Harald Bohr, 23 October 1911.
21 There is no definitive historical evidence, but it is possible that Bohr attended a lecture given by Rutherford in Cambridge about his atomic model in October.
22 Bohr (1963b), p. 31.
23 Bohr (1963c), p. 83. The official report of the first Solvay Council was published in French in 1912 and in German in 1913. Bohr read the report as soon as it became available.
24 Kay (1963), p. 131.
25 Keller (1983), quoted p. 55.
26 Nitske (1971), quoted p. 5.
27 Nitske (1971), p. 5.
28 Kragh (1999), p. 30.
29 Wilson (1983), quoted p. 127.
30 Often in textbooks and scientific histories, the French scientist Paul Villard is credited with the discovery of gamma rays in 1900. In fact Villard discovered that radium emitted gamma rays, but it was Rutherford who reported them in his first paper on uranium radiation, published in January 1899, but finished on 1 September

1898. Wilson (1983), pp. 126–8 outlines the facts and makes a convincing case for Rutherford.

31 Eve (1939), quoted p. 55.

32 Andrade (1964), quoted p. 50.

33 More accurate measurements gave a half-life of 56 seconds.

34 Howorth (1958), quoted p. 83.

35 Wilson (1983), quoted p. 225.

36 Wilson (1983), quoted p. 225.

37 Wilson (1983), quoted p. 286.

38 Wilson (1983), quoted p. 287.

39 Pais (1986), quoted p. 188.

40 Cropper (2001), quoted p. 317.

41 Wilson (1983), quoted p. 291.

42 Marsden (1948), p. 54.

43 Rhodes (1986), quoted p. 49.

44 Thomson began working on a detailed mathematical version of this model only after he came across a similar idea proposed by Kelvin in 1902.

45 Badash (1969), quoted p. 235.

46 From quoted remarks by Geiger, Wilson (1983), p. 296.

47 Rowland (1938), quoted p. 56.

48 Cropper (2001), quoted p. 317.

49 Wilson (1983), quoted p. 573.

50 Wilson (1983), quoted p. 301. Letter from William Henry Bragg to Ernest Rutherford, 7 March 1911. Received on 11 March.

51 Eve (1939), quoted p. 200. Letter from Hantaro Nagaoka to Ernest Rutherford, 22 February 1911.

52 Nagaoka had been inspired by James Clerk Maxwell's famous analysis of the stability of Saturn's rings, which had puzzled astronomers for more than 200 years. In 1855, in a bid to attract the best physicists to attack the problem, it was chosen as the topic for Cambridge University's prestigious biennial competition, the Adams Prize. Maxwell submitted the only entry to be received by the closing date in December 1857. Rather than diminish the significance of the prize and Maxwell's achievement, it only served to enhance his growing reputation by once again demonstrating the difficulty of the problem. No one else had even succeeded in completing a paper worth entering. Although when seen through telescopes they appeared to be solid, Maxwell showed conclusively that the rings would be unstable if they were either solid or liquid. In an astonishing display of mathematical virtuosity, he demonstrated that the stability of Saturn's rings was due to them being composed of an enormous number of particles revolving around the planet in concentric circles. Sir George Airy, the Astronomer Royal, declared that Maxwell's solution was 'one of the most remarkable applications of Mathematics to Physics that I have ever seen'. Maxwell was duly rewarded with the Adams Prize.

53 Rutherford (1906), p. 260.

54 Rutherford (1911a), reprinted in Boorse and Motz (1966), p. 709.

55   In their paper, published in April 1913, Geiger and Marsden argued that their data was 'strong evidence of the correctness of the underlying assumptions that an atom contains a strong charge at the centre of dimensions, small compared with the diameter of the atom'.
56   Marsden (1948), p. 55.
57   Niels Bohr, AHQP interview, 7 November 1962.
58   Niels Bohr, AHQP interview, 2 November 1962.
59   Niels Bohr, AHQP interview, 7 November 1962.
60   Rosenfeld and Rüdinger (1967), quoted p. 46.
61   Pais (1991), quoted p. 125.
62   Andrade (1964), quoted p. 210.
63   Andrade (1964), p. 209, note 3.
64   Rosenfeld and Rüdinger (1967), quoted p. 46.
65   Bohr (1963b), p. 32.
66   Niels Bohr, AHQP interview, 2 November 1962.
67   Howorth (1958), quoted p. 184.
68   Soddy (1913), p. 400. He also suggested 'isotopic elements' as an alternative.
69   Radiothorium, radioactinium, ionium and uranium-X were later identified as only four of the 25 isotopes of thorium.
70   Niels Bohr, AHQP interview, 2 November 1962.
71   Bohr (1963b), p. 33.
72   Bohr (1963b), p. 33.
73   Bohr (1963b), p. 33.
74   Niels Bohr, AHQP interview, 2 November 1962.
75   Niels Bohr, AHQP interview, 31 October 1962.
76   Niels Bohr, AHQP interview, 31 October 1962.
77   Boorse and Motz (1966), quoted p. 855.
78   Georg von Hevesy, AHQP interview, 25 May 1962.
79   Pais (1991), quoted p. 125.
80   Pais (1991), quoted p. 125.
81   Bohr (1963b), p. 33.
82   Blaedel (1985), quoted p. 48.
83   BCW, Vol. 1, p. 555. Letter from Bohr to Harald Bohr, 12 June 1912.
84   BCW, Vol. 1, p. 555. Letter from Bohr to Harald Bohr, 12 June 1912.
85   BCW, Vol. 1, p. 561. Letter from Bohr to Harald Bohr, 17 July 1912.

CHAPTER 4: THE QUANTUM ATOM

1   Margrethe Bohr, Aage Bohr and Léon Rosenfeld, AHQP interview, 30 January 1963.
2   Margrethe Bohr, Aage Bohr and Léon Rosenfeld, AHQP interview, 30 January 1963.
3   Margrethe Bohr, AHQP interview, 23 January 1963.
4   Rozental (1998), p. 34.
5   Bohr decided to delay publication of the paper until experiments being conducted in Manchester on the velocity of alpha particles became available. The paper, 'On the

Theory of the Decrease of Velocity of Moving Electrified Particles on Passing through Matter', was published in 1913 in the *Philosophical Magazine*.

6 See Chapter 3, note 6.

7 Nielson (1963), p. 22.

8 Rosenfeld and Rüdinger (1967), quoted p. 51.

9 BCW, Vol. 2, p. 577. Letter from Bohr to Ernest Rutherford, 6 July 1912.

10 Niels Bohr, AHQP interview, 7 November 1962.

11 BCW, Vol. 2, p. 136.

12 BCW, Vol. 2, p. 136.

13 Niels Bohr, AHQP interview, 1 November 1962.

14 Niels Bohr, AHQP interview, 31 October 1962.

15 BCW, Vol. 2, p. 577. Letter from Bohr to Ernest Rutherford, 4 November 1912.

16 BCW, Vol. 2, p. 578. Letter from Ernest Rutherford to Bohr, 11 November 1912.

17 Pi (π) is the numerical value of the ratio of the circumference of a circle to its diameter.

18 One electron volt (eV) was equivalent to 1.6×10^{-19} joules of energy. A 100-watt light bulb converts 100 joules of electrical energy into heat in one second.

19 BCW, Vol. 2, p. 597. Letter from Bohr to Ernest Rutherford, 31 January 1913.

20 Niels Bohr, AHQP interview, 31 October 1962.

21 In Balmer's day and well into the twentieth century, wavelength was measured in a unit named in honour of Anders Ångström. 1 Ångström = 10^{-8}cm, one hundred-millionth of a centimetre. It is equal to one-tenth of a nanometre in modern units.

22 See Bohr (1963d), with introduction by Léon Rosenfeld.

23 In 1890 the Swedish physicist Johannes Rydberg developed a more general formula than Balmer's. It contained a number, later called Rydberg's constant, which Bohr was able to calculate from his model. He was able rewrite Rydberg's constant in terms of Planck's constant, the electron's mass and the electron's charge. He was able to derive a value for Rydberg's constant that was almost an identical match for the experimentally determined value. Bohr told Rutherford that he believed it to be an 'enormous and unexpected development'. (See BCW, Vol. 2, p. 111.)

24 Heilbron (2007), quoted p. 29.

25 Gillott and Kumar (1995), quoted p. 60. Lectures delivered by Nobel Prize-winners are available at www.nobelprize.org.

26 BCW, Vol. 2, p. 582. Letter from Bohr to Ernest Rutherford, 6 March 1913.

27 Eve (1939), quoted p. 221.

28 Eve (1939), quoted p. 221.

29 BCW, Vol. 2, p. 583. Letter from Ernest Rutherford to Bohr, 20 March 1913.

30 BCW, Vol. 2, p. 584. Letter from Ernest Rutherford to Bohr, 20 March 1913.

31 BCW, Vol. 2, pp. 585–6. Letter from Bohr to Ernest Rutherford, 26 March 1913.

32 Eve (1939), p. 218.

33 Wilson (1983), quoted p. 333.

34 Rosenfeld and Rüdinger (1967), quoted p. 54.

35 Wilson (1983), quoted p. 333.

36 Blaedel (1988), quoted p. 119.

37 Eve (1939), quoted p. 223.

38 Cropper (1970), quoted p. 46.

39 Jammer (1966), quoted p. 86.

40 Mehra and Rechenberg (1982), Vol. 1, quoted p. 236.

41 Mehra and Rechenberg (1982), Vol. 1, quoted p. 236.

42 BCW, Vol. 1, p. 567. Letter from Harald Bohr to Bohr, autumn 1913.

43 Eve (1939), quoted p. 226.

44 Moseley was also able to resolve some anomalies that had arisen in the placing of three pairs of elements in the periodic table. According to atomic weight, argon (39.94) should be listed after potassium (39.10) in the periodic table. This would conflict with their chemical properties, as potassium was grouped with the inert gases and argon with the alkali metals. To avoid such chemical nonsense, the elements were placed with the atomic weights in reverse order. However, using their respective atomic numbers they are placed in the correct order. Atomic number also allowed the correct positioning of two other pairs of elements: tellurium–iodine and cobalt–nickel.

45 Pais (1991), quoted p. 164.

46 BCW, Vol. 2, p. 594. Letter from Ernest Rutherford to Bohr, 20 May 1914.

47 Pais (1991), quoted p. 164.

48 CPAE, Vol. 5, p. 50. Letter from Einstein to Arnold Sommerfeld, 14 January 1908.

49 It was discovered later that Sommerfeld's k could not be equal to zero. So k was set equal to $l+1$ where l is the orbital angular momentum number. $l = 0, 1, 2 \ldots n-1$ where n is the principal quantum number.

50 There are actually two types of Stark effect. *Linear Stark effect* is one in which splitting is proportional to the electric field and occurs in excited states of hydrogen. All other atoms exhibit the *quadratic Stark effect*, where the splitting of the lines is proportional to the square of the electric field.

51 BCW, Vol. 2, p. 589. Letter from Ernest Rutherford to Bohr, 11 December 1913.

52 BCW, Vol. 2, p. 603. Letter from Arnold Sommerfeld to Bohr, 4 September 1913.

53 In modern notation m is written m_l. For a given l there are $2l+1$ values of m_l that range from $-l$ to $+l$. If $l=1$, then there are three values of m_l: $-1,0,+1$.

54 Pais (1994), quoted p. 34. Letter from Arnold Sommerfeld to Bohr, 25 April 1921.

55 Pais (1991), quoted p. 170.

56 In 1965, when Bohr would have been 80, it was renamed the Niels Bohr Institute.

CHAPTER 5: WHEN EINSTEIN MET BOHR

1 Frank (1947), quoted p. 98.

2 CPAE, Vol. 5, p. 175. Letter from Einstein to Hendrik Lorentz, 27 January 1911.

3 CPAE, Vol. 5, p. 175. Letter from Einstein to Hendrik Lorentz, 27 January 1911.

4 CPAE, Vol. 5, p. 187. Letter from Einstein to Michele Besso, 13 May 1911.

5 Pais (1982), quoted p. 170.

6 Pais (1982), quoted p. 170.

7 CPAE, Vol. 5, p. 349. Letter from Einstein to Hendrik Lorentz, 14 August 1913.

8 Fölsing (1997), quoted p. 335.

9 CPAE, Vol. 8, p. 23. Letter from Einstein to Otto Stern, after 4 June 1914.

10 CPAE, Vol. 8, p. 10. Letter from Einstein to Paul Ehrenfest, before 10 April 1914.

11 CPAE, Vol. 5, p. 365. Letter from Einstein to Elsa Löwenthal, before 2 December 1913.

12 CPAE, Vol. 8, pp. 32–3. Memorandum from Einstein to Mileva Einstein-Maric, 18 July 1914.

13 CPAE, Vol. 8, p. 41. Letter from Einstein to Paul Ehrenfest, 19 August 1914.

14 Fromkin (2004), quoted pp. 49–50.

15 Russia, France, Britain and Serbia were joined by Japan (1914), Italy (1915), Portugal and Romania (1916), the USA and Greece (1917). The British dominions also fought with the allies. Germany and Austria-Hungary were supported by Turkey (1914) and Bulgaria (1915).

16 CPAE, Vol. 8, p. 41. Letter from Einstein to Paul Ehrenfest, 19 August 1914.

17 CPAE, Vol. 8, p. 41. Letter from Einstein to Paul Ehrenfest, 19 August 1914.

18 Heilbron (2000), quoted p. 72.

19 Fölsing (1997), quoted p. 345.

20 Fölsing (1997), quoted p. 345.

21 Gilbert (1994), quoted p. 34.

22 Fölsing (1997), quoted p. 346.

23 Fölsing (1997), quoted p. 346.

24 Large (2001), quoted p. 138.

25 CPAE, Vol. 8, p. 77. Letter from Einstein to Romain Rolland, 22 March 1915.

26 CPAE, Vol. 8, p. 422. Letter from Einstein to Hendrik Lorentz, 18 December 1917.

27 CPAE, Vol. 8, p. 422. Letter from Einstein to Hendrik Lorentz, 18 December 1917.

28 CPAE, Vol. 5, p. 324. Letter from Einstein to Arnold Sommerfeld, 29 October 1912.

29 CPAE, Vol. 8, p. 151. Letter from Einstein to Heinrich Zangger, 26 November 1915.

30 CPAE, Vol. 8, p. 22. Letter from Einstein to Paul Ehrenfest, 25 May 1914.

31 CPAE, Vol. 8, p. 243. Letter from Einstein to Michele Besso, 11 August 1916.

32 CPAE, Vol. 8, p. 243. Letter from Einstein to Michele Besso, 11 August 1916.

33 CPAE, Vol. 8, p. 246. Letter from Einstein to Michele Besso, 6 September 1916.

34 CPAE, Vol. 6, p. 232.

35 CPAE, Vol. 8, p. 613. Letter from Einstein to Michele Besso, 29 July 1918.

36 Born (2005), p. 22. Letter from Einstein to Max Born, 27 January 1920.

37 Analogy courtesy of Jim Baggott (2004).

38 Born (2005), p. 80. Letter from Einstein to Max Born, 29 April 1924.

39 Large (2001), quoted p. 134.

40 CPAE, Vol. 8, p. 300. Letter from Einstein to Heinrich Zangger, after 10 March 1917.

41 CPAE, Vol. 8, p. 88. Letter from Einstein to Heinrich Zangger, 10 April 1915.

42 In a weak gravitational field, general relativity predicts the same bending as Newton's theory.

43 Pais (1994), quoted p. 147.

44 Brian (1996), quoted p. 101.

45 In the wake of the huge interest in his work, the first English translation of *Relativity* appeared in 1920.

46 CPAE, Vol. 8, p. 412, Letter from Einstein to Heinrich Zangger, 6 December 1917.

47 Pais (1982), quoted p. 309.

48 Brian (1996), quoted p. 103.

49 Calaprice (2005), quoted p. 5. Letter from Einstein to Heinrich Zangger, 3 January 1920.

50 Fölsing (1997), quoted p. 421.

51 Fölsing (1997), quoted p. 455. Letter from Einstein to Marcel Grossmann, 12 September 1920.

52 Pais (1982), quoted p. 314. Letter from Einstein to Paul Ehrenfest, 4 December 1919.

53 Everett (1979), quoted p. 153.

54 Elon (2003), quoted pp. 359–60.

55 Moore (1966), quoted p. 103.

56 Pais (1991), quoted p. 228. Postcard from Einstein to Planck, 23 October 1919.

57 CPAE, Vol. 5, p. 20. Letter from Einstein to Conrad Habicht, sometime between 30 June and 22 September 1905.

58 CPAE, Vol. 5, pp. 20–1. Letter from Einstein to Conrad Habicht, sometime between 30 June and 22 September 1905.

59 CPAE, Vol. 5, p. 21. Letter from Einstein to Conrad Habicht, sometime between 30 June and 22 September 1905.

60 Einstein (1949a), p. 47.

61 Moore (1966), quoted p. 104.

62 Moore (1966), quoted p. 106.

63 Pais (1991) quoted p. 232.

64 CPAE, Vol. 6, p. 232.

65 Fölsing (1997), quoted p. 477. Letter from Einstein to Bohr, 2 May 1920.

66 Fölsing (1997), quoted p. 477. Letter from Einstein to Paul Ehrenfest, 4 May 1920.

67 Fölsing (1997), quoted p. 477. Letter from Bohr to Einstein, 24 June 1920.

68 Pais (1994), quoted p. 40. Letter from Einstein to Hendrik Lorentz, 4 August 1920.

69 *Arbeitsgemeinschaft deutscher Naturforscher zur Erhaltung reiner Wissenschaft.*

70 Born (2005), p. 34. Letter from Einstein to the Borns, 9 September 1920.

71 Born (2005), p. 34. Letter from Einstein to the Borns, 9 September 1920.

72 Pais (1982), quoted p. 316. Letter from Einstein to K. Haenisch, 8 September 1920.

73 Fölsing (1997), quoted p. 512. Letter from Einstein to Paul Ehrenfest, 15 March 1922.

74 BCW, Vol. 3, pp. 691–2. Letter from Bohr to Arnold Sommerfeld, 30 April 1922.

75 What Bohr was calling electron shells were really a set of electron orbits. The primary orbits were numbered from 1 to 7, with 1 being nearest to the nucleus. Secondary orbits were designated by the letters s, p, d, f (from the terms 'sharp', 'principal', 'diffuse' and fundamental', used by spectroscopists to describe the lines in atomic spectra). The orbit nearest to the nucleus is just a single orbit and is labelled 1s, the next is a pair of orbits labelled 2s and 2p, the next a trio of orbits 3s, 3p and 3d, and so on. Orbits can hold increasing numbers of electrons the further from the nucleus they are. The s can hold 2 electrons, the p ones 6, the d ones 10, and the f ones 14.

76 Brian (1996), quoted p. 138.
77 Einstein (1993), p. 57. Letter from Einstein to Maurice Solovine, 16 July 1922.
78 See Fölsing (1997), p. 520. Letter from Einstein to Marie Curie, 11 July 1922.
79 Einstein (1949a), pp. 45–7.
80 French and Kennedy (1985), quoted p. 60.
81 Mehra and Rechenberg (1982), Vol. 1, Pt. 1, p. 358. Letter from Bohr to James Franck, 15 July 1922.
82 Moore (1966), quoted p. 116.
83 Moore (1966), quoted p. 116.
84 BCW, Vol. 4, p. 685. Letter from Bohr to Einstein, 11 November 1922.
85 Pais (1982), quoted p. 317.
86 BCW, Vol. 4, p. 686. Letter from Einstein to Bohr, 11 January 1923.
87 Pais (1991), quoted p. 308.
88 Pais (1991), quoted p. 215.
89 Bohr's banquet speech is available at www.nobelprize.org.
90 Bohr (1922), p. 7.
91 Bohr (1922), p. 42.
92 Robertson (1979), p. 69.
93 Weber (1981), p. 64.
94 Bohr (1922), p. 14.
95 Stuewer (1975), quoted p. 241.
96 Stuewer (1975), quoted p. 241.
97 See Stuewer (1975).
98 Visible light does undergo the 'Compton effect'. But the difference in wavelengths for the primary and scattered visible light is so much smaller than for X-rays that the effect is not detectable by the eye, although it can be measured in the lab.
99 Compton (1924), p. 70.
100 Compton (1924), p. 70.
101 Compton (1961). A short paper by Compton recounting the experimental evidence and the theoretical considerations that led to the discovery of the 'Compton effect'.
102 The American chemist Gilbert Lewis proposed the name *photon* in 1926 for atoms of light.
103 Fölsing (1997), quoted p. 541.
104 Pais (1991), quoted p. 234.
105 Compton (1924), p. 70.
106 Pais (1982), quoted p. 414.

CHAPTER 6: THE PRINCE OF DUALITY

1 Ponte (1981), quoted p. 56.
2 Unlike Duc, Prince was not a French title. With the death of his brother, the French title took precedence and Louis became a Duc.
3 Pais (1994), quoted p. 48. Letter from Einstein to Hendrik Lorentz, 16 December 1924.
4 Abragam (1988), quoted p. 26.

5 Abragam (1988), quoted pp. 26–7.

6 Abragam (1988), quoted p. 27.

7 Abragam (1988), quoted p. 27.

8 Ponte (1981), quoted p. 55.

9 See Abragam (1988), p. 38.

10 *Corps du Génie* in French.

11 Ponte (1981), quoted pp. 55–6.

12 Pais (1991), quoted p. 240.

13 Abragam (1988), quoted p. 30.

14 Abragam (1988), quoted p. 30.

15 Abragam (1988), quoted p. 30.

16 Abragam (1988), quoted p. 30.

17 Abragam (1988), quoted p. 30.

18 Wheaton (2007), quoted p. 58.

19 Wheaton (2007, quoted pp. 54–5.

20 Elsasser (1978), p. 66.

21 Gehrenbeck (1978), quoted p. 325.

22 CPAE, Vol. 5, p. 299. Letter from Einstein to Heinrich Zangger, 12 May 1912.

23 Weinberg (1993), p. 51.

CHAPTER 7: SPIN DOCTORS

1 Meyenn and Schucking (2001), quoted p. 44.

2 Born (2005), p. 223.

3 Born (2005), p. 223.

4 Paul Ewald, AHQP interview, 8 May 1962.

5 Enz (2002), quoted p. 15.

6 Enz (2002), quoted p. 9.

7 Pais (2000), quoted p. 213.

8 Mehra and Rechenberg (1982), Vol. 1, Pt. 2, quoted p. 378.

9 Enz (2002), quoted p. 49.

10 Cropper (2001), quoted p. 257.

11 Cropper (2001), quoted p. 257.

12 Cropper (2001), quoted p. 257.

13 Mehra and Rechenberg (1982), Vol. 1, Pt. 2, p. 384.

14 Pauli (1946b), p. 27.

15 Mehra and Rechenberg (1982), Vol. 1, Pt. 1, quoted p. 281.

16 CPAE, Vol. 8, p. 467. Letter from Einstein to Hedwig Born, 8 February 1918.

17 Greenspan (2005), quoted p. 108.

18 Born (2005), p. 56. Letter from Born to Einstein, 21 October 1921.

19 Pauli (1946a), p. 213.

20 Pauli (1946a), p. 213.

21 Lorentz assumed that oscillating electrons inside atoms of the incandescent sodium gas emitted the light that Zeeman had analysed. Lorentz showed that a spectral line would split into two closely spaced lines (a doublet) or three lines (a triplet)

depending on whether the emitted light was viewed in the direction parallel or per-pendicular to that of the magnetic field. Lorentz calculated the difference in the wave-lengths of the two adjacent lines and obtained a value in agreement with Zeeman's experimental results.

22 Pais (1991), quoted p. 199.

23 Pais (2000), quoted p. 221.

24 Pauli (1946a), p. 213.

25 In 1916, 28-year-old German physicist Walther Kossel, whose father had been awarded the Nobel Prize for chemistry, was the first to establish an important con-nection between the quantum atom and the periodic table. He noticed that the dif-ference between the atomic numbers 2, 10, 18 of the first three noble gases, helium, neon, argon, was 8, and argued that the electrons in such atoms orbited in 'closed shells'. The first contained only 2 electrons, the second and third, 8 each. Bohr acknowledged the work of Kossel. But neither Kossel nor others went as far as the Dane in elucidating the distribution of electrons throughout the periodic table, the culmination of which was the correct labelling of hafnium as not a rare earth ele-ment.

26 BCW, Vol. 4, p. 740. Postcard from Arnold Sommerfeld to Bohr, 7 March 1921.

27 BCW, Vol. 4, p. 740. Letter from Arnold Sommerfeld to Bohr, 25 April 1921.

28 Pais (1991), quoted p. 205.

29 If n=3, then k=1, 2, 3.

If k=1, then m=0 and the energy state is (3,1,0).

If k=2, then m=−1, 0, 1 and the energy states are (3,2,−1), (3,2,0), and (3,2,1).

If k=3, then m=−2, −1, 0, 1, 2 and the energy states are (3,3,−2), (3,3,−1), (3,3,0), (3,3,1) and (3,3,2). The total number of energy states in the third shell n=3 is 9 and the maximum number of electrons 18. For n=4, the energy states are (4,1,0), (4,2,−1), (4,2,0), (4,2,1), (4,3,−2), (4,3,−1), (4,3,0), (4,3,1), (4,3,2), (4,4,−3), (4,4,−2), (4,4,−1), (4,4,0), (4,4,1), (4,4,2), (4,4,3).

The number of electron energy states for a given n was simply equal to n^2. For the first four shells, n=1, 2, 3 and 4, the number of energy states are 1, 4, 9, 16.

30 The first edition of *Atombau und Spektrallinien* was published in 1919.

31 Pais (2000), quoted p. 223.

32 Recall that in his model of the quantum atom, Bohr introduced the quantum into the atom through the quantisation of angular momentum ($L = nh/2\pi = mvr$). An electron moving in a circular orbit possesses angular momentum. Labelled L in calculations, the angular momentum of the electron is nothing more than the value obtained by multiplying its mass by its velocity by the radius of its orbit (in symbols, L=mvr). Only those electron orbits were permitted that had an angular momentum equal to $nh/2\pi$, where n was 1, 2, 3 and so on. All others orbits were forbidden.

33 Calaprice (2005), quoted p. 77.

34 Pais (1989b), quoted p. 310.

35 Goudsmit (1976), p. 246.

36 Samuel Goudsmit, AHQP interview, 5 December 1963.

37 Pais (1989b), quoted p. 310.

38 Pais (2000), quoted p. 222.
39 Actually, the two values are $+\frac{1}{2}(h/2\pi)$ and $-\frac{1}{2}(h/2\pi)$ or equivalently $+h/4\pi$ and $-h/4\pi$.
40 Mehra and Rechenberg (1982), Vol. 1, Pt. 2, quoted p. 702.
41 Pais (1989b), quoted p. 311.
42 George Uhlenbeck, AHQP interview, 31 March 1962.
43 Uhlenbeck (1976), p. 253.
44 BCW, Vol. 5, p. 229. Letter from Bohr to Ralph Kronig, 26 March 1926.
45 Pais (2000), quoted p. 304.
46 Robertson (1979), quoted p. 100.
47 Mehra and Rechenberg (1982), Vol. 1, Pt. 2, quoted p. 691.
48 Mehra and Rechenberg (1982), Vol. 1, Pt. 2, quoted p. 692.
49 Ralph Kronig, AHQP interview, 11 December 1962.
50 Ralph Kronig, AHQP interview, 11 December 1962.
51 Pais (2000), quoted p. 305.
52 Pais (2000), quoted p. 305.
53 Pais (2000), quoted p. 305.
54 Pais (2000), quoted p. 305.
55 Uhlenbeck (1976), p. 250.
56 Pais (2000), quoted p. 305.
57 Pais (2000), quoted p. 305.
58 Pais (2000), quoted p. 230.
59 Enz (2002), quoted p. 115.
60 Enz (2002), quoted p. 117.
61 Goudsmit (1976), p. 248.
62 Jammer (1966), p. 196.
63 Mehra and Rechenberg (1982), Vol. 2, Pt. 2, quoted p. 208. Letter from Pauli to Ralph Kronig, 21 May 1925.
64 Mehra and Rechenberg (1982), Vol. 1, Pt. 2, quoted p. 719.

CHAPTER 8: THE QUANTUM MAGICIAN

1   Mehra and Rechenberg (1982), Vol. 2, quoted p. 6.
2   Heisenberg (1971), p. 16.
3   Heisenberg (1971), p. 16.
4   Heisenberg (1971), p. 16.
5   Heisenberg (1971), p. 16.
6   Werner Heisenberg, AHQP interview, 30 November 1962.
7   Heisenberg (1971), p. 24.
8   Heisenberg (1971), p. 24.
9   Werner Heisenberg, AHQP interview, 30 November 1962.
10 Heisenberg (1971), p. 26.
11 Heisenberg (1971), p. 26.
12 Heisenberg (1971), p. 26.
13 Heisenberg (1971), p. 38.

14 Heisenberg (1971), p. 38.

15 Werner Heisenberg, AHQP interview, 30 November 1962.

16 Heisenberg (1971), p. 42.

17 Born (1978), p. 212.

18 Born (2005), p. 73. Letter from Born to Einstein, 7 April 1923.

19 Born (1978), p. 212.

20 Cassidy (1992), quoted p. 168.

21 Mehra and Rechenberg (1982), Vol. 2, quoted pp. 140–1. Letter from Heisenberg to Pauli, 26 March 1924.

22 Mehra and Rechenberg (1982), Vol. 2, quoted p. 133. Letter from Pauli to Bohr, 11 February 1924.

23 Mehra and Rechenberg (1982), Vol. 2, quoted p. 135. Letter from Pauli to Bohr, 11 February 1924.

24 Mehra and Rechenberg (1982), Vol. 2, quoted p. 142.

25 Mehra and Rechenberg (1982), Vol. 2, quoted p. 127. Letter from Born to Bohr, 16 April 1924.

26 Mehra and Rechenberg (1982), Vol. 2, quoted p. 3.

27 Mehra and Rechenberg (1982), Vol. 2, quoted p. 150.

28 Frank Hoyt, AHQP interview, 28 April 1964.

29 Mehra and Rechenberg (1982), Vol. 2, quoted p. 209. Letter from Heisenberg to Bohr, 21 April 1925.

30 Heisenberg (1971), p. 8.

31 Pais (1991), quoted p. 270.

32 Mehra and Rechenberg (1982), Vol. 2, quoted p. 196. Letter from Pauli to Bohr, 12 December 1924.

33 Cassidy (1992), quoted p. 198.

34 Pais (1991), quoted p. 275.

35 Heisenberg (1971), p. 60.

36 Heisenberg (1971), p. 60.

37 Heisenberg (1971), p. 61.

38 Heisenberg (1971), p. 61.

39 Heisenberg (1971), p. 61.

40

$$A=\begin{pmatrix} a & b \\ c & d \end{pmatrix} \quad B=\begin{pmatrix} e & f \\ g & h \end{pmatrix} \quad A\times B=\begin{pmatrix} (a\times e)+(b\times g) & (a\times f)+(b\times h) \\ (c\times e)+(d\times g) & (c\times f)+(d\times h) \end{pmatrix}$$

$$\text{If } A=\begin{pmatrix} 1 & 2 \\ 3 & 4 \end{pmatrix} \text{ If } B=\begin{pmatrix} 5 & 6 \\ 7 & 8 \end{pmatrix} \text{ then } A\times B=\begin{pmatrix} (1\times5)+(2\times7) & (1\times6)+(2\times8) \\ (3\times5)+(4\times7) & (3\times6)+(4\times8) \end{pmatrix}=\begin{pmatrix} 5+14 & 6+16 \\ 15+28 & 18+32 \end{pmatrix}=\begin{pmatrix} 19 & 22 \\ 43 & 50 \end{pmatrix}$$

$$\text{If } B=\begin{pmatrix} 5 & 6 \\ 7 & 8 \end{pmatrix} \text{ If } A=\begin{pmatrix} 1 & 2 \\ 3 & 4 \end{pmatrix} \text{ then } B\times A=\begin{pmatrix} (5\times1)+(6\times3) & (5\times2)+(6\times4) \\ (7\times1)+(8\times3) & (7\times2)+(8\times4) \end{pmatrix}=\begin{pmatrix} 5+18 & 10+24 \\ 7+24 & 14+32 \end{pmatrix}=\begin{pmatrix} 23 & 34 \\ 31 & 46 \end{pmatrix}$$

$$\text{Therefore } (A\times B)-(B\times A)=\begin{pmatrix} 19 & 22 \\ 43 & 50 \end{pmatrix}-\begin{pmatrix} 23 & 34 \\ 31 & 46 \end{pmatrix}=\begin{pmatrix} -4 & -12 \\ 12 & 4 \end{pmatrix}$$

41 Enz (2002), quoted p. 131. Letter from Heisenberg to Pauli, 21 June 1925.

42 Cassidy (1992), quoted p. 197. Letter from Heisenberg to Pauli, 9 July 1925.

43   Mehra and Rechenberg (1982), quoted p. 291.

44   Enz (2002), quoted p. 133.

45   Cassidy (1992), quoted p. 204.

46   Heisenberg (1925), p. 276.

47   Born (2005), p. 82. Letter from Born to Einstein, 15 July 1925. Born may have discovered that Heisenberg's multiplication rule was exactly the same as that for matrix multiplication by the time he wrote to Einstein. Born recalled on one occasion that Heisenberg gave him the paper on 11 or 12 July. However, on another occasion he believed the date of his identifying the strange multiplication with matrix multiplication was 10 July.

48   Born (2005), p. 82. Letter from Born to Einstein, 15 July 1925.

49   Cropper (2001), quoted p. 269.

50   Born (1978), p. 218.

51   Schweber (1994), quoted p. 7.

52   Born (2005), p. 80. Letter from Born to Einstein, 15 July 1925.

53   In 1925 and 1926, Heisenberg, Born and Jordan never used the term 'matrix mechanics'. They always spoke about the 'new mechanics' or 'quantum mechanics'. Others initially referred to 'Heisenberg's mechanics' or 'Göttingen mechanics' before some mathematicians started referring to it as '*Matrizenphysik*', 'matrix physics'. By 1927 it was routinely referred to as 'matrix mechanics', a name that Heisenberg always disliked.

54   Born (1978), p. 190.

55   Born (1978), p. 218.

56   Mehra and Rechenberg (1982), Vol. 3, quoted p. 59. Letter from Born to Bohr, 18 December 1926.

57   Greenspan (2005), quoted p. 127.

58   Pais (1986), quoted p. 255. Letter from Einstein to Paul Ehrenfest, 20 September 1925.

59   Pais (1986), quoted p. 255.

60   Pais (2000), quoted p. 224.

61   Born (1978), p. 226.

62   Born (1978), p. 226.

63   Kursunoglu and Wigner (1987), quoted p. 3.

64   Paul Dirac, AHQP interview, 7 May 1963.

65   Kragh (2002) quoted p. 241.

66   Dirac (1977), p. 116.

67   Dirac (1977), p. 116.

68   Born (2005), p. 86. Letter from Einstein to Mrs Born, 7 March 1926.

69   Bernstein (1991), quoted p. 160.

CHAPTER 9: 'A LATE EROTIC OUTBURST'

1   Moore (1989), quoted p. 191.

2   Born (1978), p. 270.

3   Moore (1989), quoted p. 23.

4   Moore (1989), quoted pp. 58–9.
5   Moore (1989), quoted p. 91.
6   Moore (1989), quoted p. 91.
7   Mehra and Rechenberg (1987) Vol. 5, Pt. 1, quoted p. 182.
8   Moore (1989), quoted p. 145.
9   Mehra and Rechenberg (1987), Vol. 5, Pt. 2, quoted p. 412.
10 Bloch (1976), p. 23. Although there is some doubt when exactly Schrödinger delivered his talk at the colloquium, 23 November is the most probable date that fits the known facts better than any alternative.
11 Bloch (1976), p. 23.
12 Bloch (1976), p. 23.
13 Abragam (1988), p. 31.
14 Bloch (1976), pp. 23–4.
15 The equation was rediscovered in 1927 by Oskar Klein and Walter Gordon and became known as the Klein-Gordon equation. It applies only to spin zero particles.
16 Moore (1989), quoted p. 196.
17 Moore (1989), quoted p. 191.
18 The title of Schrödinger's paper signalled that in his theory the quantisation of an atom's energy levels was based on the allowed values, or *eigenvalues*, of electron wavelengths. In German, *eigen* means 'proper' or 'characteristic'. The German word *eigenwert* was only half-heartedly translated into English as *eigenvalue*.
19 Cassidy (1992), quoted p. 214.
20 Moore (1989), quoted p. 209. Letter from Planck to Schrödinger, 2 April 1926.
21 Moore (1989), quoted p. 209. Letter from Einstein to Schrödinger, 16 April 1926.
22 Przibram (1967), p. 6.
23 Moore (1989), quoted p. 209. Letter from Einstein to Schrödinger, 26 April 1926.
24 Cassidy (1992), quoted p. 213.
25 Pais (2000), quoted p. 306.
26 Moore (1989), quoted p. 210.
27 Mehra and Rechenberg (1987), Vol. 5, Pt. 1, quoted p. 1. Letter from Pauli to Pascual Jordan, 12 April 1926.
28 Cassidy (1992), quoted p. 213.
29 Cassidy (1992), quoted p. 213. Letter from Heisenberg to Pascual Jordan, 19 July 1926.
30 Cassidy (1992), quoted p. 213.
31 Cassidy (1992), quoted p. 213. Letter from Born to Schrödinger, 16 May 1927.
32 Mehra and Rechenberg (1987), Vol. 5, Pt. 2, quoted p. 639. Letter from Schrödinger to Wilhelm Wien, 22 February 1926.
33 Mehra and Rechenberg (1987), Vol. 5, Pt. 2, quoted p. 639. Letter from Schrödinger to Wilhelm Wien, 22 February 1926.
34 Pauli, Dirac and the American Carl Eckhart all independently showed that Schrödinger was correct.
35 Mehra and Rechenberg (1987), Vol. 5 Pt. 2, quoted p. 639. Letter from Schrödinger to Wilhelm Wien, 22 February 1926.

36   Moore (1989), quoted p. 211.

37   Moore (1989), quoted p. 211.

38   Cassidy (1992), quoted p. 215. Letter from Heisenberg to Pauli, 8 June 1926.

39   Cassidy (1992), quoted p. 213. Letter from Heisenberg to Pascual Jordan, 8 April 1926.

40   Heisenberg's paper was received by the *Zeitschrift für Physik* on 24 July and was published on 26 October 1926.

41   Pais (2000), quoted p. 41. Letter from Born to Einstein, 30 November 1926. Not included in Born (2005).

42   Bloch (1976), p. 320. In the original German:

Gar Manches rechnet Erwin schon
Mit seiner Wellenfunktion.
Nur wissen möcht' man gerne wohl
Was man sich dabei vorstell'n soll.

43   Strictly speaking it should be the square of the 'modulus' of the wave function. Modulus is the technical term for taking the absolute value of a number regardless of whether it is positive or negative. For example, if x=−3, then the modulus of x is 3. Written as: $|x| = |{-3}| = 3$. For a complex number z=x+iy, the modulus of z is given by $|z| = \sqrt{x^2+y^2}$.

44   The square of a complex number is calculated as follows: z=4+3i, z^2 is in fact not z×z, but z×z* where z* is called the complex conjugate. If z=4+3i, then z*=4−3i.
Hence, z^2=z×z*=(4+3i)×(4−3i)=16−12i+12i−9i²=16−9(√−1)²=16−9(−1)=16+9=25.
If z=4+3i, then the modulus of z is 5.

45   Born (1978), p. 229.

46   Born (1978), p. 229.

47   Born (1978), p. 230.

48   Born (1978), p. 231.

49   Born (2005), p. 81. Letter from Born to Einstein, 15 July 1925.

50   Born (2005), p. 81. Letter from Born to Einstein, 15 July 1925.

51   Pais (2000), quoted p. 41.

52   Pais (1986), quoted p. 256.

53   Pais (2000), quoted p. 42.

54   The second paper was published in the *Zeitschrift für Physik* on 14 September.

55   Pais (1986), quoted p. 257.

56   Pais (1986), quoted p. 257.

57   Once again, technically speaking it is the absolute or modulus square of the wave function. Also, technically, rather than the 'probability', the absolute square of the wave function gives the 'probability density'.

58   Pais (1986), quoted p. 257.

59   Pais (1986), quoted p. 257.

60   Pais (2000), quoted p. 39.

61   Mehra and Rechenberg (1987), Vol. 5, Pt. 2, quoted p. 827. Letter from Schrödinger to Wien, 25 August 1926.

62   Mehra and Rechenberg (1987), Vol. 5, Pt. 2, quoted p. 828. Letter from Schrödinger to Born, 2 November 1926.

63   Heitler (1961), quoted p. 223.

64   Moore (1989), quoted p. 222.

65   Moore (1989), quoted p. 222.

66   Heisenberg (1971), p. 73.

67   Cassidy (1992), quoted p. 222. Letter from Heisenberg to Pascual Jordan, 28 July 1926.

68   Cassidy (1992), quoted p. 222. Letter from Heisenberg to Pascual Jordan, 28 July 1926.

69   Mehra and Rechenberg (1987), Vol. 5, Pt. 2, quoted p. 625. Letter from Bohr to Schrödinger, 11 September 1926.

70   Heisenberg (1971), p. 73.

71   Heisenberg (1971), p. 73.

72   See Heisenberg (1971), pp. 73–5 for the complete reconstruction of this particular exchange between Schrödinger and Bohr.

73   Heisenberg (1971), p. 76.

74   Moore (1989), p. 228. Letter from Schrödinger to Wilhelm Wien, 21 October 1926.

75   Mehra and Rechenberg (1987), Vol. 5, Pt. 2, quoted p. 826. Letter from Schrödinger to Wilhelm Wien, 21 October 1926.

76   Born (2005), p. 88. Letter from Einstein to Born, 4 December 1926.

CHAPTER 10: UNCERTAINTY IN COPENHAGEN

1   Heisenberg (1971), p. 62.

2   Heisenberg (1971), p. 62.

3   Heisenberg (1971), p. 62.

4   Heisenberg (1971), p. 62.

5   Heisenberg (1971), p. 63.

6   Heisenberg (1971), p. 63.

7   Heisenberg (1971), p. 63.

8   Werner Heisenberg, AHQP interview, 30 November 1962.

9   Heisenberg (1971), p. 63.

10   Heisenberg (1971), p. 63.

11   Heisenberg (1971), p. 64.

12   Heisenberg (1971), p. 64.

13   Heisenberg (1971), p. 64.

14   Heisenberg (1971), p. 65.

15   Cassidy (1992), quoted p. 218.

16   Pais (1991), quoted p. 296. Letter from Bohr to Rutherford, 15 May 1926.

17   Heisenberg (1971), p. 76.

18   Cassidy (1992), quoted p. 219.

19   Pais (1991), quoted p. 297.

20   Robertson (1979), quoted p. 111.

21   Pais (1991), quoted p. 300.

22 Heisenberg (1967), p. 104.

23 Mehra and Rechenberg (2000), Vol. 6, Pt. 1, quoted p. 235. Letter from Einstein to Paul Ehrenfest, 28 August 1926.

24 Werner Heisenberg, AHQP interview, 25 February 1963.

25 Werner Heisenberg, AHQP interview, 25 February 1963.

26 Werner Heisenberg, AHQP interview, 25 February 1963.

27 Heisenberg (1971), p. 77.

28 Heisenberg (1971), p. 77.

29 Heisenberg (1971), p. 77.

30 Heisenberg (1971), p. 77.

31 In another of his later writings, Heisenberg expressed the crucial switch in the question to answer: 'Instead of asking: How can one in the known mathematical scheme express a given experimental situation? The other question was put: Is it true, perhaps, that only such experimental situations can arise in nature as can be expressed in the mathematical formalism?' Heisenberg (1989), p. 30.

32 Heisenberg (1971), p. 78.

33 Heisenberg (1971), p. 78.

34 Heisenberg (1971), p. 79.

35 Momentum is preferred over velocity because it appears in fundamental equations of both classical and quantum mechanics. Both physical variables are intimately connected by the fact that momentum is just mass times velocity – even for a fast-moving electron with corrections imposed by the special theory of relativity.

36 As pointed out by Max Jammer (1974), Heisenberg used *Ungenauigkeit* (inexactness, imprecision) or *Genauigkeit* (precision, degree of precision). These two terms appear more than 30 times in his paper, whereas *Unbestimmtheit* (indeterminacy) appears only twice and *Unsicherheit* (uncertainty) three times.

37 Heisenberg in his published paper actually put it as $\Delta p \Delta q \sim h$, or Δp times Δq is approximately Planck's constant.

38 There were occasions over the years when Heisenberg seemed to suggest that it was our knowledge of the atomic world that was indeterminate: 'The uncertainty principle refers to the degree of indeterminateness in the possible present knowledge of the simultaneous values of the various quantities with which quantum theory deals ...', rather than an intrinsic feature of nature. See Heisenberg (1949), p. 20.

39 Heisenberg (1927), p. 68. An English translation can be found in Wheeler and Zurek (1983), pp. 62–84. All page references refer to this reprint.

40 Heisenberg (1927), p. 68.

41 Heisenberg (1927), p. 68.

42 Heisenberg (1989), p. 30.

43 Heisenberg (1927), p. 62.

44 Heisenberg (1989), p. 31.

45 Heisenberg (1927), p. 63.

46 Heisenberg (1927), p. 64.

47 Heisenberg (1927), p. 65.

48 Heisenberg (1989), p. 36.

49   Mehra and Rechenberg (2000), Vol. 6, Pt. 1, quoted p. 146. Letter from Pauli to Heisenberg, 19 October 1926.

50   Mehra and Rechenberg (2000), Vol. 6, Pt. 1, quoted p. 147. Letter from Pauli to Heisenberg, 19 October 1926.

51   Mehra and Rechenberg (2000), Vol. 6, Pt. 1, quoted p. 146. Letter from Pauli to Heisenberg, 19 October 1926.

52   Mehra and Rechenberg (2000), Vol. 6, Pt. 1, quoted p. 93.

53   Pais (1991), quoted p. 304. Letter from Heisenberg to Bohr, 10 March 1927.

54   Pais (1991), quoted p. 304.

55   Cassidy (1992), quoted p. 241. Letter from Heisenberg to Pauli, 4 April 1927.

56   Werner Heisenberg, AHQP interview, 25 February 1963.

57   Werner Heisenberg, AHQP interview, 25 February 1963.

58   Werner Heisenberg, AHQP interview, 25 February 1963.

59   Heisenberg (1927), p. 82.

60   The original German title was: 'Über den anschaulichen Inhalt der quanten-theoretischen Kinematik und Mechanik', *Zeitschrift für Physik*, 43, 172–98 (1927). See Wheeler and Zurek (1983), pp. 62–84.

61   Mehra and Rechenberg (2000), Vol. 6, Pt. 1, quoted p. 182. Letter from Heisenberg to Pauli, 4 April 1927.

62   Bohr (1949), p. 210.

63   There was a subtle difference between wave-particle complementarity and that involving any pair of physical observables like position and momentum. According to Bohr, the complementary wave and particle aspects of an electron or light are mutually exclusive. It is one or the other. However, only if either position or momentum of an electron, for example, is measured with pinpoint certainty are position and momentum mutually exclusive. Otherwise, the precision with which both can be measured and therefore known is given by the position-momentum uncertainty relation.

64   BCW, Vol. 6, p. 147.

65   BCW, Vol. 3, p. 458.

66   Werner Heisenberg, AHQP interview, 25 February 1963.

67   Werner Heisenberg, AHQP interview, 25 February 1963.

68   Bohr (1949), p. 210.

69   Bohr (1928), p. 53.

70   BCW, Vol. 6, p. 91.

71   Mehra and Rechenberg (2000), Vol. 6, Pt. 1, quoted p. 187. Letter from Bohr to Einstein, 13 April 1927.

72   Mehra and Rechenberg (2000), Vol. 6, Pt. 1, quoted p. 187. Letter from Bohr to Einstein, 13 April 1927.

73   BCW, Vol. 6, p. 418. Letter from Bohr to Einstein, 13 April 1927.

74   Mackinnon (1982), quoted p. 258. Letter from Heisenberg to Pauli, 31 May 1927.

75   Cassidy (1992), quoted p. 243. Letter from Heisenberg to Pauli, 16 May 1927. Heisenberg uses the symbol ≈ that means 'approximately'.

76 Mehra and Rechenberg (2000), Vol. 6, Pt. 1, quoted p. 183. Letter from Heisenberg to Pauli, 16 May 1927.
77 Heisenberg (1927), p. 83.
78 Mehra and Rechenberg (2000), Vol. 6, Pt. 1, quoted p. 184. Letter from Heisenberg to Pauli, 3 June 1927.
79 Heisenberg (1971), p. 79.
80 Pais (1991), quoted p. 309. Letter from Heisenberg to Bohr, 18 June 1927.
81 Pais (1991), quoted p. 309. Letter from Heisenberg to Bohr, 21 August 1927.
82 Cassidy (1992), quoted p. 218. Letter from Heisenberg to his parents, 29 April 1926.
83 Pais (2000), quoted p. 136.
84 Pais (1991), quoted p. 309. Letter from Heisenberg to Pauli, 16 May 1927.
85 Heisenberg (1989), p. 30.
86 Heisenberg (1989), p. 30.
87 Heisenberg (1927), p. 83.
88 Heisenberg (1927), p. 83.
89 Heisenberg (1927), p. 83.
90 Heisenberg (1927), p. 83.

CHAPTER 11: SOLVAY 1927

1 Mehra (1975), quoted p. xxiv.
2 CPAE, Vol. 5, p. 222. Letter from Einstein to Heinrich Zangger, 15 November 1911.
3 Mehra (1975), quoted p. xxiv. Lorentz's Report to the Administrative Council, Solvay Institute, 3 April 1926.
4 Mehra (1975), quoted p. xxiv.
5 Mehra (1975), quoted p. xxiii. Letter from Ernest Rutherford to B.B. Boltwood, 28 February 1921.
6 Mehra (1975), quoted p. xxii.
7 The statute of the League of Nations was drawn up in April 1919.
8 In 1936 Hitler violated the Locarno treaties when he sent German troops into the demilitarised Rhineland.
9 William H. Bragg resigned from the committee in May 1927 citing other commitments, and though invited did not attend. Edmond Van Aubel, though still on the committee, refused to attend because the Germans had been invited.
10 Mehra and Rechenberg (2000), Vol. 6, Pt. 1, quoted p. 232.
11 Mehra and Rechenberg (2000), Vol. 6, Pt. 1, quoted p. 241. Letter from Einstein to Hendrik Lorentz, 17 June 1927.
12 Mehra and Rechenberg (2000), Vol. 6, Pt. 1, quoted p. 241. Letter from Einstein to Hendrik Lorentz, 17 June 1927.
13 Bohr (1949), p. 212.
14 Bacciagaluppi and Valentini (2006), quoted p. 408.
15 Bacciagaluppi and Valentini (2006), quoted p. 408.
16 Bacciagaluppi and Valentini (2006), quoted p. 432.
17 Bacciagaluppi and Valentini (2006), quoted p. 437.
18 Mehra (1975), quoted p. xvii.

19   Bacciagaluppi and Valentini (2006), quoted p. 448.

20   Bacciagaluppi and Valentini (2006), quoted p. 448.

21   Bacciagaluppi and Valentini (2006), quoted p. 470.

22   Bacciagaluppi and Valentini (2006), quoted p. 472.

23   Bacciagaluppi and Valentini (2006), quoted p. 473.

24   Pais (1991), quoted p. 426. 'Could one not keep determinism by making it an object of belief? Must one necessarily elevate indeterminism to a principle?' (Bacciagaluppi and Valentini (2006), p. 477.)

25   Bohr (1963c), p. 91.

26   Bohr was partly to blame for the confusion, since on occasions he referred to his contribution during the general discussion as a 'report'. He did so, for example, in his lecture 'The Solvay Meetings and the Development of Quantum Physics', reprinted in Bohr (1963c).

27   Bohr (1963c), p. 91.

28   Mehra and Rechenberg (2000), Vol. 6, Pt. 1, quoted p. 240.

29   Bohr (1928), p. 53.

30   Bohr (1928), p. 54.

31   Petersen (1985), quoted p. 305.

32   Bohr (1987), p. 1.

33   Einstein (1993), p. 121. Letter from Einstein to Maurice Solovine, 1 January 1951.

34   Einstein (1949a), p. 81.

35   Heisenberg (1989), p. 174.

36   Bacciagaluppi and Valentini (2006), quoted p. 486. The translation is based on notes in the Einstein archives. The published French translation reads: 'I have to apologize for not having gone deeply into quantum mechanics. I should nevertheless want to make some general remarks.'

37   Bohr (1949), p. 213.

38   Bacciagaluppi and Valentini (2006), quoted p. 487.

39   Bacciagaluppi and Valentini (2006), quoted p. 487.

40   See Chapter 9, note 43.

41   Bacciagaluppi and Valentini (2006), quoted p. 487.

42   Bacciagaluppi and Valentini (2006), quoted p. 489.

43   Bacciagaluppi and Valentini (2006), quoted p. 489.

44   Bohr (1949).

45   Bohr (1949), p. 217.

46   Bohr (1949), p. 218.

47   Bohr (1949), p. 218.

48   Bohr (1949), p. 218.

49   Bohr (1949), p. 218.

50   Bohr (1949), p. 222.

51   De Broglie (1962), p. 150.

52   Heisenberg (1971), p. 80.

53   Heisenberg (1967), p. 107.

54   Heisenberg (1967), p. 107.

55 Heisenberg (1967), p. 107.
56 Heisenberg (1983), p. 117.
57 Heisenberg (1983), p. 117.
58 Heisenberg (1971), p. 80.
59 Bohr (1949), p. 213.
60 Mehra and Rechenberg (2000), Vol. 6, Pt. 1, pp. 251–3. Letter from Paul Ehrenfest to Samuel Goudsmit, George Uhlenbeck and Gerhard Diecke, 3 November 1927.
61 Bohr (1949), p. 218.
62 Bohr (1949), p. 218.
63 Bohr (1949), p. 206.
64 Brian (1996), p. 164.
65 Cassidy (1992), quoted p. 253. Letter from Einstein to Arnold Sommerfeld, 9 November 1927.
66 Marage and Wallenborn (1999), quoted p. 165.
67 Cassidy (1992), quoted p. 254.
68 Werner Heisenberg, AHQP interview, 27 February 1963.
69 Gamov (1966), p. 51.
70 Calaprice (2005), p. 89.
71 Fölsing (1997), quoted p. 601. Letter from Einstein to Michele Besso, 5 January 1929.
72 Brian (1996), quoted p. 168.
73 Mehra and Rechenberg (2000), Vol. 6, Pt. 1, quoted p. 256.
74 Mehra and Rechenberg (2000), Vol. 6, Pt. 1, quoted p. 266. Letter from Schrödinger to Bohr, 5 May 1928.
75 Mehra and Rechenberg (2000), Vol. 6, Pt. 1, quoted pp. 266–7. Letter from Bohr to Schrödinger, 23 May 1928.
76 Przibram (1967), p. 31. Letter from Einstein to Schrödinger, 31 May 1928.
77 Fölsing (1997), quoted p. 602. Letter from Einstein to Paul Ehrenfest, 28 August 1928.
78 Brian (1996), quoted p. 169.
79 Pais (2000), quoted p. 215. Letter from Pauli to Hermann Weyl, 11 July 1929.
80 Pais (1982), quoted p. 31.

CHAPTER 12: EINSTEIN FORGETS RELATIVITY

1 Rosenfeld (1968), p. 232.
2 Pais (2000), quoted p. 225.
3 Rosenfeld (1968), p. 232.
4 Rosenfeld (1968), p. 232.
5 Rosenfeld, AHQP interview.
6 Clark (1973) quoted p. 198.
7 'The Fabric of the Universe', *The Times*, 7 November 1919.
8 Thorne (1994), p. 100.
9 Alternatively, since the uncontrollable transfer of momentum to the light box when the pointer and scale is illuminated causes the box to move about unpredictably, the

clock inside is now moving in a gravitational field. The rate at which it ticks (the flow of time) changes unpredictably, leading to an uncertainty in the precise time when the shutter is opened and the photon escapes. Once again, the chain of uncertainties obeys the limits set by Heisenberg's uncertainty principle.

10 Pais (1982), quoted p. 449.

11 Pais (1982), quoted p. 515. Einstein had pointed out to the Swedish Academy that the achievements of Heisenberg and Schrödinger were so significant that it would not be appropriate to divide a Nobel Prize between them. However, 'who should get the prize first is hard to answer', he admitted, before suggesting Schrödinger. He had first nominated Heisenberg and Schrödinger in 1928 when he suggested that de Broglie and Davisson be given precedence. The other options he put forward involved one prize to be shared by de Broglie and Schrödinger and another by Born, Heisenberg and Jordan. The 1928 prize was deferred until 1929, when it was awarded to the British physicist Owen Richardson. As Einstein suggested, Louis de Broglie was the first of the new generation of quantum theorists to be honoured when he was awarded the 1929 prize.

12 Fölsing (1997), quoted p. 630.

13 Brian (1996), quoted p. 200.

14 Calaprice (2005), p. 323.

15 Brian (1996), quoted p. 201.

16 Brian (1996), quoted p. 201.

17 Brian (1996), quoted p. 201.

18 Henig (1998), p. 64.

19 Brian (1996), quoted p. 199.

20 Fölsing (1997), quoted p. 629.

21 Brian (1996), quoted p. 199. Letter from Sigmund Freud to Arnold Zweig, 7 December 1930.

22 Brian (1996), quoted p. 204.

23 Levenson (2003), quoted p. 410.

24 Brian (1996), quoted p. 237.

25 Fölsing (1997), quoted p. 659. Letter from Einstein to Margarete Lenbach, 27 February 1933.

26 Clark (1973), quoted p. 431.

27 Fölsing (1997), quoted p. 661 and Brian (1996), p. 244.

28 Fölsing (1997), quoted p. 662. Letter from Planck to Einstein, 19 March 1933.

29 Fölsing (1997), quoted p. 662. Letter from Planck to Einstein, 31 March 1933.

30 Friedländer (1997), quoted p. 27.

31 Physics: Albert Einstein (1921), James Franck (1925), Gustav Hertz (1925), Erwin Schrödinger (1933), Viktor Hess (1936), Otto Stern (1943), Felix Bloch (1952), Max Born (1954), Eugene Wigner (1963), Hans Bethe (1967), and Dennis Gabor (1971). Chemistry: Fritz Haber (1918), Pieter Debye (1936), Georg von Hevesy (1943), and Gerhard Hertzberg (1971). Medicine: Otto Meyerhof (1922), Otto Loewi (1936), Boris Chain (1945), Hans Krebs (1953), and Max Delbrück (1969).

32 Heilbron (2000), quoted p. 210.

33   Heilbron (2000), quoted p. 210.

34   Beyerchen (1977), quoted p. 43. This section does not appear in the account published in Heilbron (2000), pp. 210–11, which ends with: 'So saying, he hit himself hard on the knee, spoke faster and faster, and flew into such a rage that I could only remain silent and withdraw.'

35   Forman (1973), quoted p. 163.

36   Holton (2005), quoted pp. 32–3.

37   Greenspan (2005), quoted p. 175.

38   Born (1971), p. 251.

39   Greenspan (2005), quoted p. 177.

40   Born (2005), p. 114. Letter from Born to Einstein, 2 June 1933.

41   Born (2005), p. 114. Letter from Born to Einstein, 2 June 1933.

42   Born (2005), p. 111. Letter from Einstein to Born, 30 May 1933.

43   Cornwell (2003), quoted p. 134.

44   Jungk (1960), quoted p. 44.

45   Clark (1973), quoted p. 472.

46   Pais (1982), quoted p. 452. Letter from Abraham Flexner to Einstein, 13 October 1933.

47   Fölsing (1997), quoted p. 682.

48   Fölsing (1997), quoted p. 682. Letter from Einstein to the Board of Trustees of the Institute for Advanced Study, November 1933.

49   Fölsing (1997), quoted pp. 682–3. Letter from Einstein to the Board of Trustees of the Institute for Advanced Study, November 1933.

50   Moore (1989), quoted p. 280.

51   Cassidy (1992), quoted p. 325. Letter from Heisenberg to Bohr, 27 November 1933.

52   Greenspan (2005), quoted p. 191. Letter from Heisenberg to Born, 25 November 1933.

53   Born (2005), p. 200. Letter from Born to Einstein, 8 November 1953.

54   Mehra (1975), quoted p. xxvii. Letter from Einstein to Queen Elizabeth of Belgium, 20 November 1933.

CHAPTER 13: QUANTUM REALITY

1   Smith and Weiner (1980), p. 190. Letter from Robert Oppenheimer to Frank Oppenheimer, 11 January 1935.

2   Smith and Weiner (1980), p. 190. Letter from Robert Oppenheimer to Frank Oppenheimer, 11 January 1935.

3   Born (2005), quoted p. 128.

4   Bernstein (1991), quoted p. 49.

5   James Chadwick was awarded the Nobel Prize for Physics in 1935 and Enrico Fermi in 1938.

6   Brian (1996), quoted p. 251.

7   Einstein (1950), p. 238.

8   Moore (1989), quoted p. 305, Letter from Einstein to Schrödinger, 8 August 1935.

9   Jammer (1985), quoted p. 142.

10 Reprinted in Wheeler and Zurek (1983), pp. 138–41.

11 *New York Times*, 7 May 1935, p. 21.

12 Einstein et al. (1935), p. 138. References to paper reprinted in Wheeler and Zurek (1983).

13 Einstein et al. (1935), p. 138. Italics in the original.

14 Einstein et al. (1935), p. 138. Italics in the original.

15 EPR resisted the temptation to use the two-particle experiment to challenge Heisenberg's uncertainty principle. It is possible to measure the exact momentum of particle A directly and determine the momentum of particle B. While it is not possible to know the position of A, because of the measurement already performed on it, it is possible to determine the position of B directly, since no previous measurement has been directly performed on it. Therefore it may be argued that the momentum and position of particle B can be determined simultaneously, thereby circumventing the uncertainty principle.

16 Einstein et al. (1935), p. 141. Italics in the original.

17 Einstein et al. (1935), p. 141.

18 BCW, Vol. 7, p. 251. Letter from Pauli to Heisenberg, 15 June 1935.

19 BCW, Vol. 7, p. 251. Letter from Pauli to Heisenberg, 15 June 1935.

20 Fölsing (1997), quoted p. 697.

21 Rosenfeld (1967), p. 128.

22 Rosenfeld (1967), p. 128.

23 Rosenfeld (1967), p. 128.

24 Rosenfeld (1967), p. 128.

25 Rosenfeld (1967), p. 129. Also in Wheeler and Zurek (1983), quoted p. 142.

26 See Bohr (1935a).

27 See Bohr (1935b).

28 Bohr (1935b), p. 145.

29 Bohr (1935b), p. 148.

30 Heisenberg (1971), p. 104.

31 Heisenberg (1971), p. 104.

32 Heisenberg (1971), p. 104.

33 Heisenberg (1971), p. 105.

34 Bohr (1949), p. 234.

35 Bohr (1935b), p. 148.

36 Bohr (1935b), p. 148. Italics in the original.

37 Bohr (1935b), p. 148.

38 Fölsing (1997), quoted p. 699. Letter from Einstein to Cornelius Lanczos, 21 March 1942.

39 Born (2005), p. 155. Letter from Einstein to Born, 3 March 1947.

40 Petersen (1985), quoted p. 305.

41 Jammer (1974), quoted p. 161.

42 Niels Bohr, AHQP interview, 17 November 1962.

43 Moore (1989), quoted p. 304. Letter from Schrödinger to Einstein, 7 June 1935.

44 Moore (1989), quoted p. 304. Letter from Schrödinger to Einstein, 7 June 1935.

45   Schrödinger (1935), p. 161.

46   Schrödinger (1935), p. 161.

47   Fine (1986), quoted p. 68. Letter from Einstein to Schrödinger, 17 June 1935.

48   Murdoch (1987), quoted p. 173. Letter from Einstein to Schrödinger, 19 June 1935.

49   Moore (1989), quoted p. 304. Letter from Einstein to Schrödinger, 19 June 1935.

50   Fine (1986), quoted p. 78. Letter from Einstein to Schrödinger, 8 August 1935.

51   Fine (1986), quoted p. 78. Letter from Einstein to Schrödinger, 8 August 1935.

52   Schrödinger (1935), p. 157.

53   Mehra and Rechenberg (2001) Vol. 6, Pt. 2, quoted p. 743. Letter from Einstein to Schrödinger, 4 September 1935.

54   Fine (1986), quoted pp. 84–5. Letter from Einstein to Schrödinger, 22 December 1950.

55   Fine (1986), quoted pp. 84–5. Letter from Einstein to Schrödinger, 22 December 1950.

56   Moore (1989), quoted p. 314. Letter from Schrödinger to Einstein, 23 March 1936.

57   Fölsing (1997), quoted p. 688.

58   Fölsing (1997), quoted p. 688.

59   Born (2005), p. 125. Letter from Einstein to Born, undated.

60   Born (2005), p. 127.

61   Fölsing (1997), quoted p. 704.

62   Brian (1996), quoted p. 305.

63   Brian (1996), quoted p. 305.

64   Petersen (1985), quoted p. 305.

65   Einstein (1993), p. 119. Letter from Einstein to Maurice Solovine, 1 January 1951.

66   Fine (1986), quoted p. 95. Letter from Einstein to M. Laserna, 8 January 1955.

67   Einstein (1934), p. 112.

68   Einstein (1993), p. 119. Letter from Einstein to Maurice Solovine, 1 January 1951.

69   Heisenberg (1989), p. 117.

70   Heisenberg (1989), p. 117.

71   Heisenberg (1989), p. 116.

72   Einstein (1950), p. 88.

73   Heisenberg (1989), p. 44.

74   Przibram (1967), p. 31. Letter from Einstein to Schrödinger, 31 May 1928.

75   Fölsing (1997), quoted p. 704.

76   Fölsing (1997), quoted p. 705.

77   Mehra (1975), quoted p. xxvii. Letter from Einstein to Queen Elizabeth of Belgium, 9 January 1939.

78   Pais (1994), quoted p. 218. Letter from Einstein to Roosevelt, 7 March 1940.

79   Clark (1973), quoted p. 29.

80   Heilbron (2000), quoted p. 195.

81   Heilbron (2000), quoted p. 195.

82   Fölsing (1997), quoted p. 729. Letter from Einstein to Marga Planck, October 1947.

83   Pais (1967), p. 224.

84   Pais (1967), p. 225.

85 Heisenberg (1983), p. 121.
86 Holton (2005), quoted p. 32.
87 Einstein (1993), p. 85. Letter from Einstein to Solovine, 10 April 1938.
88 Brian (1996), quoted p. 400.
89 Nathan and Norden (1960), pp. 629–30. Letter from Einstein to Bohr, 2 March 1955.
90 Pais (1982), quoted p. 477. Letter from Helen Dukas to Abraham Pais, 30 April 1955.
91 Overbye (2001), quoted p. 1.
92 Clark (1973), quoted p. 502.
93 Bohr (1955), p. 6.
94 Pais (1994), quoted p. 41.

CHAPTER 14: FOR WHOM BELL'S THEOREM TOLLS

1   Born (2005), p. 146. Letter from Einstein to Born, 7 September 1944.
2   Stapp (1977), p. 191.
3   Petersen (1985), quoted p. 305.
4   Przibam (1967), p. 39. Letter from Einstein to Schrödinger, 22 December 1950.
5   Goodchild (1980), quoted p. 162.
6   Bohm (1951), pp. 612–13.
7   Bohm (1951), p. 622.
8   Bohm (1951), p. 611.
9   Bohm (1952a), p. 382.
10 Bohm (1952a), p. 369.
11 Bell (1987), p. 160.
12 Bell (1987), p. 160.
13 The German title of von Neumann's book was *Mathematische Grundlagen der Quantenmechanik.*
14 Von Neumann (1955), p. 325.
15 Maxwell (1860), p. 19.
16 Maxwell (1860), p. 19.
17 Von Neumann (1955), pp. 327–8.
18 Bernstein (1991), quoted p. 12.
19 Bernstein (1991), quoted p. 15.
20 Bernstein (1991), quoted p. 64.
21 Bell (1987), quoted p. 159.
22 Bell (1987), quoted p. 159.
23 Bell (1987), quoted p. 159.
24 Bernstein (1991), quoted p. 65.
25 Bell (1987), p. 160.
26 Bell (1987), p. 167.
27 Beller (1999), quoted p. 213.
28 Born (2005), p. 189. Letter from Einstein to Born, 12 May 1952.
29 Bernstein (1991), quoted p. 66.

30   Bernstein (1991), quoted p. 72.
31   Bernstein (1991), quoted p. 72.
32   Bernstein (1991), quoted p. 73.
33   Born (2005), p. 153. Letter from Einstein to Born, 3 March 1947.
34   Bohm's modification of EPR appeared in chapter 22 of his book *Quantum Theory*. It involved a molecule with a spin of zero that disintegrates into two atoms, one with spin-up ($+\frac{1}{2}$) and the other with spin-down ($-\frac{1}{2}$), whose combined spin remains zero. Since its inception it has become standard practice to replace the atoms with a pair of electrons.
35   The mutually perpendicular axes x, y, and z are chosen only for convenience and because they are most familiar. Any set of three axes serves just as well for measuring the components of quantum spin.
36   Bell (1987), p. 139.
37   Bell (1987), p. 143.
38   Bell (1987), p. 143.
39   Also known as 'Bell's inequalities'.
40   Bell (1964). Reprinted in Bell (1987) and Wheeler and Zurek (1983).
41   Bell (1966), p. 447. Reprinted in Bell (1987) and Wheeler and Zurek (1983).
42   Bell (1966), p. 447.
43   Born (2005), p. 218. Letter from Pauli to Born, 31 March 1954.
44   Born (2005), p. 218. Letter from Pauli to Born, 31 March 1954.
45   Bell (1964), p. 199.
46   Clauser (2002), p. 71.
47   Clauser (2002), p. 70.
48   Redhead (1987), p. 108, table 1.
49   Aczel (2003), quoted p. 186.
50   Aczel (2003), quoted p. 186.
51   Aspect et al. (1982), p. 94.
52   Davies and Brown (1986), p. 50.
53   Davies and Brown (1986), p. 51.
54   Davies and Brown (1986), p. 47.

CHAPTER 15: THE QUANTUM DEMON

1   Pais (1982), quoted p. 9.
2   Einstein (1950), p. 91.
3   Pais (1982), quoted p. 460.
4   Pais (1982), p. 9.
5   Feynman (1965), p. 129.
6   Feynman (1965), p. 129.
7   Bernstein (1991), p. 42.
8   Born (2005), p. 162. Comment on manuscript from Einstein to Born, 18 March 1948.
9   Heisenberg (1983), p. 117. One example of Einstein using his famous phrase.
10   Born (2005), p. 216. Letter from Pauli to Born, 31 March 1954.

11   Born (2005), p. 216. Letter from Pauli to Born, 31 March 1954.

12   Born (2005), p. 216. Letter from Pauli to Born, 31 March 1954.

13   Born (2005), p. 216. Letter from Pauli to Born, 31 March 1954.

14   Stachel (2002), quoted p. 390. Letter from Einstein to Georg Jaffe, 19 January 1954.

15   Born (2005), p. 88. Letter from Einstein to Born, 4 December 1926.

16   Born (2005), p. 219. Letter from Pauli to Born, 31 March 1954.

17   Isaacson (2007), quoted p. 460. Letter from Einstein to Jerome Rothstein, 22 May 1950.

18   Rosenthal-Schneider (1980), quoted p. 70. Postcard from Einstein to Ilse Rosenthal, 31 March 1944.

19   Aspect (2007), p. 867.

20   Einstein et al. (1935), p. 141.

21   Einstein (1949b), p. 666.

22   Fine (1986), quoted p. 57. Letter from Einstein to Aron Kupperman, 10 November 1954.

23   Isaacson (2007), quoted p. 466.

24   Heisenberg (1971), p. 81.

25   Heisenberg (1971), p. 80.

26   Born (2005), p. 69.

27   Born (1949), pp. 163–4.

28   Clauser (2002), p. 72.

29   Blaedel (1988), p. 11.

30   Clauser (2002), p. 61.

31   Wolf (1988), quoted p. 17.

32   Pais (2000), quoted p. 55.

33   Gell-Mann (1979), p. 29.

34   Tegmark and Wheeler (2001), p. 61.

35   Among the 30 there were those who supported the 'consistent histories' approach that has its origins in the many worlds interpretation. It is based on the idea that out of all possible means by which an observed experimental result may have been caused, only a few make sense under the rules of quantum mechanics.

36   Buchanan (2007), quoted p. 37.

37   Buchanan (2007), quoted p. 38.

38   Stachel (1998), p. xiii.

39   French (1979), quoted p. 133.

40   Pais (1994) quoted p. 57.

BIBLIOGRAPHY

The Collected Papers of Albert Einstein (CPAE), published by Princeton University Press:

Volume 1 – The Early Years: 1879–1902. Edited by John Stachel (1987)

Volume 2 – The Swiss Years: Writings, 1900–1909. Edited by John Stachel (1989)

Volume 3 – The Swiss Years: Writings, 1909–1911. Edited by Martin J. Klein, A.J. Kox, Jürgen Renn and Robert Schulmann (1993)

Volume 4 – The Swiss Years: Writings, 1912–1914. Edited by Martin J. Klein, A.J. Kox, Jürgen Renn and Robert Schulmann (1996)

Volume 5 – The Swiss Years: Correspondence, 1902–1914. Edited by Martin J. Klein, A.J. Kox and Robert Schulmann (1994)

Volume 6 – The Berlin Years: Writings, 1914–1917. Edited by Martin J. Klein, A.J. Kox and Robert Schulmann (1997)

Volume 7 – The Berlin Years: Writings, 1918–1921. Edited by Michel Janssen, Robert Schulmann, Jozsef Illy, Christop Lehner and Diana Kormos Buchwald (2002)

Volume 8 – The Berlin Years: Correspondence, 1914–1918. Edited by Robert Schulmann, A.J. Kox, Michel Janssen and Jozsef Illy (1998)

Volume 9 – The Berlin Years: Correspondence, January 1919–April 1920. Edited by Diana Kormos Buchwald and Robert Schulman (2004)

CPAE note: Volumes 1 to 5 translated by Anna Beck, Volumes 6 and 7 by Alfred Engel, and Volumes 8 and 9 by Ann Hentschel. Publication dates are for the English translations.

Niels Bohr Collected Works (BCW), published by North-Holland, Amsterdam:

Volume 1 – Early work, 1905–1911. Edited by J. Rud Nielsen, general editor Léon Rosenfeld (1972)

Volume 2 – Work on atomic physics, 1912–1917. Edited by Ulrich Hoyer, general editor Léon Rosenfeld (1981)

Volume 3 – The correspondence principle, 1918–1923. Edited by J. Rud Nielsen, general editor Léon Rosenfeld (1976)

Volume 4 – The periodic system, 1920–1923. Edited by J. Rud Nielsen (1977)

Volume 5 – The emergence of quantum mechanics, mainly 1924–1926. Edited by
  Klaus Stolzenburg, general editor Erik Rüdinger (1984)
Volume 6 – Foundations of quantum physics I, 1926–1932. Edited by Jørgen
  Kalckar, general editor Erik Rüdinger (1985)
Volume 7 – Foundations of quantum physics II, 1933–1958. Edited by Jørgen
  Kalckar, general editors Finn Aaserud and Erik Rüdinger (1996)

Abragam, A. (1988), 'Louis Victor Pierre Raymond de Broglie', *Biographical
  Memoirs of Fellows of the Royal Society*, **34**, 22–41 (London: Royal Society)
Aczel, Amir D. (2003), *Entanglement* (Chichester: John Wiley)
Albert, David Z. (1992), *Quantum Mechanics and Experience* (Cambridge, MA:
  Harvard University Press)
Andrade, E.N. da C. (1964), *Rutherford and the Nature of the Atom* (Garden
  City, NY: Doubleday Anchor)
Ashton, Francis W. (1940), 'J.J. Thomson', *The Times*, London, 4 September
Aspect, Alain, Philippe Grangier, and Gérard Roger (1981), 'Experimental tests
  of realistic local theories via Bell's theorem', Physical Review Letters, **47**,
  460–463
Aspect, Alain, Philippe Grangier, and Gérard Roger (1982), 'Experimental
  realization of Einstein-Podolsky-Rosen-Bohm *Gedankenexperiment*: A new
  violation of Bell's inequalities', *Physical Review Letters*, **49**, 91–94
Aspect, Alain (2007), 'To be or not to be local', *Nature*, **446**, 866

Bacciagaluppi, Guido and Anthony Valentini (2006), *Quantum Theory at
  the Crossroads: Reconsidering the 1927 Solvay Conference*, arXiv:quant-ph/
  0609184v1, 24 September. To be published by Cambridge University Press
  in December 2008
Badash, Lawrence (1969), *Rutherford and Boltwood* (New Haven, CT: Yale
  University Press)
Badash, Lawrence (1987), 'Ernest Rutherford and Theoretical Physics', in
  Kargon and Achinstein (1987)
Baggott, Jim (2004), *Beyond Measure* (Oxford: Oxford University Press)
Baierlein, Ralph (2001), *Newton to Einstein: The Trail of Light* (Cambridge:
  Cambridge University Press)
Ballentine, L.E. (1972), 'Einstein's Interpretation of Quantum Mechanics',
  American Journal of Physics, **40**, 1763–1771
Barkan, Diana Kormos (1993), 'The Witches' Sabbath: The First International
  Solvay Congress in Physics', in Beller et al. (1993)

Bell, John S. (1964), 'On The Einstein Podolsky Rosen Paradox', *Physics*, **1**, 3, 195–200. Reprinted in Bell (1987) and Wheeler and Zurek (1983)

Bell, John S. (1966), 'On the Problem of Hidden Variables in Quantum Mechanics', *Review of Modern Physics*, **38**, 3, 447–452. Reprinted in Bell (1987) and Wheeler and Zurek (1983)

Bell, John S. (1982), 'On the Impossible Pilot Wave', *Foundations of Physics*, **12**, 989–999. Reprinted in Bell (1987)

Bell, John S. (1987), *Speakable and Unspeakable in Quantum Mechanics* (Cambridge: Cambridge University Press)

Beller, Mara (1999), *Quantum Dialogue: The Making of a Revolution* (Chicago: University of Chicago Press)

Beller, Mara, Jürgen Renn, and Robert S. Cohen (eds) (1993), *Einstein in Context*. Special issue of *Science in Context*, **6**, no. 1 (Cambridge: Cambridge University Press)

Bernstein, Jeremy (1991), *Quantum Profiles* (Princeton, NJ: Princeton University Press)

Bertlmann, R.A. and A. Zeilinger (eds) (2002), *Quantum [Un]speakables: From Bell to Quantum Information* (Berlin: Springer)

Beyerchen, Alan D. (1977), *Scientists under Hitler: Politics and the Physics Community in the Third Reich* (New Haven, CT: Yale University Press)

Blaedel, Niels (1985), *Harmony and Unity: The Life of Niels Bohr* (Madison, WI: Science Tech Inc.)

Bloch, Felix (1976), 'Reminiscences of Heisenberg and the Early Days of Quantum Mechanics', *Physics Today*, **29**, December, 23–7

Bohm, David (1951), *Quantum Theory* (Englewood Cliffs, NJ: Prentice-Hall)

Bohm, David (1952a), 'A Suggested Interpretation of the Quantum Theory in Terms of "Hidden" Variables I', reprinted in Wheeler and Zurek (1983), 369–382

Bohm, David (1952b), 'A Suggested Interpretation of the Quantum Theory in Terms of "Hidden" Variables II', reprinted in Wheeler and Zurek (1983), 383–396

Bohm, David (1957), *Causality and Chance in Modern Physics* (London: Routledge)

Bohr, Niels (1922), 'The structure of the atom', Nobel lecture delivered on 11 December. Reprinted in *Nobel Lectures* (1965), 7–43

Bohr, Niels (1928), 'The Quantum Postulate and the Recent Development of Atomic Theory', in Bohr (1987)

Bohr, Niels (1935a), 'Quantum Mechanics and Physical Reality', *Nature*, **136**, 65. Reprinted in Wheeler and Zurek (1983), 144

Bohr, Niels (1935b), 'Can Quantum-Mechanical Description of Physical Reality Be Considered Complete?', *Physical Review*, **48**, 696–702. Reprinted in Wheeler and Zurek (1983), 145–151

Bohr, Niels (1949), 'Discussion with Einstein on Epistemological Problems in Atomic Physics', in Schilpp (1969)

Bohr, Niels (1955), 'Albert Einstein: 1879–1955', *Scientific American*, **192**, June, 31–33

Bohr, Niels (1963a), *Essays 1958–1962 on Atomic Physics and Human Knowledge* (New York: John Wiley)

Bohr, Niels (1963b), 'The Rutherford Memorial Lecture 1958: Reminiscences of the Founder of Nuclear Science and of Some Developments Based on his Work', in Bohr (1963a)

Bohr, Niels (1963c), 'The Solvay Meetings and the Development of Quantum Physics', in Bohr (1963a)

Bohr, Niels (1963d), *On the Constitution of Atoms and Molecules: Papers of 1913 reprinted from the* Philosophical Magazine *with an Introduction by L. Rosenfeld* (Copenhagen: Munksgaard Ltd; New York: W.A. Benjamin)

Bohr, Niels (1987), *The Philosophical Writings of Niels Bohr: Volume 1 – Atomic Theory and the Description of Nature* (Woodbridge, CT: Ox Bow Press)

Boorse, Henry A. and Lloyd Motz (eds) (1966), *The World of the Atom*, 2 vols (New York: Basic Books)

Born, Max (1948), 'Max Planck', *Obituary Notices of Fellows of the Royal Society*, **6**, 161–88 (London: Royal Society)

Born, Max (1949), 'Einstein's Statistical Theories', in Schilpp (1969)

Born, Max (1954), 'The statistical interpretation of quantum mechanics', Nobel lecture delivered on 11 December. Reprinted in *Nobel Lectures* (1964), 256–267

Born, Max (1970), *Physics in My Generation* (London: Longman)

Born, Max (1978), *My Life: Recollections of a Nobel Laureate* (London: Taylor and Francis)

Born, Max (2005), *The Born–Einstein Letters 1916–1955: Friendship, Politics and Physics in Uncertain Times* (New York: Macmillan)

Brandstätter, Christian (ed.) (2005), *Vienna 1900 and the Heroes of Modernism* (London: Thames and Hudson)

Brian, Denis (1996), *Einstein: A Life* (New York: John Wiley)

Broglie, Louis de (1929), 'The wave nature of the electron', Nobel lecture delivered on 12 December. Reprinted in *Nobel Lectures* (1965), 244–256

Broglie, Louis de (1962), *New Perspectives in Physics* (New York: Basic Books)

Brooks, Michael (2007), 'Reality Check', *New Scientist*, 23 June, 30–33

Buchanan, Mark (2007), 'Quantum Untanglement', *New Scientist*, 3 November, 36–39

Burrow, J.W. (2000), *The Crisis of Reason: European Thought, 1848–1914* (New Haven, CT: Yale University Press)

Cahan, David (1985), 'The Institutional Revolution in German Physics, 1865–1914', *Historical Studies in the Physical Sciences*, **15**, 1–65

Cahan, David (1989), *An Institute for an Empire: The Physikalisch-Technische Reichsanstalt 1871–1918* (Cambridge: Cambridge University Press)

Cahan, David (2000), 'The Young Einstein's Physics Education: H.F. Weber, Hermann von Helmholtz and the Zurich Polytechnic Physics Institute', in Howard and Stachel (2000)

Calaprice, Alice (ed.) (2005), *The New Quotable Einstein* (Princeton, NJ: Princeton University Press)

Cassidy, David C. (1992), *Uncertainty: The Life and Science of Werner Heisenberg* (New York: W.H. Freeman and Company)

Cercignani, Carlo (1998), *Ludwig Boltzmann: The Man Who Trusted Atoms* (Oxford: Oxford University Press)

Clark, Christopher (2006), *Iron Kingdom: The Rise and Downfall of Prussia, 1600–1947* (London: Allen Lane)

Clark, Roland W. (1973), *Einstein: The Life and Times* (London: Hodder and Stoughton)

Clauser, John F. (2002), 'The Early History of Bell's Theorem', in Bertlmann and Zeilinger (2002)

Cline, Barbara Lovett (1987), *Men Who Made a New Physics* (Chicago: University of Chicago Press)

Compton, Arthur H. (1924), 'The Scattering of X-Rays', *Journal of the Franklin Institute*, **198**, 57–72

Compton, Arthur H. (1961), 'The Scattering of X-Rays as Particles', reprinted in Phillips (1985)

Cornwell, John (2003), *Hitler's Scientists: Science, War and the Devil's Pact* (London: Viking)

Cropper, William H. (2001), *Great Physicists: The Life and Times of Leading Physicists from Galileo to Hawking* (Oxford: Oxford University Press)

Cropper, William H. (1970), *The Quantum Physicists* (New York: Oxford University Press)

Cushing, James T. (1994), *Quantum Mechanics: Historical Contingency and the Copenhagen Hegemony* (Chicago: University of Chicago Press)

Cushing, James T. (1998), *Philosophical Concepts in Physics: The Historical Relation Between Philosophy and Scientific Theories* (Cambridge: Cambridge University Press)

Cushing, James T. and Ernan McMullin (eds) (1989), *Philosophical Consequences of Quantum Theory: Reflections on Bell's Theorem* (Notre Dame, IN: University of Notre Dame Press)

Davies, Paul C.W. and Julian Brown (1986), *The Ghost in the Atom* (Cambridge: Cambridge University Press)

De Hass-Lorentz, G.L. (ed.) (1957), *H.A. Lorentz: Impressions of his Life and Work* (Amsterdam: North-Holland Publishing Company)

Dirac, P.A.M. (1927), 'The physical interpretation of quantum dynamics', *Proceedings of the Royal Society* A, **113**, 621–641

Dirac, P.A.M. (1933), 'Theory of electrons and positrons', Nobel lecture delivered on 12 December. Reprinted in *Nobel Lectures* (1965), 320–325

Dirac, P.A.M (1977), 'Recollections of an exciting era', in Weiner (1977)

Dresden, M. (1987), *H.A. Kramers* (New York: Springer)

Einstein, Albert (1905a), 'On a Heuristic Point of View Concerning the Production and Transformation of Light', reprinted in Stachel (1998)

Einstein, Albert (1905b), 'On the Electrodynamics of Moving Bodies', reprinted in Einstein (1952)

Einstein, Albert (1934), *Essays in Science* (New York: Philosophical Library)

Einstein, Albert, Boris Podolsky, and Nathan Rosen (1935), 'Can Quantum-Mechanical Description of Physical Reality Be Considered Complete?', Physical Review, **47**, 777–780. Reprinted in Wheeler and Zurek (1983), pp. 138–41.

Einstein, Albert (1949a), 'Autobiographical Notes', in Schilpp (1969)

Einstein, Albert (1949b), 'Reply to Criticism', in Schilpp (1969)

Einstein, Albert (1950), *Out of My Later Years* (New York: Philosophical Library)

Einstein, Albert (1952), *The Principle of Relativity: A Collection of original papers on the special and general theory of relativity* (New York: Dover Publications)

Einstein, Albert (1954), *Ideas and Opinions* (New York: Crown)

Einstein, Albert (1993), *Letters to Solovine, with an introduction by Maurice Solovine* (New York: Citadel Press)

Elitzur, A., S. Dolev, and N. Kolenda (eds) (2005), *Quo Vadis Quantum Mechanics?* (Berlin: Springer)

Elon, Amos (2002), *The Pity of it All: A Portrait of Jews in Germany 1743–1933* (London: Allen Lane)

Elsasser, Walter (1978), *Memoirs of a Physicist* (New York: Science History Publications)

Emsley, John (2001), *Nature's Building Blocks: An A–Z Guide to the Elements* (Oxford: Oxford University Press)

Enz, Charles P. (2002), *No Time to be Brief: A scientific biography of Wolfgang Pauli* (Oxford: Oxford University Press)

Evans, James and Alan S. Thorndike (eds) (2007), *Quantum Mechanics at the Crossroads* (Berlin: Springer-Verlag)

Evans, Richard J. (2003), *The Coming of the Third Reich* (London: Allen Lane)

Eve, Arthur S. (1939), *Rutherford: Being the Life and Letters of the Rt. Hon. Lord Rutherford, O.M.* (Cambridge: Cambridge University Press)

Everdell, William R. (1997), *The First Moderns* (Chicago: University of Chicago Press)

Everett, Susanne (1979), *Lost Berlin* (New York: Gallery Books)

Feynman, Richard P. (1965), *The Character of Physical Law* (London: BBC Publications)

Fine, Arthur (1986), *The Shaky Game: Einstein, Realism and the Quantum Theory* (Chicago: University of Chicago Press)

Forman, Paul (1971), 'Weimar Culture, Causality, and Quantum Theory, 1918–1927: Adaptation by German Physicists and Mathematicians to a Hostile Intellectual Environment', *Historical Studies in the Physical Sciences*, **3**, 1–115

Forman, Paul (1973), 'Scientific internationalism and the Weimar physicists: The ideology and its manipulation in Germany after World War I', *Isis*, **64**, 151–178

Forman, Paul, John L. Heilbron, and Spencer Weart (1975), 'Physics circa 1900: Personnel, Funding, and Productivity of the Academic Establishments', *Historical Studies in the Physical Sciences*, **5**, 1–185

Fölsing, Albrecht (1997), *Albert Einstein: A Biography* (London: Viking)

Frank, Philipp (1947), *Einstein: His Life and Times* (New York: DaCapo Press)

Franklin, Allan (1997), 'Are There Really Electrons? Experiment and Reality', *Physics Today*, October, 26–33

French, A. P. (ed.) (1979), *Einstein: A Centenary Volume* (London: Heinemann)

French, A. P. and P.J. Kennedy (eds) (1985), *Niels Bohr: A Centenary* (Cambridge, MA: Harvard University Press)

Friedländer, Saul (1997), *Nazi Germany and The Jews: Volume 1 – The Years of Persecution 1933–39* (London: Weidenfeld and Nicolson)

Frisch, Otto (1980), *What Little I Remember* (Cambridge: Cambridge University Press)

Fromkin, David (2004), *Europe's Last Summer: Why the World Went to War in 1914* (London: William Heinemann)

Fulbrook, Mary (2004), *A Concise History of Germany*, 2nd edn (Cambridge: Cambridge University Press)

Gamov, George (1966), *Thirty Years That Shocked Physics* (New York: Dover Publications)

Gay, Ruth (1992), *The Jews of Germany: A Historical Portrait* (New Haven, CT: Yale University Press)

Gehrenbeck, Richard K. (1978), 'Electron Diffraction: Fifty Years Ago', reprinted in Weart and Phillips (1985)

Gell-Mann, Murray (1979), 'What are the Building Blocks of Matter?', in Huff and Prewett (1979)

Gell-Mann, Murray (1981), 'Questions for the Future', in Mulvey (1981)

Gell-Mann, Murray (1994), *The Quark and the Jaguar* (London: Little Brown)

Geiger, Hans and Ernest Marsden (1913), 'The Laws of Deflection of α-Particles through Large Angles', *Philosophical Magazine*, Series 6, **25**, 604–623

German Bundestag (1989), *Questions on German History: Ideas, forces, decisions from 1800 to the present*, 3rd edition, updated (Bonn: German Bundestag Publications Section)

Gilbert, Martin (1994), *The First World War* (New York: Henry Holt and Co.)

Gilbert, Martin (2006), *Kristallnacht: Prelude to Destruction* (London: HarperCollins)

Gillispie, Charles C. (ed.-in-chief) (1970–1980), *Dictionary of Scientific Biography*, 16 vols (New York: Scribner's)

Gillott, John and Manjit Kumar (1995), *Science and the Retreat from Reason* (London: Merlin Press)

Goodchild, Peter (1980), *J. Robert Oppenheimer: Shatterer of Worlds* (London: BBC)

Goodman, Peter (ed.) (1981), *Fifty Years of Electron Diffraction* (Dordrecht, Holland: D. Reidel)

Goudsmit, Samuel A. (1976), 'It might as well be spin', *Physics Today*, June, reprinted in Weart and Phillips (1985)

Greenspan, Nancy Thorndike (2005), *The End of The Certain World: The Life and Science of Max Born* (Chichester: John Wiley)

Greenstein, George and Arthur G. Zajonc (2006), *The Quantum Challenge: Modern Research on the Foundations of Quantum Mechanics*, 2nd edition (Sudbury, MA: Jones and Bartlett Publishers)

Gribbin, John (1998), *Q is for Quantum: Particle Physics from A to Z* (London: Weidenfeld and Nicolson)

Gröblacher, Simon et al. (2007) 'An experimental test of non-local realism', *Nature*, **446**, 871–875

Grunberger, Richard (1974), *A Social History of the Third Reich* (London: Penguin Books)

Haar, Dirk ter (1967), *The Old Quantum Theory* (Oxford: Pergamon)

Harman, Peter M. (1982), *Energy, Force and Matter: The Conceptual Development of Nineteenth Century Physics* (Cambridge: Cambridge University Press)

Harman, Peter M. and Simon Mitton (eds) (2002), *Cambridge Scientific Minds* (Cambridge: Cambridge University Press)

Heilbron, John L. (1977), 'Lectures on the History of Atomic Physics 1900–1922', in Weiner (1977)

Heilbron, John L. (2000), *The Dilemmas of An Upright Man: Max Planck and the Fortunes of German Science* (Cambridge, MA: Harvard University Press)

Heilbron, John L. (2007), 'Max Planck's compromises on the way to and from the Absolute', in Evans and Thorndike (2007)

Heilbron, John L. and Thomas S. Kuhn (1969), 'The Genesis of the Bohr Atom', *Historical Studies in the Physical Sciences*, **1**, 211–90

Heisenberg, Werner (1925), 'On a Quantum-Theoretical Reinterpretation of Kinematics and Mechanical Relations', reprinted and translated in Van der Waerden (1967)

Heisenberg, Werner (1927), 'The Physical Content of Quantum Kinematics and Mechanics', reprinted and translated in Wheeler and Zurek (1983)

Heisenberg, Werner (1933), 'The development of quantum mechanics', Nobel lecture delivered on 11 December. Reprinted in *Nobel Lectures* (1965), 290–301

Heisenberg, Werner (1949), *The Physical Principles of Quantum Theory* (New York: Dover Publications)

Heisenberg, Werner (1967), 'The Quantum Theory and its Interpretation', in Rozental (1967)

Heisenberg, Werner (1971), *Physics and Beyond: Encounters and Conversations* (London: George Allen and Unwin)

Heisenberg, Werner (1983), *Encounters with Einstein: And Other Essays on People, Places, and Particles* (Princeton, NJ: Princeton University Press)

Heisenberg, Werner (1989), *Physics and Philosophy* (London: Penguin Books)

Heitler, Walter (1961), 'Erwin Schrödinger', *Biographical Memoirs of Fellows of the Royal Society*, **7**, 221–228

Henig, Ruth (1998), *The Weimar Republic 1919–1933* (London: Routledge)

Hentschel, Anna M. and Gerd Grasshoff (2005), *Albert Einstein: 'Those Happy Bernese Years'* (Bern: Staempfli Publishers)

Hentschel, Klaus (ed.) (1996), *Physics and National Socialism: An Anthology of Primary Sources* (Basel: Birkhäuser)

Hermann, Armin (1971), *The Genesis of Quantum Theory* (Cambridge, MA: MIT Press)

Hiebert, Erwin N. (1990), 'The Transformation of Physics', in Teich and Porter (1990)

Highfield, Roger and Paul Carter (1993), *The Private Lives of Albert Einstein* (London: Faber and Faber)

Holton, Gerald (2005), *Victory and Vexation in Science: Einstein, Bohr, Heisenberg and Others* (Cambridge, MA: Harvard University Press)

Honner, John (1987), *The Description of Nature: Niels Bohr and the Philosophy of Quantum Physics* (Oxford: Clarendon Press)

Howard, Don and John Stachel (eds)(2000), *Einstein: The Formative Years 1879–1909* (Boston, MA: Birkhäuser)

Howorth, Muriel (1958), *The Life of Frederick Soddy* (London: New World)

Huff, Douglas and Omer Prewett (eds) (1979), *The Nature of the Physical Universe* (New York: John Wiley)

Isaacson, Walter (2007), *Einstein: His Life and Universe* (London: Simon and Schuster)

Isham, Chris J. (1995), *Lectures on Quantum Theory* (London: Imperial College Press)

Jammer, Max (1966), *The Conceptual Development of Quantum Mechanics* (New York: McGraw-Hill)

Jammer, Max (1974), *The Philosophy of Quantum Mechanics: The Interpretations of Quantum Mechanics in Historical Perspective* (New York: Wiley-Interscience)

Jammer, Max (1985), 'The EPR problem and its historical development', in Lahti and Mittelstaedt (1985)

Jordan, Pascual (1927), 'Philosophical Foundations of Quantum Theory', *Nature*, **119**, 566

Jungk, Robert (1960), *Brighter Than A Thousand Suns: A Personal History of The Atomic Scientists* (London: Penguin)

Jungnickel, Christa and Russell McCormmach (1986), *Intellectual Mastery of Nature: Theoretical physics from Ohm to Einstein*, 2 vols (Chicago: University of Chicago Press)

Kangro, Hans (1970), 'Max Planck', *Dictionary of Scientific Biography*, 7–17 (New York: Scribner)

Kangro, Hans (1976), *Early History of Planck's Radiation Law* (London: Taylor and Francis)

Kargon, Robert and Peter Achinstein (eds) (1987), *Kelvin's Baltimore Lectures and Modern Theoretical Physics: Historical and Philosophical Perspectives* (Cambridge, MA: MIT Press)

Kay, William A. (1963), 'Recollections of Rutherford: Being the Personal Reminiscences of Lord Rutherford's Laboratory Assistant Here Published for the First Time', *The Natural Philosopher*, 1, 127–155

Keller, Alex (1983), *The Infancy of Atomic Physics: Hercules in his Cradle* (Oxford: Clarendon Press)

Kelvin, Lord (1901), 'Nineteenth Clouds Over the Dynamical Theory of Heat and Light', *Philosophical Magazine*, 2, 1–40

Klein, Martin J. (1962), 'Max Planck and the Beginnings of Quantum Theory', *Archive for History of Exact Sciences*, 1, 459–479

Klein, Martin J. (1965), 'Einstein, Specific Heats, and the Early Quantum Theory', *Science*, 148, 173–180

Klein, Martin J. (1966), 'Thermodynamics and Quanta in Planck's Work', *Physics Today*, November

Klein, Martin J. (1967), 'Thermodynamics in Einstein's Thought', *Science*, 157, 509–516

Klein, Martin J. (1970), 'The First Phase of the Einstein-Bohr Dialogue', *Historical Studies in the Physical Sciences*, 2, 1–39

Klein, Martin J. (1985), *Paul Ehrenfest: the making of a theoretical physicist*, Vol. 1 (Amsterdam: North-Holland)

Knight, David (1986), *The Age of Science* (Oxford: Blackwell)

Kragh, Helge (1990), *Dirac: A Scientific Biography* (Cambridge: Cambridge University Press)

Kragh, Helge (1999), *Quantum Generations: A History of Physics in the Twentieth Century* (Princeton, NJ: Princeton University Press)

Kragh, Helge (2002), 'Paul Dirac: A Quantum Genius', in Harman and Mitton (2002)

Kuhn, Thomas S. (1987), *Blackbody Theory and the Quantum Discontinuity, 1894–1912, with new afterword* (Chicago: University of Chicago Press)

Kursunoglu, Behram N. and Eugene P. Wigner (eds) (1987), *Reminiscences about a Great Physicist: Paul Adrien Maurice Dirac* (Cambridge: Cambridge University Press)

Lahti, P. and P. Mittelstaedt (eds) (1985), *Symposium on the Foundations of Modern Physics* (Singapore: World Scientific)

Laidler, Keith J. (2002), *Energy and the Unexpected* (Oxford: Oxford University Press)

Large, David Clay (2001), *Berlin: A Modern History* (London: Allen Lane)

Levenson, Thomas (2003), *Einstein in Berlin* (New York: Bantam Dell)

Levi, Hilde (1985), *George de Hevesy: Life and Work* (Bristol: Adam Hilger Ltd)

Lindley, David (2001), *Boltzmann's Atom: The Great Debate That Launched a Revolution in Physics* (New York: The Free Press)

MacKinnon, Edward M. (1982), *Scientific Explanation and Atomic Physics* (Chicago: University of Chicago Press)

Magris, Claudio (2001), *Danube* (London: The Harvill Press)

Mahon, Basil (2003), *The Man Who Changed Everything: The Life of James Clerk Maxwell* (Chichester: John Wiley)

Marage, Pierre and Grégoire Wallenborn (eds) (1999), *The Solvay Councils and the Birth of Modern Physics* (Basel: Birkhäuser)

Marsden, Ernest (1948), 'Rutherford Memorial Lecture', in Rutherford (1954)

Maxwell, James Clerk (1860), 'Illustrations of the Dynamical Theory of Gases', *Philosophical Magazine*, **19**, 19–32. Reprinted in Niven (1952)

Mehra, Jagdish (1975), *The Solvay Conferences on Physics: Aspects of the Development of Physics since 1911* (Dordrecht, Holland: D. Reidel)

Mehra, Jagdish and Helmut Rechenberg (1982), *The Historical Development of Quantum Theory*, Vol. 1, Parts 1 and 2: *The Quantum Theory of Planck, Einstein, Bohr, and Sommerfeld: Its Foundations and the Rise of Its Difficulties 1900–1925* (Berlin: Springer)

Mehra, Jagdish and Helmut Rechenberg (1982), *The Historical Development of Quantum Theory*, Vol. 2: *The Discovery of Quantum Mechanics* (Berlin: Springer)

Mehra, Jagdish and Helmut Rechenberg (1982), *The Historical Development of Quantum Theory*, Vol. 3: *The Formulation of Matrix Mechanics and Its Modifications 1925–1926* (Berlin: Springer)

Mehra, Jagdish and Helmut Rechenberg (1982), *The Historical Development of Quantum Theory*, Vol. 4: *The Fundamental Equations of Quantum Mechanics 1925–1926* and *The Reception of the New Quantum Mechanics 1925–1926* (Berlin: Springer)

Mehra, Jagdish and Helmut Rechenberg (1987), *The Historical Development of Quantum Theory*, Vol. 5, Parts 1 and 2: *Erwin Schrödinger and the Rise of Wave Mechanics* (Berlin: Springer)

Mehra, Jagdish and Helmut Rechenberg (2000), *The Historical Development of Quantum Theory*, Vol. 6, Part 1: *The Completion of Quantum Mechanics 1926–1941* (Berlin: Springer)

Mehra, Jagdish and Helmut Rechenberg (2001), *The Historical Development of Quantum Theory*, Vol. 6, Part 2: *The Completion of Quantum Mechanics 1926–1941* (Berlin: Springer)

Mendelssohn, Kurt (1973), *The World of Walther Nernst: The Rise and Fall of German Science* (London: Macmillan)

Metzger, Rainer (2007), *Berlin in the Twenties: Art and Culture 1918–1933* (London: Thames and Hudson)

Meyenn, Karl von and Engelbert Schucking (2001), 'Wolfgang Pauli', *Physics Today*, February, 43–48

Millikan, Robert A. (1915), 'New tests of Einstein's photoelectric equation', *Physical Review*, **6**, 55

Moore, Ruth (1966), *Niels Bohr: The Man, His Science, and The World They Changed* (New York: Alfred A. Knopf)

Moore, Walter (1989), *Schrödinger: Life and Thought* (Cambridge: Cambridge University Press)

Mulligan, Joseph F. (1994), 'Max Planck and the "black year" of German Physics', *American Journal of Physics*, **62**, 12, 1089–1097

Mulligan, Joseph F. (1999), 'Heinrich Hertz and Philipp Lenard: Two Distinguished Physicists, Two Disparate Men', *Physics in Perspective*, **1**, 345–366

Mulvey, J. H. (ed.) (1981), *The Nature of Matter* (Oxford: Oxford University Press)

Murdoch, Dugald (1987), *Niels Bohr's Philosophy of Physics* (Cambridge: Cambridge University Press)

Nathan, Otto and Heinz Norden (eds) (1960), *Einstein on Peace* (New York: Simon and Schuster)

Neumann, John von (1955), *Mathematical Foundations of Quantum Mechanics* (Princeton, NJ: Princeton University Press)

Nielsen, J. Rud (1963), 'Memories of Niels Bohr', *Physics Today*, October

Nitske, W. Robert (1971), *The Life of Wilhelm Conrad Röntgen: Discoverer of the X-Ray* (Tucson, AZ: University of Arizona Press)

Niven, W. D. (ed.) (1952), *The Scientific Papers of James Clerk Maxwell*, 2 vols (New York: Dover Publications)

Nobel Lectures (1964), *Physics 1942–1962* (Amsterdam: Elsevier)

Nobel Lectures (1965), *Physics 1922–1941* (Amsterdam: Elsevier)

Nobel Lectures (1967), *Physics 1901–1921* (Amsterdam: Elsevier)

Norris, Christopher (2000), *Quantum Theory and the Flight from Reason: Philosophical Responses to Quantum Mechanics* (London: Routledge)

Nye, Mary Jo (1996), *Before Big Science: The Pursuit of Modern Chemistry and Physics 1800–1940* (Cambridge, MA: Harvard University Press)

Offer, Avner (1991), *The First World War: An Agrarian Interpretation* (Oxford: Oxford University Press)

Omnès, Roland (1999), *Quantum Philosophy: Understanding and Interpreting Contemporary Science* (Princeton, NJ: Princeton University Press)

Overbye, Dennis (2001), *Einstein in Love* (London: Bloomsbury)

Ozment, Steven (2005), *A Mighty Fortress: A New History of the German People, 100 BC to the 21st Century* (London: Granta Books)

Pais, Abraham (1967), 'Reminiscences of the Post-War Years', in Rozental (1967)

Pais, Abraham (1982), *'Subtle is the Lord …': The Science and the Life of Albert Einstein* (Oxford: Oxford University Press)

Pais, Abraham (1986), *Inward Bound: Of Matter and Forces in the Physical World* (Oxford: Clarendon Press)

Pais, Abraham (1989a), 'Physics in the Making in Bohr's Copenhagen', in Sarlemijn and Sparnaay (1989)

Pais, Abraham (1989b), 'George Uhlenbeck and the Discovery of Electron Spin', *Physics Today*, December. Reprinted in Phillips (1992)

Pais, Abraham (1991), *Niels Bohr's Times, in Physics, Philosophy, and Polity* (Oxford: Clarendon Press)

Pais, Abraham (1994), *Einstein Lived Here* (Oxford: Clarendon Press)

Pais, Abraham (2000), *The Genius of Science: A portrait gallery of twentieth-century physicists* (New York: Oxford University Press)

Pais, Abraham (2006), *J. Robert Oppenheimer: A Life* (Oxford: Oxford University Press)

Park, David (1997), *The Fire Within The Eye: A Historical Essay on the Nature and Meaning of Light* (Princeton, NJ: Princeton University Press)

Pauli, Wolfgang (1946a), 'Remarks on the History of the Exclusion Principle', *Science*, **103**, 213–215

Pauli, Wolfang (1946b), 'Exclusion principle and quantum mechanics', Nobel lecture delivered on 13 December. Reprinted in *Nobel Lectures* (1964), 27–43

Penrose, Roger (1990), *The Emperor's New Mind* (London: Vintage)

Penrose, Roger (1995), *Shadows of the Mind* (London: Vintage)

Penrose, Roger (1997), *The large, the small and the human mind* (Cambridge: Cambridge University Press)

Petersen, Aage (1985), 'The Philosophy of Niels Bohr', in French and Kennedy (1985)

Petruccioli, Sandro (1993), *Atoms, Metaphors and Paradoxes: Niels Bohr and the construction of a new physics* (Cambridge: Cambridge University Press)

Phillips, Melba Newell (ed.) (1985), *Physics History from AAPT Journals* (College Park, MD: American Association of Physics Teachers)

Phillips, Melba (ed.) (1992), *The Life and Times of Modern Physics: History of Physics II* (New York: American Institute of Physics)

Planck, Max (1900a), 'On An Improvement of Wien's Equation for the Spectrum', reprinted in Haar (1967)

Planck, Max (1900b), 'On the Theory of the Energy Distribution Law of the Normal Spectrum', reprinted in Haar (1967)

Planck, Max (1949), *Scientific Autobiography and Other Papers* (New York: Philosophical Library)

Planck, Max (1993), *A Survey of Physical Theory* (New York: Dover Publications)

Ponte, M.J.H. (1981), 'Louis de Broglie', in Goodman (1981)

Powers, Jonathan (1985), *Philosophy and the New Physics* (London: Methuen)

Przibram, Karl (ed.) (1967), *Letters on Wave Mechanics*, translation and introduction by Martin Klein (New York: Philosophical Library)

Purrington, Robert D. (1997), *Physics in the Nineteenth Century* (New Brunswick, NJ: Rutgers University Press)

Redhead, Michael (1987), *Incompleteness, Nonlocality and Realism* (Oxford: Clarendon Press)

Rhodes, Richard (1986), *The Making of the Atomic Bomb* (New York: Simon and Schuster)

Robertson, Peter (1979), *The Early Years: The Niels Bohr Institute 1921–1930* (Copenhagen: Akademisk Forlag)

Robinson, Andrew (2006), *The Last Man Who Knew Everything* (New York: Pi Press)

Rosenkranz, Ze'ev (2002), *The Einstein Scrapbook* (Baltimore, MD: Johns Hopkins University Press)

Rowland, John (1938), *Understanding the Atom* (London: Gollancz)

Rosenfeld, Léon (1967), 'Niels Bohr in the Thirties. Consolidation and extension of the Conception of Complementarity', in Rozental (1967)

Rosenfeld, Léon (1968), 'Some Concluding Remarks and Reminiscences', in Solvay Institute (1968)

Rosenfeld, Léon and Erik Rüdinger (1967), 'The Decisive Years: 1911–1918', in Rozental (1967)

Rosenthal-Schneider, Ilse (1980), *Reality and Scientific Truth: Discussions with Einstein, von Laue, and Planck,* edited by Thomas Braun (Detroit, MI: Wayne State University Press)

Rozental, Stefan (ed.) (1967), *Niels Bohr: His Life and Work as seen by his Friends and Colleagues* (Amsterdam: North-Holland)

Rozental, Stefan (1998), *Niels Bohr: Memoirs of a Working Relationship* (Copenhagen: Christian Ejlers)

Ruhla, Charles (1992), *The Physics of Chance: From Blaise Pascal to Niels Bohr* (Oxford: Oxford University Press)

Rutherford, Ernest (1906), *Radioactive Transformations* (London: Constable)

Rutherford, Ernest (1911a), 'The Scattering of Alpha and Beta Particles by Matter and the Structure of the Atom', *Philosophical Magazine,* **21**, 669–688. Reprinted in Boorse and Motz (1966), Vol. 1

Rutherford, Ernest (1911b), 'Conference on the Theory of Radiation', *Nature,* **88**, 82–83

Rutherford, Ernest (1954), *Rutherford By Those Who Knew Him. Being the Collection of the First Five Rutherford Lectures of the Physical Society* (London: The Physical Society)

Rutherford, Ernest and Hans Geiger (1908a), 'An Electrical Method for Counting the Number of Alpha Particles from Radioactive Substances', *The Proceedings of the Royal Society* A, **81**, 141–161

Rutherford, Ernest and Hans Geiger (1908b), 'The Charge and Nature of the Alpha Particle', *The Proceedings of the Royal Society* A, **81**, 162–173

Sarlemijn, A. and M.J. Sparnaay (eds) (1989), *Physics in the Making* (Amsterdam: Elsevier)

Schilpp, Paul A. (ed.) (1969), *Albert Einstein: Philosopher-Scientist* (New York: MJF Books). Collection first published in 1949 as Vol. VII in the series *The Library of Living Philosophers* by Open Court, La Salle, IL

Schrödinger, Erwin (1933), 'The fundamental idea of wave mechanics', Nobel lecture delivered on 12 December. Reprinted in *Nobel Lectures* (1965), 305–316

Schrödinger, Erwin (1935), 'The Present Situation in Quantum Mechanics', reprinted and translated in Wheeler and Zurek (1983), 152–167

Schweber, Silvan S. (1994), *QED and the Men Who Made It: Dyson, Feynman, Schwinger, and Tomonaga* (Princeton, NJ: Princeton University Press)

Segrè, Emilio (1980), *From X-Rays to Quarks: Modern Physicists and Their Discoveries* (New York: W.H. Freeman and Company)

Segrè, Emilio (1984), *From Falling Bodies to Radio Waves* (New York: W.H. Freeman and Company)

Sime, Ruth Lewin (1996), *Lise Meitner: A Life in Physics* (Berkeley, CA: University of California Press)

Smith, Alice Kimball and Charles Weiner (eds) (1980), *Robert Oppenheimer: Letters and Recollections* (Cambridge, MA: Harvard University Press)

Snow, C. P. (1969), *Variety of Men: Statesmen, Scientists, Writers* (London: Penguin)

Snow, C. P. (1981), *The Physicists* (London: Macmillan)

Soddy, Frederick (1913), 'Intra-Atomic Charge', *Nature*, **92**, 399–400

Solvay Institute (1968), *Fundamental Problems in Elementary Particle Physics: proceedings of the 14th Solvay Council held in Brussels in 1967* (New York: Wiley Interscience)

Stachel, John (ed.) (1998), *Einstein's Miraculous Year: Five Papers That Changed the Face of Physics* (Princeton, NJ: Princeton University Press)

Stachel, John (2002), *Einstein from 'B to Z'* (Boston, MA: Birkhäuser)

Stapp, Henry P. (1977), 'Are superluminal connections necessary?', *Il Nuovo Cimento*, **40B**, 191–205

Stürmer, Michael (1999), *The German Century* (London: Weidenfeld and Nicolson)

Stürmer, Michael (2000), *The German Empire* (London: Weidenfeld and Nicolson)

Stuewer, Roger H. (1975), *The Compton Effect: Turning Point in Physics* (New York: Science History Publications)

Susskind, Charles (1995), *Heinrich Hertz: A Short Life* (San Francisco: San Francisco Press)

Tegmark, Max and John Wheeler (2001), '100 Years of Quantum Mysteries', *Scientific American*, February, 54–61

Teich, Mikulas and Roy Porter (eds) (1990), *Fin de Siècle and its Legacy* (Cambridge: Cambridge University Press)

Teichmann, Jürgen, Michael Eckert, and Stefan Wolff (2002), 'Physicists and Physics in Munich', *Physics in Perspective*, **4**, 333–359

Thomson, George P. (1964), *J.J. Thomson and the Cavendish Laboratory in his day* (London: Nelson)

Thorne, Kip S. (1994), *Black Holes and Time Warps: Einstein's Outrageous Legacy* (London: Picador)

Trigg, Roger (1989), *Reality at Risk: A Defence of Realism in Philosophy and the Sciences* (Hemel Hempstead: Harvester Wheatsheaf)

Treiman, Sam (1999), *The Odd Quantum* (Princeton, NJ: Princeton University Press)

Tuchman, Barbara W. (1966), *The Proud Tower: A Portrait of the World Before the War 1890–1914* (New York: Macmillan)

Uhlenbeck, George E. (1976), 'Personal reminiscences', *Physics Today*, June. Reprinted in Weart and Phillips (1985)

Van der Waerden, B.L. (1967), *Sources of Quantum Mechanics* (New York: Dover Publications)

Weart, Spencer R. and Melba Phillips (eds) (1985), *History of Physics: Readings from Physics Today* (New York: American Institute of Physics)

Weber, Robert L. (1981), *Pioneers of Science: Nobel Prize Winners in Physics* (London: The Scientific Book Club)

Wehler, Hans-Ulrich (1985), *The German Empire* (Leamington Spa: Berg Publishers)

Weinberg, Steven (1993), *Dreams of a Final Theory: The Search for the Fundamental Laws of Nature* (London: Hutchinson)

Weinberg, Steven (2003), *The Discovery of Subatomic Particles* (Cambridge: Cambridge University Press)

Weiner, Charles (ed.) (1977), *History of Twentieth Century Physics* (New York: Academic)

Wheaton, Bruce R. (2007), 'Atomic Waves in Private Practice', in Evans and Thorndike (2007)

Wheeler, John A. (1994), *At Home in the Universe* (Woodbury, NY: AIP Press)

Wheeler, John A. and Wojciech H. Zurek (eds) (1983), *Quantum Theory and Measurement* (Princeton, NJ: Princeton University Press)

Whitaker, Andrew (2002), 'John Bell in Belfast: Early Years and Education', in Bertlmann and Zeilinger (2002)

Wilson, David (1983), *Rutherford: Simple Genius* (London: Hodder and Stoughton)

Wolf, Fred Alan (1988), *Parallel Universes: The Search for Other Worlds* (London: The Bodley Head)

ACKNOWLEDGEMENTS

For many years the photograph of those who gathered in Brussels in October 1927 for the fifth Solvay conference hung on my wall. Occasionally I would pass it and think that it was the perfect starting point for a narrative history of the quantum. When I finally wrote a proposal for *Quantum*, I had the great good fortune of submitting it to Patrick Walsh. His enthusiasm was instrumental in getting the project off the ground. I was lucky a second time when the talented science editor and publisher Peter Tallack joined Conville & Walsh and became my agent. To Pete I offer my heartfelt thanks for being both friend and agent through the years that this book took to write and for handling with grace all the difficulties that arose from my prolonged bouts of ill-health. Together with Pete, Jake Smith-Bosanquet has served as my point man for the foreign language publishers of *Quantum*; I would like to express my thanks to him and the rest of the team at Conville & Walsh, particularly Claire Conville and Sue Armstrong, for their steadfast support and help. It is a pleasure to have this opportunity to thank Michael Carlisle and especially Emma Parry for their work on my behalf in the USA.

I owe a great deal to the studies of scholars cited in the notes and listed in the bibliography; however, I am particularly indebted to Denis Brian, David C. Cassidy, Albrecht Fölsing, John L. Heilbron, Martin J. Klein, Jagdish Mehra, Walter Moore, Dennis Overbye, Abraham Pais, Helmut Rechenberg, and John Stachel. I would like to thank Guido Bacciagaluppi and Anthony Valentini for making available the first English-language translation of the proceedings of the fifth Solvay conference and their commentary prior to its publication.

Pandora Kay-Kreizman, Ravi Bali, Steven Böhm, Jo Cambridge, Bob Cormican, John Gillott, and Eve Kay all read drafts of the book. Thanks go to each and every one of them for their astute criticism and suggestions.

Mitzi Angel was at one time my editor and her insightful comments on an earlier draft of the book were invaluable. Christopher Potter was an early champion of *Quantum* and I am deeply grateful to him for having been so. Simon Flynn, my publisher at Icon Books, has been indefatigable in bringing the book to press. He has done much that was beyond the call of duty and I thank him for it. Duncan Heath has been an astonishingly eagle-eyed copy-editor; every writer should be so fortunate. I am grateful to Andrew Furlow and Najma Finlay of Icon for their enthusiasm and work on behalf of *Quantum* and to Nicholas Halliday for producing the fine diagrams that illustrate the text. Thanks also go to Neal Price and his team at Faber & Faber.

This book would not have been possible without the unfailing support over many years of Lahmber Ram, Gurmit Kaur, Rodney Kay-Kreizman, Leonora Kay-Kreizman, Rajinder Kumar, Santosh Morgan, Eve Kay, John Gillott, and Ravi Bali.

Finally, I wish to thank, with all my heart, my wife Pandora and my sons Ravinder and Jasvinder. Words are not enough to convey what I owe to the three of you.

Manjit Kumar
London, August 2008

INDEX